VOLUME SIX HUNDRED AND THIRTEEN

METHODS IN ENZYMOLOGY

Enzymes of Energy Technology

METHODS IN ENZYMOLOGY

VOLUME SIX HUNDRED AND THIRTEEN

METHODS IN ENZYMOLOGY

Enzymes of Energy Technology

Edited by

FRASER ARMSTRONG
Department of Chemistry
University of Oxford
Oxford, United Kingdom

ELSEVIER

ACADEMIC PRESS
An imprint of Elsevier

Academic Press is an imprint of Elsevier
50 Hampshire Street, 5th Floor, Cambridge, MA 02139, United States
525 B Street, Suite 1650, San Diego, CA 92101, United States
The Boulevard, Langford Lane, Kidlington, Oxford OX5 1GB, United Kingdom
125 London Wall, London, EC2Y 5AS, United Kingdom

First edition 2018

ISBN: 978-0-12-816361-0
ISSN: 0076-6879

For information on all Academic Press publications
visit our website at https://www.elsevier.com/books-and-journals

Publisher: Fiona Pattison
Acquisition Editor: Ashlie Jackman
Editorial Project Manager: Shellie Bryant
Production Project Manager: Divya Krishna Kumar
Cover Designer: Mark Rogers

Typeset by SPi Global, India

Contents

Contributors

Michael W.W. Adams
Department of Biochemistry and Molecular Biology, University of Georgia, Athens, GA, United States

Stephen E. Beaton
Inorganic Chemistry Laboratory, University of Oxford, Oxford, United Kingdom

Rodney Burton
Department of Biological Chemistry, University of Michigan Medical School, Ann Arbor, MI, United States

Julea N. Butt
Centre for Molecular and Structural Biochemistry, School of Chemistry and School of Biological Sciences, University of East Anglia, Norwich Research Park, Norwich, United Kingdom

Mehmet Can
Department of Biological Chemistry, University of Michigan Medical School, Ann Arbor, MI, United States

Valerie L. Cash
Department of Biochemistry, Virginia Tech, Blacksburg, VA, United States

Yolande Charmasson
Aix Marseille Université, CNRS, Centrale Marseille, iSm2, Marseille, France

Luisa Ciano
Department of Chemistry, University of York, York, United Kingdom

Thomas A. Clarke
Centre for Molecular and Structural Biochemistry, School of Chemistry and School of Biological Sciences, University of East Anglia, Norwich Research Park, Norwich, United Kingdom

David A. Collins
Department of Biology, San Diego State University, San Diego, CA, United States

Gideon J. Davies
Department of Chemistry, University of York, York, United Kingdom

Antonio L. De Lacey
Instituto de Catálisis y Petroleoquímica, CSIC, Madrid, Spain

Dennis R. Dean
Department of Biochemistry, Virginia Tech, Blacksburg, VA, United States

Elise Courvoisier Dezord
Aix Marseille Université, CNRS, Centrale Marseille, iSm2, Marseille, France

Marcus J. Edwards
Centre for Molecular and Structural Biochemistry, School of Chemistry and School of Biological Sciences, University of East Anglia, Norwich Research Park, Norwich, United Kingdom

Vera Engelbrecht
AG Photobiotechnologie, Fakultät für Biologie und Biotechnologie, Ruhr-Universität Bochum, Universitätsstraße, Bochum, Germany

Daniel Esckilsen
Department of Biological Chemistry, University of Michigan Medical School, Ann Arbor, MI, United States

Rhiannon M. Evans
Inorganic Chemistry Laboratory, University of Oxford, Oxford, United Kingdom

Stefan Frielingsdorf
Department of Chemistry, Technische Universität Berlin, Berlin, Germany

Dominik K. Haja
Department of Biochemistry and Molecular Biology, University of Georgia, Athens, GA, United States

Thomas Happe
AG Photobiotechnologie, Fakultät für Biologie und Biotechnologie, Ruhr-Universität Bochum, Universitätsstraße, Bochum, Germany

Glyn R. Hemsworth
School of Molecular and Cellular Biology, Faculty of Biological Sciences; Astbury Centre for Structural Molecular Biology, University of Leeds, Leeds, United Kingdom

Russ Hille
Department of Biochemistry, University of California, Riverside, CA, United States

Emilio Jiménez-Vicente
Department of Biochemistry, Virginia Tech, Blacksburg, VA, United States

Marina G. Kalyuzhnaya
Department of Biology, San Diego State University, San Diego, CA, United States

Keisuke Kawakami
The OCU Advanced Research Institute for Natural Science & Technology (OCARINA), Osaka City University, Osaka, Japan

Lars Lauterbach
Department of Chemistry, Technische Universität Berlin, Berlin, Germany

Oliver Lenz
Department of Chemistry, Technische Universität Berlin, Berlin, Germany

Colin W.J. Lockwood
Centre for Molecular and Structural Biochemistry, School of Chemistry and School of Biological Sciences, University of East Anglia, Norwich Research Park, Norwich, United Kingdom

Julia Sanchez Martin Del Campo
Department of Biochemistry, Virginia Tech, Blacksburg, VA, United States

Pedro M. Matias
Instituto de Tecnologia Química e Biológica António Xavier, Universidade Nova de Lisboa; iBET, Instituto de Biologia Experimental e Tecnológica, Oeiras, Portugal

Yasmina Mekmouche
Aix Marseille Université, CNRS, Centrale Marseille, iSm2, Marseille, France

William W. Metcalf
Carl R. Woese Institute for Genomic Biology; Department of Microbiology, University of Illinois, Urbana, IL, United States

Dipti D. Nayak
Carl R. Woese Institute for Genomic Biology; Department of Microbiology, University of Illinois, Urbana, IL, United States

Simone Newton-Payne
Centre for Molecular and Structural Biochemistry, School of Chemistry and School of Biological Sciences, University of East Anglia, Norwich Research Park, Norwich, United Kingdom

Dimitri Niks
Department of Biochemistry, University of California, Riverside, CA, United States

Inês A.C. Pereira
Instituto de Tecnologia Química e Biológica António Xavier, Universidade Nova de Lisboa, Oeiras, Portugal

Samuel E.H. Piper
Centre for Molecular and Structural Biochemistry, School of Chemistry and School of Biological Sciences, University of East Anglia, Norwich Research Park, Norwich, United Kingdom

Marcos Pita
Instituto de Catálisis y Petroleoquímica, CSIC, Madrid, Spain

Stephen W. Ragsdale
Department of Biological Chemistry, University of Michigan Medical School, Ann Arbor, MI, United States

Lu Ren
Aix Marseille Université, CNRS, Centrale Marseille, iSm2, Marseille, France

David J. Richardson
Centre for Molecular and Structural Biochemistry, School of Chemistry and School of Biological Sciences, University of East Anglia, Norwich Research Park, Norwich, United Kingdom

Viviane Robert
Aix Marseille Université, CNRS, Centrale Marseille, iSm2, Marseille, France

Pierre Rousselot-Pailley
Aix Marseille Université, CNRS, Centrale Marseille, iSm2, Marseille, France

Lance C. Seefeldt
Department of Chemistry and Biochemistry, Utah State University, Logan, UT, United States

Jian-Ren Shen
Graduate School of Natural Science and Technology, Okayama University, Okayama, Japan

Thierry Tron
Aix Marseille Université, CNRS, Centrale Marseille, iSm2, Marseille, France

Jessica H. van Wonderen
Centre for Molecular and Structural Biochemistry, School of Chemistry and School of Biological Sciences, University of East Anglia, Norwich Research Park, Norwich, United Kingdom

Marisela Vélez
Instituto de Catálisis y Petroleoquímica, CSIC, Madrid, Spain

Paul H. Walton
Department of Chemistry, University of York, York, United Kingdom

Gaye F. White
Centre for Molecular and Structural Biochemistry, School of Chemistry and School of Biological Sciences, University of East Anglia, Norwich Research Park, Norwich, United Kingdom

Seth Wiley
Department of Biological Chemistry, University of Michigan Medical School, Ann Arbor, MI, United States

Chang-Hao Wu
Department of Biochemistry and Molecular Biology, University of Georgia, Athens, GA, United States

Zhi-Yong Yang
Department of Chemistry and Biochemistry, Utah State University, Logan, UT, United States

Sónia Zacarias
Instituto de Tecnologia Química e Biológica António Xavier, Universidade Nova de Lisboa, Oeiras, Portugal

Simeng Zhou
Aix Marseille Université, CNRS, Centrale Marseille, iSm2, Marseille, France

Preface

On a per-active-site basis, enzymes are the most active catalysts for several chemical reactions that are crucial for future renewable energy—the storage and usage of solar fuels as well as other green, sustainable technologies. The most widely studied processes are referred to (and abbreviated) as the hydrogen evolution reaction (HER), important in water splitting by electricity or sunlight; the hydrogen oxidation reaction (HOR), used in fuel cells; the four-electron oxygen evolution reaction (OER), required in water splitting; the four-electron oxygen reduction reaction (ORR) used in fuel cells; the nitrogen reduction reaction (NRR), which produces ammonia for agriculture; and direct carbon dioxide reduction reactions (CO_2RR) for use in CO_2 capture and utilization (CCU). The acronyms foster familiarity and bring together scientists from many different fields.

Many of these enzymes have been studied using protein film electrochemistry, which has enabled their activities to be quantified in unusual detail. Those catalyzing hydrogen and CO_2 activation are already identified as special electrocatalysts: when attached to a suitable electrode, they require only a minimal overpotential (above the thermodynamic equilibrium value) to drive oxidation or reduction in either direction. Only one such process is well established in industry—the interconversion of protons and hydrogen at platinum, which still remains the catalyst employed in fuel cells used in transport, but at considerable cost. Biology has exploited abundant elements to optimize rates and minimize energy wastage, even evolving "wires" to conduct electrons in and out of living cells. It may be stated with confidence that enzymes represent the best catalysts, in terms of activity, that industry could hope to achieve, so it is worth knowing why. All the enzymes addressed in this volume are oxidoreductases and require a tightly bound cofactor: in some cases, such as nitrogenase, hydrogenases, methane-evolving F430, and carbon monoxide dehydrogenases, the cofactor is particularly complex and requires several genes to produce special proteins for its synthesis. Many of the enzymes are large, multisubunit assemblies and are often membrane-bound. This volume is devoted to the practical details of obtaining the enzymes and handling the organisms that produce them.

Chapter 1 leads with an account by Kawakami and Shen, who explain how to produce highly active, pure samples of Photosystem II for crystallization and structure determination at high resolution. Efforts of the Shen group and others are now providing an unprecedented level of structural detail on intermediates formed at the [4MnCa-5O] active site which carries out the biological OER. Through considerable skill, it has recently proved possible to obtain time-resolved structures of short-lived species by using an X-ray free-electron laser.

The biological ORR is represented by the "blue" Cu oxidases produced by fungi - enzymes set to play an important role in industry once they become cheap and abundant. Laccases perform clean, four-electron reduction of O_2 to H_2O while oxidizing organics or taking up electrons at electrodes. In Chapter 2, Mekmouche and coworkers describe strategies they have developed to obtain large yields of laccase variants using *Saccharomyces cerevisiae* and *Aspergillus niger* as hosts.

Lytic polysaccharide monooxygenases (LPMOs) are Cu-containing enzymes which use O_2 to convert recalcitrant polysaccharides into smaller, usable units, greatly expanding the scope for biofuel production, e.g., from lignin. Walton and his colleagues discuss the production and characterization of LPMOs in Chapter 3.

The inspirational topic of biological H_2 activation (HOR and HER) is taken up by five groups of authors who describe procedures for producing some key representatives of hydrogenases, well known to be O_2-sensitive enzymes. In Chapter 4, Evans and Beaton explain how they solved the problem of the poor expression of an important [NiFe]-hydrogenase known as Hyd-2, found in the common bacterium *Escherichia coli*: in its full form, Hyd-2 is a membrane-bound proton translocator and has long defied structural characterization. In Chapter 5, Lenz, Lauterbach, and Frielingsdorf describe the production, isolation, and characterization of an O_2-tolerant [NiFe]-hydrogenase from *Ralstonia eutropha*, which can grow using H_2 as the only fuel source and O_2 as the only oxidant—a "living fuel cell." Chapter 6, written by Adams and coworkers, deals with the robust [NiFe]-hydrogenases produced by *Pyrococcus furiosus* which grows optimally at 100°C. The [NiFeSe]-hydrogenases, in which one cysteine ligand to the active-site Ni is replaced by selenocysteine, are interesting because they have higher activities than their [NiFe]-counterparts and are less inactivated by O_2: in Chapter 7, Matias and colleagues describe the procedures used to express, purify, and study the [NiFeSe]-hydrogenase from a sulfate-reducing bacterium. The best HER catalysts are [FeFe]-hydrogenases, and their presence in photosynthetic green algae, notably *Chlamydomonas reinhardtii*, has inspired scientists to consider the possibility of "hydrogen farms," in which H_2 gas is produced continually instead of biomass. The [FeFe]-hydrogenases are also present in fermentative anaerobes. The active site, known as the H-cluster, is particularly instructive from a mechanistic viewpoint. In Chapter 8, Engelbrecht and Happe explain how to express, purify, and manipulate [FeFe]-hydrogenases produced by green algae.

The NRR is a particularly challenging process. Unlike the industrial Haber–Bosch process which operates at high temperature and high pressure, microbes produce ammonia under ambient conditions through the action of nitrogenase. The fact that so many ATP molecules (16) are sacrificed to produce each pair of ammonia molecules suggests that even Biology has struggled to evolve a catalyst to activate the very strong N≡N bond. Three questions have driven research for over 60 years: (1) how does nitrogenase work? (2) could we (assisted by our knowledge of the mechanism) synthesize alternative catalysts to the Haber–Bosch process? (3) can we engineer organisms to produce more ammonia in soils? In Chapter 9, Dean and colleagues outline the quaternary organization of nitrogenases and describe powerful, affinity-based purification protocols that are greatly enhancing mechanistic investigations.

Electrogenic bacteria live by making electrical connections across their outer cell membranes to tap the energy available from extracellular redox sources such as minerals. These organisms can be wired to an electrode, eventually forming the basis of new green technologies such as wastewater remediation. Determining the mechanism by which electron transport occurs across the outer membrane requires thorough investigations of the proteins and their cofactors that are involved. In Chapter 10, Butt and her colleagues outline approaches for obtaining and characterizing membrane-bound and soluble components of MR-1 from *Shewanella oneidensis*.

Direct reduction of molecular CO_2 is mainly carried out by two classes of enzyme: formate dehydrogenases catalyze the interconversion of CO_2 and formate ($HCOO^-$), while carbon monoxide dehydrogenases (CODHs) catalyze the interconversion of CO_2 and CO, the latter eventually leading to the formation of acetyl-CoA. There are different types of each class. In Chapter 11, Niks and Hille describe methods for obtaining metal-dependent (Mo or W) formate dehydrogenases. In Chapter 12, Ragsdale and coworkers focus on CODHs that contain a NiFe cofactor and are proficient at producing CO.

Methane is both a relatively clean-burning fuel (a carbon-neutral asset) and a greenhouse gas (an environmental hazard): its management and exploitation are therefore of great importance in sustainability. In

Chapter 13, Nayak and Metcalf focus on the final stage of methanogenesis. They describe genetic tools and purification procedures that are greatly facilitating the investigation of methyl coenzyme M reductase (MCR), the enzyme responsible for completing the synthesis of CH_4, which could be beneficial if controlled. Finally, in Chapter 14, Collins and Kalyuzhnaya provide a helpful practical guide for working with methanotrophs, methane-consuming bacteria, and the enzymes that are involved—leading to insight that may help curtail uncontrolled methane production.

FRASER ARMSTRONG
Oxford
October 2018

CHAPTER ONE

Purification of fully active and crystallizable photosystem II from thermophilic cyanobacteria

Keisuke Kawakami[a], Jian-Ren Shen[b,*]
[a]The OCU Advanced Research Institute for Natural Science & Technology (OCARINA), Osaka City University, Osaka, Japan
[b]Graduate School of Natural Science and Technology, Okayama University, Okayama, Japan
*Corresponding author: e-mail address: shen@cc.okayama-u.ac.jp

Contents

Abstract

Photosystem II (PSII) is a membrane protein complex which functions to catalyze light-induced water oxidation in oxygenic photosynthesis. Through the water-splitting reaction of PSII, light energy is converted into biologically useful chemical energy, and molecular oxygen is formed which transformed the atmosphere into an aerobic one and sustained aerobic life on the Earth. The PSII core complex from cyanobacteria consists of 17 transmembrane subunits and 3 extrinsic subunits with a total molecular mass of approximately 350 kDa per monomer, and PSII exists predominately in a dimeric form in vivo. This chapter describes the purification procedures leading to highly pure, homogenous, and highly active PSII core dimers from a thermophilic cyanobacterium, *Thermosynechococcus vulcanus* (*T. vulcanus*), that are used for successful crystallization and diffraction at atomic resolution. The purity and homogeneity of the PSII dimers thus obtained are characterized by absorption spectra, low-temperature fluorescence spectra, SDS-PAGE, clear native PAGE, blue native PAGE, gel filtration chromatography, and oxygen-evolving activity measurements. Finally, high-quality crystals obtained from the purified PSII dimers are shown.

Methods in Enzymology, Volume 613
ISSN 0076-6879
https://doi.org/10.1016/bs.mie.2018.10.002

1. Introduction

Photosystem II (PSII) catalyzes light-induced charge separation and water oxidation, leading to the evolution of electrons, protons, and molecular oxygen. The former two products are used for the synthesis of NADPH and ATP, whereas the latter product is indispensable for sustaining aerobic life on Earth. Most of the PSII components are conserved, from the prokaryotic oxygenic photosynthetic organisms cyanobacteria to eukaryotic higher plants (Shen, 2015; Shen, Henmi, & Kamiya, 2008; Wydrzynski & Satoh, 2006); however, the types and numbers of light-harvesting antenna complexes, the composition of extrinsic proteins involved in the oxygen-evolving reaction, and that of some low-molecular weight subunits vary among different organisms (Enami et al., 2008; Wydrzynski & Satoh, 2006). Both PSII-enriched membrane fragments and PSII core complexes devoid of their antenna proteins have been purified from higher plants (Berthold, Babcock, & Yocum, 1981; Enami, Kamino, Shen, Satoh, & Katoh, 1989; Ikeuchi & Inoue, 1986), green algae (Sugiura, Inoue, & Minagawa, 1998; Suzuki et al., 2003), euglena (Suzuki et al., 2004), diatoms (Nagao et al., 2007; Okumura et al., 2008), red algae (Adachi et al., 2009), cyanobacteria (Kuhl et al., 2000; Shen, Ikeuchi, & Inoue, 1992; Shen & Inoue, 1993), and the purification method of PSII varies depending on the species used. On the other hand, only PSII core complexes from thermophilic cyanobacteria have been used successfully in producing PSII crystals (Shen & Kamiya, 2000; Zouni et al., 2001), and crystal quality and resolution have improved remarkably in recent years (Ferreira, Iverson, Maghlaoui, Barber, & Iwata, 2004; Kamiya & Shen, 2003; Loll, Kern, Saenger, Zouni, & Biesiadka, 2005; Umena, Kawakami, Shen, & Kamiya, 2011). The improvement is largely due to a higher stability of the PSII core complexes purified from thermophilic cyanobacteria (Sugiura & Inoue, 1999), since they grow at temperatures of 45–60°C, which are much higher than the mesophilic organisms. This property is especially important for the handling of protein complexes such as PSII, since some of its subunits, in particular some of the extrinsic proteins, bind weakly within the complex and are easily lost during solubilization and purification, making the purification of intact and highly active PSII core complexes from mesophilic organisms difficult.

Two closely related thermophilic cyanobacteria, *Thermosynechococcus elongatus* and *T. vulcanus*, have been used for the purification and crystallization of PSII core complexes (Ferreira et al., 2004; Kamiya & Shen, 2003;

Umena et al., 2011; Zouni et al., 2001). Both of these microorganisms were collected from Japanese hot springs at two different locations. Two forms of the PSII core are obtained from the cyanobacteria, one is a monomer and the other one is a dimer: the latter exists in a much larger amount than the monomer, and it is believed to be the functional form of PSII in vivo. Both PSII dimers and monomers have been crystallized (Broser et al., 2010; Ferreira et al., 2004; Kamiya & Shen, 2003; Umena et al., 2011; Zouni et al., 2001), although the resolution of the PSII dimer crystals is consistently much higher than that of the monomers; hence, we will focus here on the purification and characterization of the PSII dimers.

Each PSII core of cyanobacteria contains 17 membrane-spanning transmembrane subunits (D1, D2, CP43, CP47, PsbE, PsbF, PsbH, PsbI, PsbJ, PsbK, PsbL, PsbM, PsbT, PsbX, PsbY, PsbZ, and Psb30), 3 extrinsic proteins (PsbO, PsbU, and PsbV), and over 100 cofactors including chlorophylls (Chls), carotenoids (Cars), Mn, Ca, Fe, plastoquinones (PQs), which give rise to a total molecular mass of 350 kDa (Shen, 2015; Shen et al., 2008). Among the protein subunits, the extrinsic proteins contribute to maintaining the stability and activity of the oxygen-evolving complex; they bind weakly to the PSII intrinsic proteins and are therefore easily lost during purification (Shen et al., 1992; Shen & Inoue, 1993). In the following sections, we describe the purification procedure for obtaining PSII dimers with a high oxygen-evolving activity that has been used to obtain high-quality crystals and provide a wealth of insight.

2. Growth and harvest of cells

A small-scale cell culture (50 mL) of *T. vulcanus* is grown in a sterilized tube in a growth chamber (LH-410S, NK System, Japan) at 50°C under a light intensity of 20–30 μmol photons m^2 s^{-1}, provided either by LED light with wavelengths dominating at the region for plant growth or with white fluorescent light. The growth medium used is described in Shen, Kawakami, and Koike (2011), but the BG11 medium can be used equally effectively. The medium needs to be changed every 1–2 weeks before the cells become too dense, since a high cell density affects the viability of the cells. For a large-scale culture, 1 L of the mature cells are diluted in a tank containing 50 L of the medium maintained at 50°C, with illumination from two sides and continuous, strong stirring with a water circulating pump. Alternatively, the culture can also be divided into several 10-L bottles or 5-L flasks, and incubated in the same growth chamber (LH-410S, NK System, Japan). In either case, the cells are grown for 7–10 days under continuous illumination, with the light intensity adjusted to 20–30 μmol photons m^2 s^{-1} at the beginning and gradually increased to a final intensity of 100–150 μmol photons m^{-2} s^{-1} 1–2 days before harvest. To accelerate the growth of the cells, the liquid culture is bubbled with CO_2-enriched air containing 3%–5% (v/v) CO_2. Following the growth, cells are collected by centrifugation at 8000 × *g* for 5 min, or with a continuous centrifugation rotor. Typically, 50–60 g (wet weight) of cells are obtained from a culture of 50 L for 7–10 days. The harvested cells are washed with 1.0 L of 40 m*M* KH_2PO_4–KOH (pH 6.8) (buffer-A) to replace the culture medium, and resuspended in 1.0 L of the same buffer.

3. Isolation of thylakoid membranes

Cells are broken by a combination of enzyme treatment and freeze–thawing to isolate the thylakoid membranes. First, the cells in buffer-A are changed to 1.0 L of buffer-B containing 40 m*M* KH_2PO_4–KOH (pH 6.8) and 0.4 mannitol by centrifugation, and transferred to a 2.0-L flask, to which EDTA, 6-aminocaproic acid, and benzamidine are added to give final concentrations of 1.0 m*M* in each component, from their stock solutions. Subsequently, 200 μL of DNase I (1.0 mg/mL) and 1.4 g lysozyme are added to the flask and mixed gently. The flask is then incubated at 38°C for 90 min with gentle shaking in the dark. Following the lysozyme treatment, the cells are precipitated by centrifugation (8000 × *g*, 10 min, 4°C), and, unless otherwise stated, the following procedures are performed on ice or at low temperature to obtain the thylakoid membranes.

The precipitate from the centrifugation described above is suspended in a small amount of buffer-C containing 30 m*M* HEPES–NaOH (pH 7.0), 10 m*M* $MgCl_2$, 25% (w/v) glycerol, and the suspension is stored at −80°C. The frozen cells are thawed in a water bath kept at 20–25°C for 30–40 min to disrupt the cells to be used for the subsequent isolation of thylakoid membranes and crude PSII particles. The thylakoid membranes are collected by centrifugation at 10,000 × *g* for 10 min, resuspended in buffer-C, and used immediately for the subsequent isolation of crude PSII particles or stored at −80°C until use.

4. Isolation of crude PSII particles

To obtain pure PSII dimers, the thylakoids are first solubilized with a detergent, *N*,*N*-dimethyldodecylamine-*N*-oxide (LDAO), to remove a large amount of phycobiliproteins, PSI, and other contaminating proteins. At this step, PSII is selectively solubilized from the membrane by appropriate concentrations of LDAO, so that most of PSI are not solubilized and can be removed as precipitates by centrifugation. All of the procedures are performed on ice or at 4°C under dim light conditions to preserve the intactness of the PSII complexes and avoid possible photo damage.

The thylakoid membranes obtained from the previous step are washed with buffer-C and resuspended with buffer-D containing 30 m*M* HEPES–NaOH, pH 7.0, 10 m*M* $MgCl_2$ to remove phycobiliproteins and other soluble proteins. Since large amounts of phycobilisomes are dissociated from the membrane when the cells are disrupted, they should be removed as far as possible by the wash step (multiple washing steps may be needed if a large amount of phycobilisomes remains). The concentration and total amount of Chl *a* of the thylakoid membranes are determined by extraction with 100% methanol based on the method of Porra, Thompson, and Kriedemann (1989).

From the thylakoid membranes obtained, aliquots containing 320 mg Chl *a* are washed with 1.0 L of buffer-D and separated by centrifugation (8000 × *g*, 10 min, 4°C). The precipitates are then resuspended with a small amount of buffer-E containing 30 m*M* HEPES–NaOH (pH 7.0), 10 m*M* $MgCl_2$, 5% (w/v) glycerol to a total weight of 130 g. Prior to the following procedures, 0.86 g LDAO solution (with a concentration of approximately 30%, from Sigma-Aldrich) is added to 30 g of fresh buffer-E and kept on ice. Subsequently, this buffer-E solution containing LDAO is mixed with the thylakoid membrane sample prepared above, and the mixture is stirred gently on ice for 5 min in the dark to allow solubilization. This step of LDAO treatment does not solubilize the thylakoid membranes but dissociates most of the phycobilisomes and other soluble proteins from the membrane. The mixture is subsequently centrifuged at 38,900 × *g* for 60 min at 4°C to remove phycobilisomes and other soluble proteins in the supernatant. If the color of the supernatant is green, there is a possibility that PSII is solubilized. In this case, the LDAO concentration should be reduced by 5%–15%.

The precipitated membranes obtained in the previous step are resuspended with buffer-C to a total weight of 170 g, to which 30 g of ice-cooled, premixed buffer-C containing 1.65 g of LDAO solution (concentration approximately 30%, Sigma-Aldrich) is added. The mixture is stirred gently on ice for 5 min in the dark to solubilize the membranes. This step of LDAO treatment selectively solubilizes the PSII cores from the membrane, and the solution is centrifuged at 100,000 × *g* for 60 min at 4°C to recover the PSII particles in the supernatant. The precipitates contain unsolubilized PSI and other membrane proteins, and are discarded. The supernatant is collected, then 50% (w/v) PEG1,500 is added with gentle mixing on ice to give a final concentration of 13% (w/v) PEG1,500, and the PSII particles are collected by centrifugation at 100,000 × *g* for 60 min at 4°C. The precipitate is suspended in a small amount of buffer-C and its Chl *a* concentration is determined by the method of Porra et al. (1989). Finally, the crude PSII particles are stored at −80°C or in liquid nitrogen until further usage.

The success of the above procedures for isolation of the crude PSII particles can be gauged by spectroscopy. Fig. 1 shows the absorption spectra of

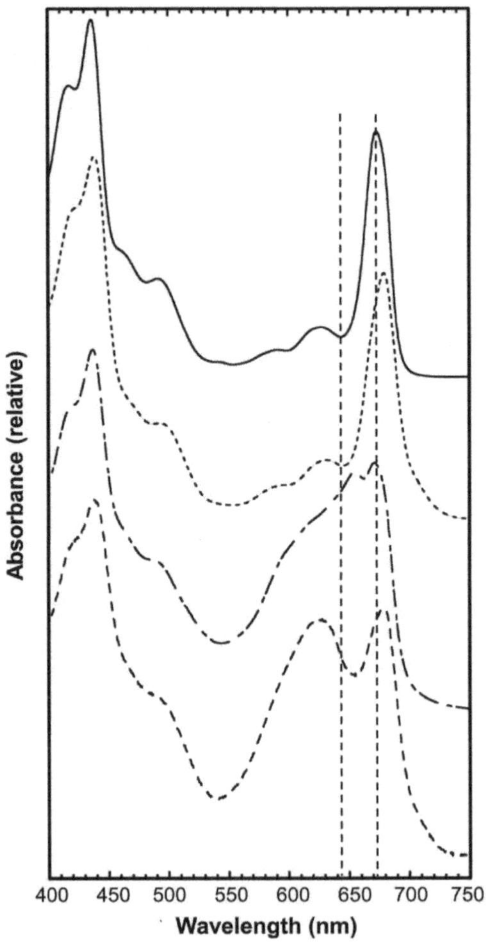

Fig. 1 Room-temperature absorption spectra of (from bottom to top) thylakoid membranes (*dashed line*), crude PSII particles (*dash-dotted line*), PSI trimers (*dotted line*), and the purified PSII dimers (*solid line*), respectively. The two vertical dashed lines indicate the trough (643 nm) and peak (673 nm) positions that are characteristic of purified PSII.

thylakoid membranes (dashed line), crude PSII particles (dash-dotted line), the PSI trimer (dotted line), and the purified PSII dimer (see later) (solid line), respectively. Since thylakoid membranes scatter light, the spectrum is measured by the opal glass method. The dominant Qy absorption peak of the crude PSII appears at 673 nm, similar to that of purified PSII dimers, whereas for those of PSI and thylakoid membranes it appears at 680 nm. If the PSII particles contain a large amount of PSI, the Qy absorption peak shifts to a longer wavelength. Other methods that can be used to assess the purity of the crude PSII particles are high-resolution clear native PAGE, blue native PAGE, 77 K fluorescence spectra, and SDS-PAGE (see later). Once optimal LDAO treatment is achieved, almost no bands derived from PSI trimer are detected. The amount of PSI in the sample can easily be estimated by measuring absorption and fluorescence spectra; however, these measurements cannot give an estimate regarding the amount of PSII monomers in the sample.

It should be noted that the amount of LDAO used for the second solubilization may vary depending on the sample conditions among different batches of the samples. Based on the amount of Chl *a* in the starting sample of thylakoid membranes (320 mg Chl *a*), the Chl *a* yield of the crude PSII particles obtained should be in the range of 10%–15%, and the oxygen-evolving activity of the sample should fall in the range 1000–2000 μmol O_2 mg Chl a^{-1} h^{-1}. A higher Chl *a* yield of the crude PSII particles (over 20%) suggests oversolubilization of the membrane and higher contamination by PSI; therefore, the amount of LDAO used for the second LDAO treatment should be reduced by 5%–15%. On the contrary, if the yield of the crude PSII particles is too low, PSII is not solubilized from the membrane effectively, so that the LDAO amount used for the second LDAO treatment should be increased by 5%–15% to achieve the optimal yield of crude PSII particles.

5. Purification of PSII dimers

Since the crude PSII particles still contain a large amount of contaminating components (mainly PSI and phycobiliproteins), it is necessary to perform further purifications in order to obtain pure PSII dimers capable of yielding high-quality crystals. This is carried out through an additional step of solubilization with another detergent, *n*-dodecyl-β-D-maltoside (β-DDM), followed by separation of PSII dimers from other components by anion exchange column chromatography. These procedures should be carried out on ice or at 4°C under dim light conditions.

The crude PSII particles in buffer-C are solubilized with 1.0% (w/v) β-DDM at a Chl *a* concentration of 1.0 mg/mL for 30 min at 4°C with gentle mixing. The solubilized sample is then diluted threefold with buffer-C, followed by addition of PEG1,500 from a 50% (w/v) stock solution to a final concentration of 15% (w/v), and the solution is stirred gently for 3 min on ice. Subsequently, the sample is centrifuged at 100,000 × *g* for 30 min at 4°C to collect the precipitate which is resuspended with a small amount of buffer-C. The solubilization step is repeated with 1.0% (w/v) β-DDM. These two steps of β-DDM solubilization appear to be helpful in increasing the yield of PSII dimers obtainable in the following purification steps.

Following the second step of β-DDM solubilization, the sample is centrifuged at 18,000 × *g* for 3–5 min at 4°C to remove a small amount of unsolubilized materials. The supernatant is then filtered with a 0.45- or 0.2-μm disc filter, and the sample is loaded onto a Q-Sepharose High-Performance column (GE Healthcare) preequilibrated with a column buffer containing 30 m*M* MES–NaOH, pH 6.0, 3.0 m*M* $CaCl_2$, 5% (w/v) glycerol, and 0.03% (w/v) β-DDM. The column is washed with 170 m*M* NaCl in the same column buffer (more than 10 column volumes), until the absorbance at 280 nm decreases to a sufficiently low constant level. The proteins eluted by this washing step are mainly PSI monomers and phycobilisomes. Following this washing step, the column is eluted with a linear gradient from 170 to 300 m*M* NaCl in the same column buffer at a flow rate of 2 mL/min. The elution pattern is shown in Fig. 2, where three main fractions are eluted in the order of PSII monomers, PSII dimers, and PSI trimers. The PSII dimer fraction is collected, diluted threefold with the column buffer, and then PEG1,500 is added from a 50% stock solution to a final concentration of 13% (w/v) with gentle stirring for 3 min on ice. Subsequently, the sample is centrifuged at 38,900 × *g* for 30 min at 4°C to collect the PSII dimers, and the precipitate is resuspended in a small amount of the supernatant. After further centrifugation at 38,900 × *g* for 30 min at 4°C, the precipitates are taken and resuspended with a small amount of buffer-F containing 20 m*M* MES–NaOH, pH 6.0, 3.0 m*M* $CaCl_2$, 20 m*M* NaCl, 5% (w/v) glycerol, to ensure that the Chl concentration of the final sample is around 8.0 mg Chl *a*/mL or higher. The Chl *a* concentration of the purified PSII dimers is determined using the method of Porra et al. (1989). Finally, the samples are stored under liquid nitrogen until further usage.

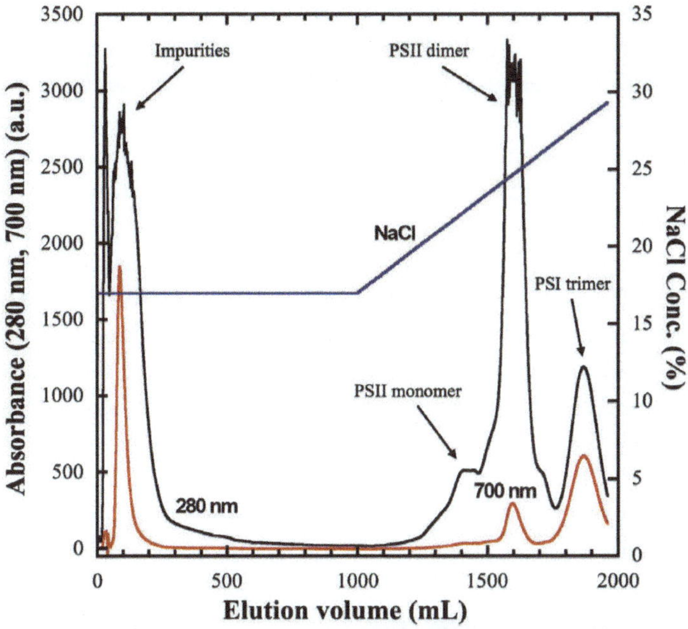

Fig. 2 Elution pattern of the β-DDM solubilized crude PSII particles from a Q-Sepharose-High Performance column. PSII monomers, PSII dimers, and PSI trimers are separated by a linear gradient of 170–300 m*M* NaCl (indicated by 17%–30% of 1 *M* NaCl in the right *Y*-axis). The elution was monitored at 280 nm (*black line*) and 700 nm (*red line*). Monitoring at 700 nm is useful for monitoring the PSI fraction because PSI absorbs at longer wavelengths than P700.

The typical yields of PSII dimers obtained using the above purification procedures range from 3% to 5% from the starting thylakoid membranes, based on the amount of Chl *a*. A lower yield may imply some failure in certain steps of the purification procedure, whereas a higher yield may imply the presence of impurities in the final sample.

6. Characterization of the purified PSII core dimers

The purity, homogeneity, and activity of the PSII samples can be assessed by absorption and 77 K fluorescence spectra, gel filtration chromatography, clear native PAGE, blue native PAGE, SDS-PAGE, etc. The room temperature absorption spectra of the purified PSII dimers are shown in Fig. 1, for comparison with those of the thylakoid membranes, crude PSII particles, and PSI trimers. The absorption band of the purified PSII dimers is located at 673 nm, in contrast to the 680 nm observed in the thylakoid membrane and PSI trimer samples. Furthermore, there is a deep trough at around 643 nm, indicating the complete removal of phycobiliproteins. For comparison, crude PSII particles show a strong absorption at around 643 nm, indicating a significant contamination by phycobiliproteins. A deep trough at 643 nm is also observed in the absorption spectrum of PSI trimers; however, its absorption is located at 680 nm.

The protein components and purity of the PSII dimers can be confirmed by SDS-PAGE analysis, as shown in Fig. 3, in which all the PSII components are present and virtually no other contaminating bands are observed except two weak bands from phycobiliproteins. The presence of the trace amounts of phycobiliproteins was proven not to affect the crystallization, and these bands are removed from the PSII dimers after crystallization (Shen & Kamiya, 2000). Fluorescence spectra measured at 77 K can also be used to monitor the purity of PSII, where two characteristic peaks (F685, F695)

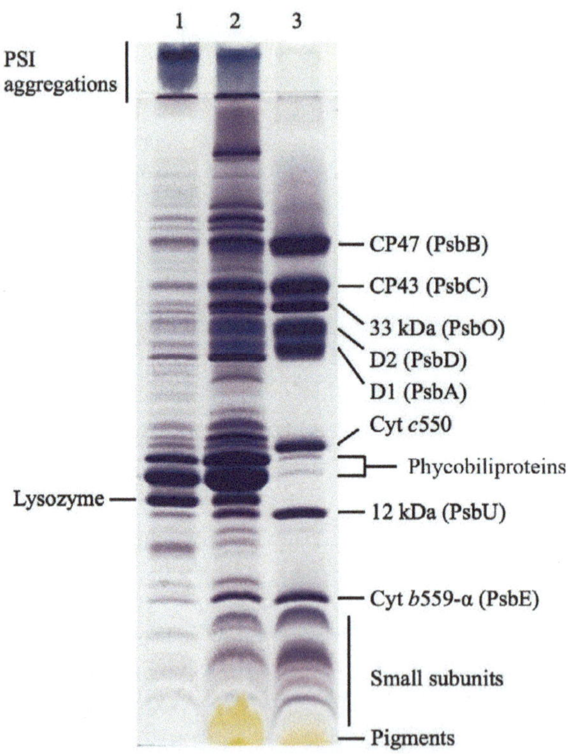

Fig. 3 Polypeptide compositions of thylakoid membranes (*lane 1*), crude PSII particles (*lane 2*), and the purified PSII dimers (*lane 3*) analyzed by SDS-PAGE.

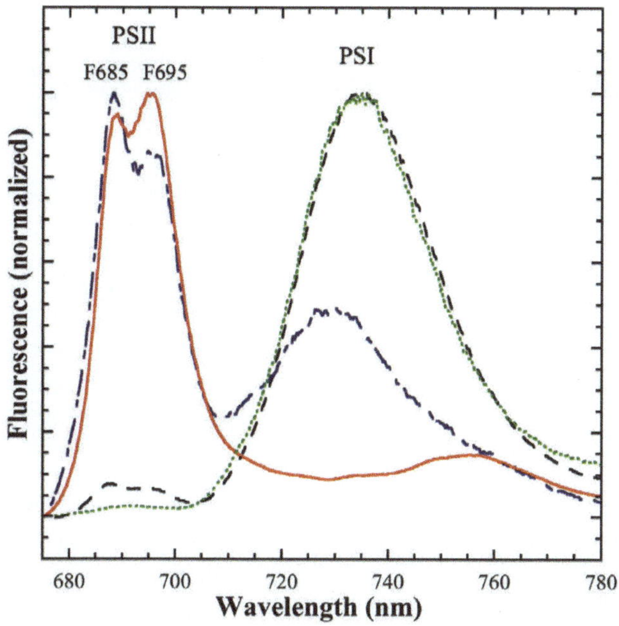

Fig. 4 Fluorescence emission spectra of thylakoid membranes (*dashed line*), crude PSII particles (*dash-dotted line*), PSI trimers (*dotted line*), and the purified PSII dimers (*solid line*), respectively, measured at 77 K.

originating from PSII are observed in the purified PSII dimers, and the characteristic peak of PSI at around 730–735 nm completely disappears (Fig. 4). On the other hand, a small peak at around 730 nm was still observed in the crude PSII particles purified after solubilization with LDAO, indicating contamination by PSI in the crude PSII particles.

Homogenous distribution of the protein particles in solution is another important factor for successful crystallization, in addition to the purity. The homogeneity of proteins in solution is often assessed by light-scattering measurements; however, in the case of membrane proteins and their complexes, light-scattering measurements may give results that represent a mixture of the proteins and the detergent micelles, which may make the results difficult to interpret. Since photosynthetic proteins such as PSII, PSI, and light-harvesting antenna proteins bind various pigments, their size distribution can be assessed by clear native PAGE and/or blue native PAGE (Wittig, Braun, & Schägger, 2006; Wittig, Karas, & Schägger, 2007). High-resolution clear native PAGE analysis shown in Fig. 5A demonstrated that the purified PSII dimers are devoid of PSI trimers, phycobiliproteins, and most of the PSII monomers.

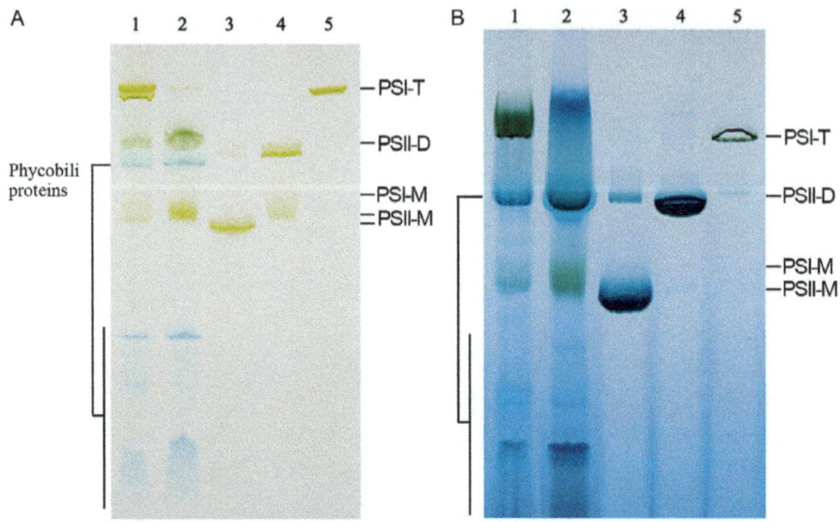

Fig. 5 Clear native PAGE (A) and blue native PAGE (B) analysis of thylakoid membranes (*lane 1*), crude PSII particles (*lane 2*), purified PSII monomers (*lane 3*), purified PSII dimers (*lane 4*), and PSI trimers (*lane 5*), respectively.

In the clear native PAGE, however, a weak green band is visible at the position equivalent to that of PSI monomer or PSII monomer, in the purified PSII dimers (Fig. 5A). Since there is no PSI contamination in the purified PSII dimer sample based on the SDS-PAGE analysis, this band is due to a partial decomposition of the PSII dimers during electrophoresis. In fact, blue native PAGE analysis of the same sample clearly showed that the PSII dimers are highly pure, and there are no PSII monomers present (Fig. 5B). The size distribution of the purified PSII dimers is further confirmed by gel filtration chromatography with a Superdex 200 10/30 column (GE Healthcare) (Fig. 6), which showed a single sharp peak, indicating that the PSII dimers are homogenous and there are no other contaminating components present in the preparation.

Finally, the purified PSII dimers typically exhibit an oxygen-evolving activity in the range of 3000–6000 μmol O_2 mg Chl a^{-1} h^{-1} measured with electron acceptors 0.4 m*M* phenyl-*p*-benzoquinone (PPBQ) and 1 m*M* potassium ferricyande (Fecy) (Table 1). A lower activity may reflect some damage to the PSII core during the purification procedures, although PSII dimers with an oxygen-evolving activity close to 2000 μmol O_2 mg Chl a^{-1} h^{-1} can also be used successfully for crystallization. The activity when measured with PPBQ alone is only slightly lower, but is remarkably lower when measured with Fecy alone, suggesting limited accessibility of Fecy

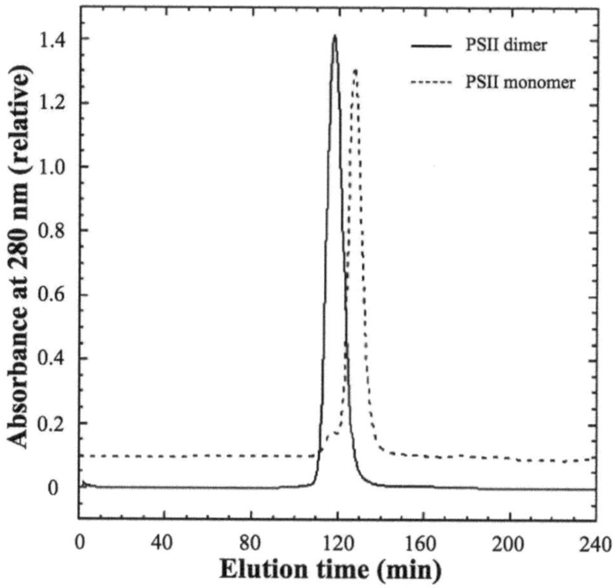

Fig. 6 Elution pattern of gel filtration chromatography of purified PSII monomers and dimers obtained with a prepacked Superdex 200 10/30 column (GE Healthcare).

Table 1 Oxygen-evolving activity of the PSII dimers purified from *T. culcanus*

	Oxygen-evolving activity (μmol O_2 mg Chl a^{-1} h^{-1})		
Electron acceptors	−DCMU	+DCMU (10 μ*M*)	Inhibition by DCMU (%)
PPBQ (0.4 m*M*)	3970 ± 260 (4)	315 ± 30 (3)	92
Fecy (1.0 m*M*)	1180 ± 20 (3)	620 ± 20 (3)	48
PPBQ (0.4 m*M*) + Fecy (1.0 m*M*)	5610 ± 270 (4)	600 ± 30 (3)	89

Numbers in parentheses indicate the number of independent measurements.

to the Q_B-binding site and thus an intact Q_B-binding site in the purified sample. This is supported by an almost complete inhibition of the activity by 3-(3,4-dichlorophenyl)-1, 1-dimethylurea (DCMU) with the acceptor PPBQ (Table 1).

7. Crystallization of purified PSII dimers

Using the purified PSII dimer sample, crystals can be grown with a hanging drop or sitting drop vapor diffusion or an oil batch method at

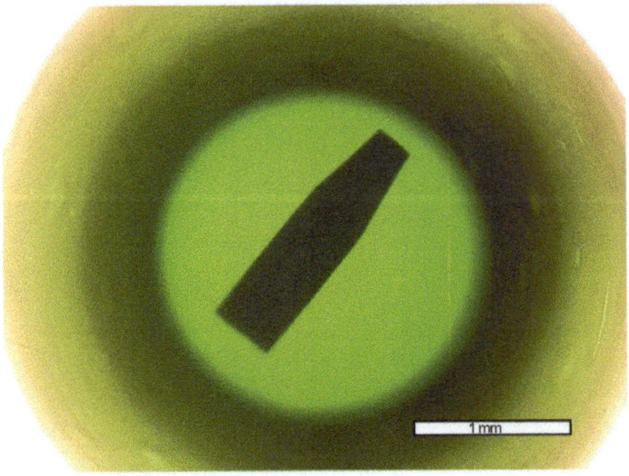

Fig. 7 A crystal obtained from the purified PSII dimers that diffracted to resolution above 2.0 Å. The crystal was grown at 12°C for 3–5 days and has dimensions of $1.5 \times 0.5 \times 0.4$ mm. The scale bar represents 1.0 mm.

4–20°C as described earlier (Kamiya & Shen, 2003; Shen & Kamiya, 2000; Umena et al., 2011). To improve the sample purity and homogeneity, and thus the crystal quality, a recrystallization step is introduced in which small crystals are grown first at a higher concentration of the precipitant for one night; the crystals thus obtained are collected, solubilized in the crystallization buffer without the precipitant, and then used for the second step crystallization at a lower concentration of the precipitant for 3–5 days. This recrystallization step removes some contaminating phycobiliproteins and PSII monomers present in the initial dimer sample (Shen & Kamiya, 2000), and results in high-quality crystals that can be used for structural analysis at a 1.9 Å resolution (Fig. 7) (Suga et al., 2015, 2017; Umena et al., 2011).

Acknowledgments

We thank Prof. Nobuo Kamiya and Ms. Rie Uno for their supports during the experiments.

References

Adachi, H., Umena, Y., Enami, I., Henmi, T., Kamiya, N., & Shen, J.-R. (2009). Towards structural elucidation of eukaryotic photosystem II: Purification, crystallization and preliminary X-ray diffraction analysis of photosystem II from a red alga. *Biochimica et Biophysica Acta—Bioenergetics*, *1787*, 121–128.

Berthold, D. A., Babcock, G. T., & Yocum, C. F. (1981). A highly resolved, oxygen-evolving photosystem II preparation from spinach thylakoid membranes. *FEBS Letters*, *134*, 231–234.

Broser, M., Gabdulkhakov, A., Kern, J., Guskov, A., Müh, F., Saenger, W., et al. (2010). Crystal structure of monomeric photosystem II from *Thermosynechococcus elongatus* at 3.6-Å resolution. *Journal of Biological Chemistry*, *285*(34), 26255–26262.

Enami, I., Kamino, K., Shen, J.-R., Satoh, K., & Katoh, S. (1989). Isolation and characterization of photosystem-II complexes which lack light-harvesting chlorophyll-a/b proteins but retain 3 extrinsic proteins related to oxygen evolution from spinach. *Biochimica et Biophysica Acta—Bioenergetics*, *977*, 33–39.

Enami, I., Okumura, A., Nagao, R., Suzuki, T., Iwai, M., & Shen, J.-R. (2008). Structures and functions of the extrinsic proteins of photosystem II from different species. *Photosynthesis Research*, *98*, 349–363.

Ferreira, K. N., Iverson, T. M., Maghlaoui, K., Barber, J., & Iwata, S. (2004). Architecture of the photosynthetic oxygen-evolving center. *Science*, *303*(5665), 1831–1838.

Ikeuchi, M., & Inoue, Y. (1986). Characterization of O_2 evolution by a wheat photosystem II reaction center complex isolated by a simplified method: Disjunction of secondary acceptor quinone and enhanced Ca^{2+} demand. *Archives of Biochemistry and Biophysics*, *247*, 97–107.

Kamiya, N., & Shen, J.-R. (2003). Crystal structure of oxygen-evolving photosystem II from *Thermosynechococcus vulcanus* at 3.7-Å resolution. *Proceedings of the National Academy of Sciences of the United States of America*, *100*(1), 98–103.

Kuhl, H., Kruip, J., Seidler, A., Krieger-Liszkay, A., Bunker, M., Bald, D., et al. (2000). Towards structural determination of the water-splitting enzyme. Purification, crystallization, and preliminary crystallographic studies of photosystem II from a thermophilic cyanobacterium. *Journal of Biological Chemistry*, *275*, 20652–20659.

Loll, B., Kern, J., Saenger, W., Zouni, A., & Biesiadka, J. (2005). Towards complete cofactor arrangement in the 3.0 Å resolution structure of photosystem II. *Nature*, *438*(7070), 1040–1044.

Nagao, R., Ishii, A., Tada, O., Suzuki, T., Dohmae, N., Okumura, A., et al. (2007). Isolation and characterization of oxygen-evolving thylakoid membranes and photosystem II particles from a marine diatom *Chaetoceros gracilis*. *Biochimica et Biophysica Acta—Bioenergetics*, *1767*, 1353–1362.

Okumura, A., Nagao, R., Suzuki, T., Yamagoe, S., Iwai, M., Nakazato, K., et al. (2008). A novel protein in photosystem II of a diatom *Chaetoceros gracilis* is one of the extrinsic proteins located on lumenal side and directly associates with PSII core components. *Biochimica et Biophysica Acta—Bioenergetics*, *1777*, 1545–1551.

Porra, R. J., Thompson, W. A., & Kriedemann, P. E. (1989). Determination of accurate extinction coefficients and simultaneous-equations for assaying chlorophyll a and chlorophyll b extracted with 4 different solvents—Verification of the concentration of chlorophyll standards by atomic absorption spectroscopy. *Biochimica et Biophysica Acta—Bioenergetics*, *975*, 384–394.

Shen, J.-R. (2015). Structure of photosystem II and the mechanism of water oxidation in photosynthesis. *Annual Review of Plant Biology*, *66*, 23–48.

Shen, J.-R., Henmi, T., & Kamiya, N. (2008). Structure and function of photosystem II. In F. Fromme (Ed.), *Photosynthetic protein complexes: A structural approach* (pp. 83–106). Wiley-VCH.

Shen, J.-R., Ikeuchi, M., & Inoue, Y. (1992). Stoichiometric association of extrinsic cytochrome *c*-550 and 12 kDa protein with a highly purified oxygen-evolving photosystem II core complex from *Synechococcus vulcanus*. *FEBS Letters*, *301*, 145–149.

Shen, J.-R., & Inoue, Y. (1993). Binding and functional properties of two new extrinsic components, cytochrome *c*-550 and a 12 kDa protein, in cyanobacterial photosystem II. *Biochemistry*, *32*, 1825–1832.

Shen, J.-R., & Kamiya, N. (2000). Crystallization and the crystal properties of the oxygen-evolving photosystem II from *Synechococcus vulcanus*. *Biochemistry*, *39*, 14739–14744.

Shen, J.-R., Kawakami, K., & Koike, H. (2011). Purification and crystallization of oxygen-evolving photosystem II core complex from thermophilic cyanobacteria. In R. Carpentier (Ed.), *Photosynthesis research protocols* (pp. 41–51). New York/ Dordrecht/ Heidelberg/London: Humana Press/Springer.
Suga, M., Akita, F., Hirata, K., Ueno, G., Murakami, H., Nakajima, Y., et al. (2015). Native structure of photosystem II at 1.95 Å resolution viewed by femtosecond X-ray pulses. *Nature, 517*, 99–103.
Suga, M., Akita, F., Sugahara, M., Kubo, M., Nakajima, Y., Nakane, T., et al. (2017). Light-induced structural changes and the site of O=O bond formation in PSII caught by XFEL. *Nature, 543*, 131–135.
Sugiura, M., & Inoue, Y. (1999). Highly purified thermo-stable oxygen-evolving photosystem II core complex from the thermophilic cyanobacterium *Synechococcus elongatus* having his-tagged CP43. *Plant and Cell Physiology, 40*, 1219–1231.
Sugiura, M., Inoue, Y., & Minagawa, J. (1998). Rapid and discrete isolation of oxygen-evolving His-tagged photosystem II core complex from *Chlamydomonas reinhardtii* by Ni^{2+} affinity column chromatography. *FEBS Letters, 426*(1), 140–144.
Suzuki, T., Minagawa, J., Tomo, T., Sonoike, K., Ohta, H., & Enami, I. (2003). Binding and functional properties of the extrinsic proteins in oxygen-evolving photosystem II particle from a green alga, *Chlamydomonas reinhardtii* having His-tagged CP47. *Plant and Cell Physiology, 44*, 76–84.
Suzuki, T., Tada, O., Makimura, M., Tohri, A., Ohta, H., Yamamoto, Y., et al. (2004). Isolation and characterization of oxygen-evolving photosystem II complexes retaining the PsbO, P and Q proteins from *Euglena gracilis*. *Plant and Cell Physiology, 45*, 1168–1175.
Umena, Y., Kawakami, K., Shen, J.-R., & Kamiya, N. (2011). Crystal structure of oxygen-evolving photosystem II at a resolution of 1.9 Å. *Nature, 473*, 55–60.
Wittig, I., Braun, H.-P., & Schägger, H. (2006). Blue native PAGE. *Nature Protocols, 1*, 418–428.
Wittig, I., Karas, M., & Schägger, H. (2007). High resolution clear native electrophoresis for in-gel functional assays and fluorescence studies of membrane protein complexes. *Molecular & Cellular Proteomics, 6*, 1215–1225.
Wydrzynski, T., & Satoh, K. (Eds.), (2006). In *Photosystem II: The light-driven water: Plastoquinone oxidoreductase*, . , Vol. 22. *Advances in photosynthesis and respirationSpringer.*
Zouni, A., Witt, H. T., Kern, J., Fromme, P., Krauss, N., Saenger, W., et al. (2001). Crystal structure of photosystem II from *Synechococcus elongatus* at 3.8 Å resolution. *Nature, 409*(6821), 739–743.

CHAPTER TWO

Production and manipulation of blue copper oxidases for technological applications

Simeng Zhou, Pierre Rousselot-Pailley, Lu Ren, Yolande Charmasson, Elise Courvoisier Dezord, Viviane Robert, Thierry Tron, Yasmina Mekmouche*

Aix Marseille Université, CNRS, Centrale Marseille, iSm2, Marseille, France
*Corresponding author: e-mail address: y.mekmouche@univ-amu.fr

Contents

Abstract

Fungal laccases are robust multicopper oxidoreductases. Perfectly amenable to synthetic evolution, the fungal laccase scaffold is a potential generic for the production of tailored biocatalysts, which, in principle, can be secreted at substantial levels in industrially relevant organisms. In this chapter, the strategy we have developed for the rapid production of hundreds of milligram of laccase variants is detailed. It is based on the use of two heterologous expression hosts: the yeast *Saccharomyces cerevisiae* for a rapid upstream screening and the fungus *Aspergillus niger* for downstream production.

Methods in Enzymology, Volume 613
ISSN 0076-6879
https://doi.org/10.1016/bs.mie.2018.10.015

Methods for screening active and nonactive laccase variants, convenient setups for enzyme production in both organisms as well as a methodology for efficient purification of large amounts of recombinant enzymes are given. The general procedure for developing new materials for artificial catalysis is also described.

Abbreviations

AA	amino acid
Ab-LAC	polyclonal anti-lac3 antibody
ABTS	2,2′-azino-bis(3-ethylbenzothiazoline-6-sulfonic acid)
ABTS$^{\bullet+}$	ABTS stable radical cation
AlP	alkaline phosphatase
ATA	aurin tricarboxylic acid ammonium salt
Bp	base pair
BSA	bovine serum albumin
DEAE	diethylaminoethanol
DNA	deoxyribonucleic acid
DTT	dithiothreitol
ELISA	enzyme-linked immunosorbent assay
FPLC	fast protein liquid chromatography
GRAS	generally regarded as safe
MWCN	multiwall carbon nanotube
MWCO	molecular weight cut off
PBST	phosphate buffer saline with tween 20
PCR	polymerase chain reaction
PEG	polyethylene glycol
PES	polyethersulfone
pNPP	4-nitrophenyl phosphate disodium salt hexahydrate
PPD	*p*-phenylenediamine
rpm	rotation per minute
RT	room temperature
SD	synthetic defined
SDS-PAGE	sodium dodecyl sulfate–polyacrylamide gel electrophoresis
SGZ	syringaldazine
SN	supernatant
sp	signal peptide
ss	single strand
STC	sodium tris chloride
TFA	trifluoroacetic acid
TFF	tangential flow filtration
TNC	trinuclear cluster
UF	ultrafiltration
UNIK	single lysine LAC3 variant
UV–vis	ultraviolet–visible
YPD	yeast extract-peptone-dextrose
ε	molar extinction coefficient

1. Introduction

New technologies for a more sustainable society are urgently needed (Kates, 2017). Considering the needed change of paradigm from fossil to renewable feedstock, a selective and efficient conversion of bioresources is crucial and will have a real impact on the sustainability of our developed society (Armstrong & Blundell, 2007). A number of current industrial processes feature catalytic methods for aerobic oxidation. However, their long-term use is condemned as reagents such as metal-oxide or chlorine-based oxidants are often required, generating organic and heavy metal wastes with safety concerns (Anastas & Warner, 1998). As an oxidant, dioxygen (O_2) is available at no cost and produces no environmentally hazardous by-products. Therefore, catalytic systems capable of selective oxidation reactions driven by clean dioxygen reduction with no by-products represent an important asset to progress toward sustainability.

Nowadays, the principle of using enzymes for large-scale industrial processes is well established (Kirk, Borchert, & Fuglsang, 2002; Meyer et al., 2013; Nestl, Hammer, Nebel, & Hauer, 2014). Limitations to the use of these excellent (enantio) selective and renewable catalysts, such as good operational stability or high production costs remain, but are progressively overcome thanks to the availability of constantly improving molecular evolution, fermentation, and immobilization methodologies.

Laccase (EC 1.10.3.2 from plant, fungi, or bacteria) is a particularly robust and well-studied biocatalyst. This copper-containing enzyme (four copper ions) couples the oxidation of a wide range of substrates to the neat reduction of one molecule of dioxygen to two water molecules. In the past 2 decades, a significant number of reports have illustrated the potential of this enzyme in technological and bioremediation processes (e.g., bleaching, clarification of beverages, detoxification, sensors) (Zucca, Cocco, Sollai, & Sanjust, 2016), focusing on its stability and efficiency (Alcalde, Ferrer, Plou, & Ballesteros, 2006; Couto & Toca-Herrera, 2006, 2007), as well as on its adaptability (plasticity) to different reaction conditions (e.g., tolerance to different organic solvents) (Kunamneni et al., 2008). Recent reviews address the use of laccases in these different fields (Cannatelli & Ragauskas, 2017; Mate & Alcalde, 2017) as well as in organic synthesis (Cracknell, Vincent, & Armstrong, 2008; Mikolasch & Schauer, 2009; Riva, 2006; Witayakran & Ragauskas, 2009) and in biofuel cells, in which they are wired on solid support and serve as the cathodic catalyst reducing dioxygen to water (Fernández-Fernández, Sanromán, & Moldes, 2013; Gutiérrez-Sánchez, Pita, Vaz-Domínguez, Shleev, & De Lacey, 2012; Jones & Solomon, 2015; Lalaoui et al., 2016; Le Goff, Holzinger, & Cosnier, 2015; Milton & Minteer, 2017; Pistone et al., 2016; Pita et al., 2011, 2014; Qiu et al., 2009; Shleev et al., 2005; Shleev, Pita, Yaropolov, Ruzgas, & Gorton, 2006). Recently, the potential of laccases for the development of molecular hybrid catalysts has been demonstrated (Lazarides et al., 2013; Mekmouche et al., 2015; Schneider et al., 2015; Simaan et al., 2011; Tarrago et al., 2018).

Genes encoding laccases from different origins have been heterologously expressed with various successes in bacteria, yeasts, filamentous ascomycetes, basidiomycetes, or plants reaching expression levels of up to a gram per liter (Alves et al., 2004; Kiiskinen et al., 2004; Record et al., 2002). Both variation of the cultivation conditions and mutagenesis in a given host can be used to tune production (Bulter et al., 2003; Mate et al., 2010; Rodgers et al., 2010). In addition, well-established protein engineering methodologies are available for modulating enzyme properties such as redox potential, pH tolerance, and thermostability (Mate & Alcade, 2015). However, it is noteworthy that a complex posttranslational maturation of the enzyme is influencing expression of laccases genes as well as the diversity obtainable among synthetic variants. The two copper-containing redox centers of laccases—a blue copper site located near the surface (oxidation of substrates) and an embedded trinuclear copper center (O_2 reduction) (Fig. 1) (Jones & Solomon, 2015; Riva, 2006; Solomon et al., 2014; Solomon, Szilagyi, DeBeer George, & Basumallick, 2004; Tron, 2013)—need to be precisely inserted within the three cupredoxin domains that are typical of proteins from the multicopper oxidase family. Moreover, fungal laccases are secreted, a process involving both enzymatic decorations of the enzyme surface with sugar moieties (glycosylation) and of the signal sequence by specific peptidase(s) during secretion. Thus, maximizing the production of fungal laccase variants requires a careful selection of the initial sequences, highly secreting hosts and a simple and straightforward selection procedure. Since the "perfect" host (allowing both selection and production at the same time) does not exist, best performances are obtained by splitting the process into two steps: (1) selection of variants (from sequences initially propagated in *Escherichia coli*) and (2) production of selected variants, involving each a specialized heterologous expression host (e.g., yeasts for step 1 and filamentous fungi for step 2).

We have forged our own experience in large-scale production of recombinant laccases from the study of isoenzymes originating from the fungus

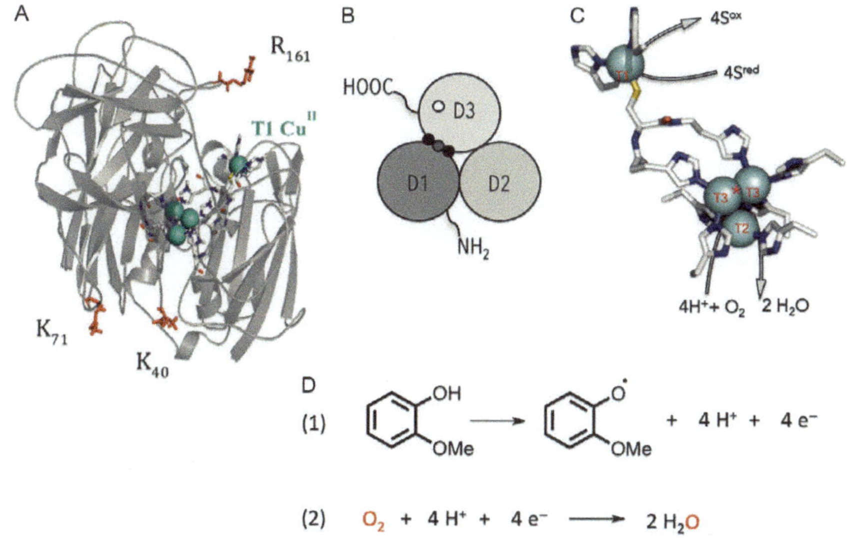

Fig. 1 Structure and function of laccase. (A) Structural model (*ribbon*) of LAC3 from *Trametes* sp. C30 with amino acids subjected to mutagenesis colored in *red*; Cu(II) ions are depicted as *blue spheres.* (B) Cartoon of the three cupredoxin domains (D1, D2, and D3) in the laccase structure; *white circle,* T1 copper ion; *gray circle,* T2 copper ion, *black circle,* T3 copper ions. (C) Close-up of the copper-active sites; *blue spheres*: copper ions annotated T1, T2, and T3. * This sign stands for the presence of a hidden histidine residue coordinated to the T3 copper site. (D) Overall outcome of the catalytic reaction.

Trametes sp. C30 (Robert, Mekmouche, Rousselot-Pailley, & Tron, 2011). The isoenzyme LAC3, which is the enzyme pivotal to all of our studies since 2005, has been selected from a functional screen of the yeast *Saccharomyces cerevisiae* transformants. LAC3, and several of its variants are produced in the yeast *S. cerevisiae* in the milligram per liter range (Balland et al., 2008; Cusano, Mekmouche, Meglecz, & Tron, 2009; Klonowska et al., 2005; Liu et al., 2014). Further, recombinant laccase production has recently been scaled up to levels of nearly $1\,g\,L^{-1}$, through the transformation and subsequent optimization of fermentation conditions in the GRAS ascomycete *Aspergillus niger* (Mekmouche et al., 2014). Altogether, these works highlight the potential of combining a generic laccase sequence with yeast and fungal expression systems for the generation and large-scale production of variants. Demonstrations of the practical use of this combination can be found in our most recent works on the production and use of robust laccase-based hybrids catalysts (Lalaoui et al., 2016; Mekmouche et al., 2015; Robert et al., 2017; Schneider et al., 2015).

In this chapter, we will detail our methodologies to produce large-scale synthetic laccases with desired properties for the purpose of developing new

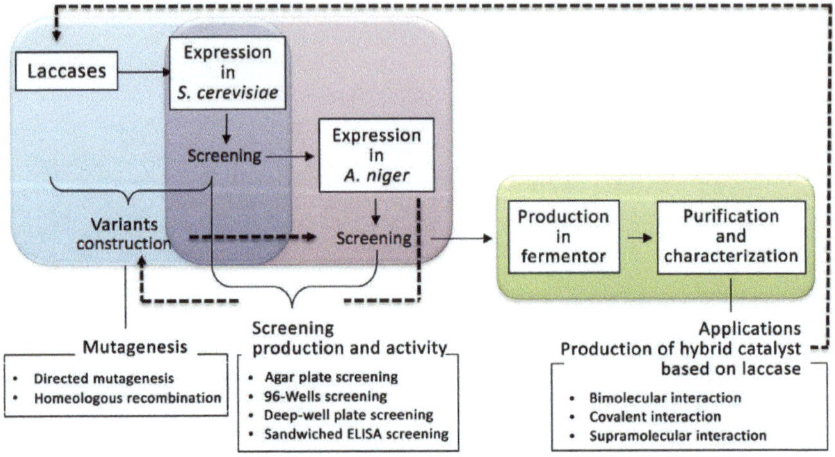

Fig. 2 General strategy for large-scale production of laccase synthetic variants covering the construction of variants (mutagenesis), the transformation an screening of laccase secreting transformants (screening production and activity) and finally the large-scale production, purification and targeted chemical functionalization (applications, production of hybrid catalyst based on laccase).

molecular hybrids for catalysis. The general strategy is depicted in Fig. 2. We will cover: (i) *the construction of variants* based on rational (structure guided) or random (molecular evolution) available mutagenesis methodologies; (ii) *the transformation procedures*; (iii) *the screening of laccase secreting transformants* directly on transformation plates containing a chromogenic substrate (e.g., guaiacol) or in various plate formats and, for nonfunctional variants (i.e., copper-depleted variants), with a sandwiched ELISA test; (iv) *the large-scale production* of laccase variants from reactor-optimized fermentation of *A. niger*; and (v) *the downstream purification* of laccase variants. We will then describe *selective chemical functionalization of the laccase surface* with synthetic effectors (transition metal complexes, macrocycles, organic probes, materials) via robust protein chemistry methodologies. We will discuss these different steps in depth by exploring the heterologous production of fungal laccases and their purification and characterization. Finally, we will provide an overview of possible applications.

2. Heterologous production of fungal laccases

Engineering of recombinant laccases for targeted oxidative transformations at desired conditions is a challenging issue.

Despite high production capabilities of fungal strains naturally producing laccases, heterologous expression of fungal laccase genes is popular because it offers possibilities: (i) to overcome the production of several isoenzymes resulting from the coexpression of multiple laccase-encoding genes usually found in parental organisms; (ii) to produce laccases from genes poorly expressed in parental organisms; (iii) to engineer a laccase by random or rational mutagenesis to improve its properties (e.g., stability vs pH, *T*°C, solvents). Recent reviews covering laccase engineering are available (Alcalde, 2015; Mate & Alcade, 2015; Pardo & Camarero, 2015; Rodgers et al., 2010).

As mentioned, our reference enzyme is LAC3, a laccase from *Trametes* sp. C 30. Cloning and heterologous expression of laccase *Trametes* sp. C 30 cDNAs has been undertaken to select enzymes with high oxidative capacities. In particular, *clac3* (GenBank accession number AY397783), which encodes a laccase isoform previously undetected in fungal cultures, has been isolated and successfully expressed in the yeast *S. cerevisiae* (Klonowska et al., 2005).

2.1 Cloning

Laccase-encoding sequences (cDNA, synthetic genes) are fused to bacteria/yeast shuttle vectors that are either commercially available (ex. Gateway™ pYES-DEST52 Vector) or custom (ex. pAK145, a constitutive yeast expression vector the construction of which is detailed in Klonowska et al., 2005) and are propagated in *E. coli* strain DH5α (F^- Δ80lacZ ΔM15 Δ(lacZYA-argF)U169 deoR recA1 endA1 hsdR17(rk^-, mk^+) phoA supE44 thi-1gyrA96 relA1 λ^-) using standard techniques (Sambrook, Fritsch, & Maniatis, 1989). Note that the secretion signal (encoding the signal peptide) to be fused to the sequence encoding the mature peptide is a matter of tuning. Heterologous peptides like the yeast alpha factor sp (Camarero et al., 2012; Mate et al., 2010) or that of the yeast invertase (Klonowska et al., 2005) are often chosen to replace the native laccase sp although in our hand the latter was found to lead to LAC3 secretion as well (Klonowska, unpublished results). In addition, improvements in the secretion (and therefore the production) of laccases have been linked to the accumulation of mutations in the sp during molecular evolution rounds (Bulter et al., 2003; Camarero et al., 2012; Mate et al., 2010). We sought both cloning methods to produce recombinant laccases and variants, i.e., a rational and a random method (Fig. 3).

2.1.1 Site-directed mutagenesis

Site-directed mutagenesis (Fig. 3A) is the basis for structure and function studies (Mate & Alcade, 2015). In this view, a 3D structure of the favorite

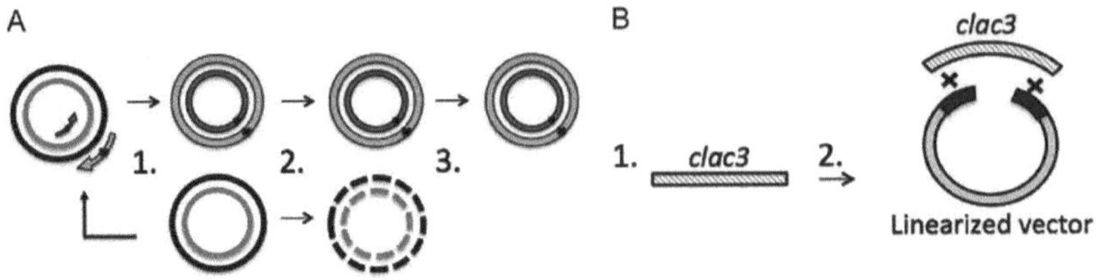

Fig. 3 Selected methods for producing variants of laccase. (A) Rational-directed mutagenesis: (1) mutant strand synthesis containing the targeted mutation; (2) digestion of template; and (3) transformation of mutated molecule into competent cells. (B) Gap repair technique: (1) gene of interest and (2) yeast transformed with vector and insert and homologous recombination to repair the gap.

enzyme or, failing that, a robust model (easily obtainable from web-based protein modeling platforms like SwissProt, PMP, Modeller, etc.) is highly recommended. Through site-directed mutagenesis, we have challenged the plasticity of our favorite enzyme (Klonowska et al., 2005; Liu et al., 2014), as well as its robustness in electrocatalysis (Balland et al., 2008; Lalaoui et al., 2016) or in chemo-enzymatic catalysis involving electron transfer from a transition metal catalyst (Mekmouche et al., 2015; Schneider et al., 2015) or from photosensitizers (Robert et al., 2017) (see Table 1). Replacement of selected sequences or codons in the *lac3* original coding sequence was obtained using a commercial kit (QuikChange®, Stratagene™). Mutated laccase sequences were constructed by designing the appropriate primers and following the recommendation of the manufacturer (Stratagene).

Recently, we started the construction of single lysine laccase variants of LAC3 with the intention of obtaining genetically defined, single surface-located, primary amino group variants amenable for posttranslationally site-directed chemical modification. A three-step semiempirical strategy was used in order to achieve the following aims: (i) To replace the only two lysine residues (K_{40} and K_{71}) naturally present in the mature sequence of LAC3 (Fig. 1A). Libraries of sequences coding K_{40}Q, M or I and K_{71}H, A, R, E (representing natural variations found at positions 40 and 71) were constructed from which the variant K_0 (K_{40}M, K_{71}H) devoid of lysine was selected: (ii) To replace a chosen surface accessible residue by a lysine, for example, in the UNIK_{161} (R_{161}K) variant (Lalaoui et al., 2016; Robert et al., 2017).

2.1.2 Yeast recombination

In vivo gene cloning or gap repair is a very efficient and fast method by which to assemble recombinant DNA in the yeast *S. cerevisiae* (Fig. 3B). This method produces, in a single step, propagatable circular DNA by homologous recombination between a linearized vector (gap) and a linear

Table 1 Variants of LAC3 engineered by site-directed mutagenesis

Study	Characterization	Main Results	References
Functional expression in yeast	E° T1; EPR spectroscopy; activity	Characterization of recombinant enzyme and comparison with homologous LAC1 and LAC2 from *Trametes* sp. C30	Klonowska et al. (2005)
Immobilization of LACCASE on electrode	Electrocatalysis; activity; spectroscopy	Characterization of a self-assembled monolayer immobilized laccase. Redox-mediated electrocatalysis	Balland et al. (2008)
Functional expression of C- and N-terminus modified laccases	Activity	Characterization and comparison of C-terminal modified laccases: depleted or/and His-tagged. Effect of glycosylation pattern on activity	Liu et al. (2014)
Oriented immobilization of LAC on electrodes to maximize direct electron transfer	Molecular biology; spectroscopy; activity; electrocatalysis	Efficient direct electron transfer for reduction of dioxygen using nanostructured electrodes modified with a site-specific surface modification laccase (single covalently bound pyrene group close to the T1 center)	Lalaoui et al. (2016)
Controlled orientation of a graft at the surface of a fungal laccase	Spectroscopy; activity; molecular biology; covalent docking; molecular dynamics simulations	Engineering of laccase variants with unique surface accessible lysine residues and orientated covalent grafting of a photosensitizer to promote photocatalysis. Simulation studies fit well with observed reactivity	Robert et al. (2017)

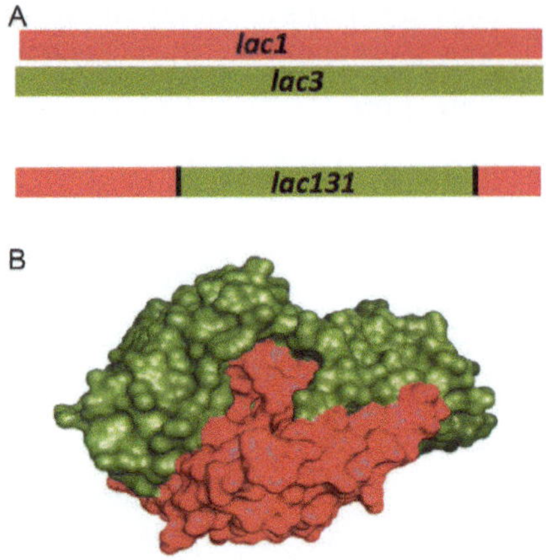

Fig. 4 Construction of laccase chimeras through yeast homologous recombination of *Trametes* sp. strain C30 laccase cDNAs sharing 65%–71% identity. (A) Schematic view of intermolecular recombination events between the *clac1* and *clac3* parental genes and the resulting hybrid *lac131* gene. Genes are represented by *rectangles* of variable lengths and recombinant junctions are represented by *vertical black bars*. (B) Molecular models of LAC131 hybrids constructed using the structure of the laccase 2HRG from *Trametes trogii* as template. The surface of LAC3 is colored in *green*; that of LAC1 is colored in *pink*. *Adapted from Cusano, A. M., Mekmouche, Y., Meglecz, E., & Tron, T. (2009). Plasticity of laccase generated by homeologous recombination in yeast.* FEBS Journal, 276, *5471–5480.*

overlapping DNA fragment (insert) cotransformed in yeast (Oldenburg, Vo, Michaelis, & Paddon, 1997; Orr-Weaver, Szostak, & Rothstein, 1983). The recombination occurs within short homologous sequences between the vector and the insert. Laccase-encoding sequences sharing 65%–71% identity (*clac1*, *clac2*, *clac3*, and *clac5*) have been shuffled in vivo by homologous recombination in yeast using a *clac3* PCR fragment as the repair sequence. In recombinant sequences, junctions have been found to map within short stretches of identity varying from 5 to 45 base pairs. Three chimeric laccases (LAC131, LAC232, and LAC535) have been isolated and characterized (Fig. 4). This single random mutagenesis step has led to the generation of variants having extended pH tolerance toward alkaline conditions (Cusano et al., 2009).

2.2 Transformations

2.2.1 S. cerevisiae

The yeast *S. cerevisiae* W303-1A (*MATa* {*leu2-3,112 trp1-1 can1-100 ura3-1 ade2-1 his3-11,15*}) is transformed with recombinant DNA materials using either a lithium acetate/ss DNA carrier/PEG protocol (Gietz & Woods,

2002) or electroporation (Becker & Guarente, 1991). After transformation, cells are spread on selective media and plates are incubated at 28°C for 4 days. Below is a brief description of: (i) the preparation of electrocompetent cells; (ii) electroporation; and (iii) plating.

Equipment

1. Temperature-controlled orbital shaker set at 30°C.
2. Stationary incubator at 30°C.
3. Electroporation cuvette (sterile).
4. Electroporator (e.g., Eppendorf Eporator®).
5. Floor centrifuge.
6. UV–vis spectrometer.

Buffers and reagents

Sterilization is performed by autoclaving unless otherwise specified. Standard recipes can be found in *Methods in Enzymology*.

1. YPD (1% yeast extract, 2% peptone, 2% dextrose) medium.
2. 1 *M* sorbitol in sterile distilled water.
3. Yeast strain (e.g., *S. cerevisiae* W303-1A).
4. Synthetic dextrose (SD) medium agar.

Example: YNB without AA, 0.67%; dextrose, 2%; adenine, 30 mg L^{-1}; casein hydrolysate, 0.5%; tryptophan, 40 mg L^{-1}; $CuSO_4$, 100 μ*M*; 50 m*M* succinate buffer pH 5.3; 15 g L^{-1} Agar. Sterile (filtered through Ø 0.2 μm filter) adenine, tryptophan, and guaiacol solutions are added postautoclave. Note that the composition of this medium has been optimized for the selection of uracyl prototrophs (*URA3* marker as part of the expression plasmid) secreting laccase.

Procedure

- Electrocompetent cells
  1. A fresh culture is prepared by growing 5 mL of *S. cerevisiae* strain in YPD in a 50 mL conical tube at 30°C, at 160 rpm overnight in order to reach an OD_{600} = 1.3–1.5. The preparation can be started directly from a plate.
  2. The cells are centrifuged at 1500 × *g* for 5 min at 4°C in a precooled rotor, the pellet is retained and the supernatant is discarded. This step allows for concentrating the cells.
  3. The pellet is resuspended with 5 mL of *ice-cold* sterile water, vortexing the cells if necessary.
  4. The cells are centrifuged as in step 2, the pellet is resuspended with 2.5 mL of *ice-cold* sterile water.
  5. The cells are centrifuged as in step 2, the pellet is resuspended in 1.5 mL of *ice-cold* 1 *M* sorbitol.

6. The cells are centrifuged as in step 2, then the pellet is resuspended in 0.1 mL of *ice-cold* 1 *M* sorbitol to a final volume of approximately 0.15 mL. *The cells are kept on ice* for immediate use, not stored.

- Electroporation
  7. Yeast suspension (50 μL) from step 6 is mixed with 2 μg of plasmid (in 5 μL sterile water) in a sterile Eppendorf tube.
  8. The mixture is transferred to an ice-cold 0.2 cm sterile electroporation cuvette.
  9. The cuvette is placed on ice for 5 min.
  10. The cells are pulsed according to the parameters for *S. cerevisiae* as suggested by the manufacturer of the specific electroporation device used.
  11. Ice-cold 1 *M* sorbitol (1 mL) is added immediately to the cuvette using a sterile pipette. After mixing gently, the cuvette contents are transferred to a 15 mL sterile tube.
  12. The tube is incubated at 30°C without shaking for 1 h.
- Plating
  13. Aliquots (200 μL) of the transformation (either LiAc or electroporation) are plated on SD plates. Transformants are selected for the appropriate prototrophy resulting from the complementation of one of the auxotrophies of *S. cerevisiae* W303-1A after transformation (corresponding to the selection marker present on the plasmid).
  14. Plates are incubated for 2–3 days at 30°C until colonies form.
  15. 10–20 colonies are chosen and streaked onto fresh SD plates. It is possible to screen for laccase activity/extracellular production by plating on an activity plate (i.e., containing guaiacol or ABTS), as will be further discussed in Section 2.3.2.

2.2.2 A. niger

Efficient expression in *A. niger* requires multiple integrations of the heterologous sequence to the genome. We are routinely using the strain D15#26 [(*pyrG*$^{-}$) (*prtT*) (*phmA*)] (Gordon et al., 2000) and plasmids pAN52-4 (EMBL accession number Z32750) for laccase sequence subcloning and pAB4-1 harboring pyrG as selection marker (van Hartingsveldt, Mattern, van Zeijl, Pouwels, & van den Hondel, 1987) all obtained from Peter Punt's laboratory. The following procedures have been adapted from previously published works (Mekmouche et al., 2014; Punt & van den Hondel, 1992; Record et al., 2002).

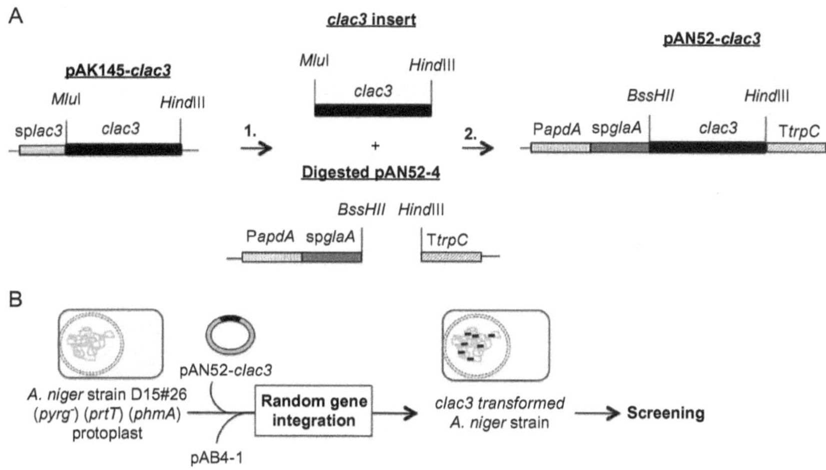

Fig. 5 (A) Description of the gene expression vectors for LAC3 plasmid pAK145-*clac3* for *S. cerevisiae* expression and plasmid pAN52-*clac3* for *A. niger* expression. The cloning restriction sites are annotated and signal peptides are indicated as sp (splac3 for pAK145-*clac3* and *spgla*A for pAN52-*clac3*). The constitutive promotor is the untranslated region P*gdp*A. The terminator gene is *trp*C. (B) Schematic description of pAN52-*clac3* and pAB4-1 transformation of *A. niger* [strain D15#26 ($pyrg^-$) (prtT) (phmA)]. (i) The host strain is sensitive to the antibiotic; (ii) preparation of protoplasts competent for DNA integration and subsequent regeneration; and (iii) selection procedures for optimal transformation frequencies. *From Punt, P. J., & van den Hondel, C.A. (1992). Transformation of filamentous fungi based on hygromycin B and phleomycin resistance markers.* Methods in Enzymology, 216, *447–457.*

a. *Construction of a pAN52-laccase expression plasmid*

The sequence encoding a laccase of interest (a *MluI/HindIII* PCR fragment encompassing the sequence coding for the mature LAC3 in our case) is easily inserted at the *BssHII/HindIII* sites of the pAN52-4 vector and is known as pAN52-*clac3* (Fig. 5). In the recombinant molecule, the sequence encoding the mature laccase is fused to a synthetic signal sequence (from the glucoamylase gene) and flanked by the 5′ untranslated region of the glyceraldehyde-3-phosphate dehydrogenase gene (constitutive promoter) and the terminator region of the *trpC* gene from *A. nidulans*. The plasmid is propagated in *E. coli* (ampicillin resistance) (Fig. 5).

Site-directed mutagenesis can be directly performed on this expression vector. K_0, a variant of LAC3 devoid of lysine was constructed starting from pAN52-*clac3* expression plasmid (unpublished).

b. *Transformation of A. niger*

A. niger D15#26 is transformed by the pAN52-*lac* plasmid along with the pAB4-1 plasmid that houses the auxotrophic marker *pyrG* (cotransformation) as described by Punt and van den Hondel (Punt & van den Hondel, 1992; van Hartingsveldt et al., 1987). Stable $pyrG^+$ transformants resulting from recombination at the pyrG locus are selected for uridine prototrophy. Cotransformants with multiple random integration of the laccase-encoding sequence in the *A. niger* genome are selected for laccase activity.

Equipment

1. Malassez cell counting chamber.
2. Microscope.
3. Temperature-controlled orbital shaker.
4. ∅90 mm Petri dish plates.

Buffers and reagents

1. *A. niger* [strain D15#26 ($pyrg^-$) (prtT) (phmA)].
2. Trace elements (1000×) solution: 76 m*M* $ZnSO_4$, 178 m*M* H_3BO_3, 25 m*M* $MnCl_2$, 18 m*M* $FeSO_4$, 7.1 m*M* $CoCl_2$, 6.4 m*M* $CuSO_4$, 6.2 m*M* Na_2MoO_4, 174 m*M* EDTA.
3. Complete medium: 70 m*M* $NaNO_3$; 3.7 m*M* KCl, 11 m*M* KH_2HPO_4, 2 m*M* $MgSO_4$, glucose 1% (w/v), trace elements (1×), 10 m*M* uridine; 2 gL^{-1} casaminoacid; 5 gL^{-1} yeast extract.
4. Solid minimum medium: 70 m*M* $NaNO_2$, 3.7 m*M* KCl, 11 m*M* KH_2HPO_4, 2 m*M* $MgSO_4$, glucose 1% (w/v), trace elements (1×) 15 gL^{-1} agar. Make the final solution 50 mgL^{-1} $CuSO_4$, $5H_2O$, and 0.2 m*M* ABTS from sterile stock solutions prior to gelification (hardening) if a laccase activity detection is desired.
5. Vector pAB4-1 containing the pyrG selection marker.
6. STC buffer: 1.2 *M* sorbitol; 10 m*M* Tris–HCl, pH 7.5, 50 m*M* $CaCl_2 \cdot 2H_2O$, 35 m*M* NaCl.
7. $CaCl_2$ buffer: 0.27 *M* $CaCl_2 \cdot 2H_2O$; 0.6 *M* NaCl.
8. PEG buffer: 60% PEG 4000; 10 m*M* Tris–HCl, pH 7.5; 50 m*M* $CaCl_2 \cdot 2H_2O$.
9. pAN-52-*lac*.
10. Glucanex from Sigma (L1412).
11. ATA (aurin tricarboxylic acid ammonium salt, 10–20 m*M*, MERCK).

Trace elements stock solution: 76 m*M* $ZnSO_4$, 178 m*M* H_3BO_3, 25 m*M* $MnCl_2$, 18 m*M* $FeSO_4$, 7.1 m*M* $CoCl_2$, 6.4 m*M* $CuSO_4$, 6.2 m*M* Na_2MoO_4, 174 m*M* EDTA.

12. Minimum medium: 70 mM $NaNO_3$, 7 mM KCl, 200 mM Na_2HPO_4, 2 mM $MgSO_4$, 50 gL^{-1} glucose, 50 mgL^{-1} $CuSO_4$, $5H_2O$. Then add 1‰ of trace element stock solution and adjusted to pH 5 with a 1 M citric acid solution.
13. Tween 80–water: 10 drops of tween 80 in 10 mL of water.
14. Sterile 1 M citric acid.

Procedure

- Generation of protoplast

  Protoplasts generation is generally the basis for fungal transformation: it is followed by a PEG/$CaCl_2$-mediated DNA uptake.

  1. 250 mL of complete medium is inoculated in a 1 L conical flask at a final concentration of 2×10^6 spores mL^{-1} (described in Section 2.2.2, Mass production of spores) and incubated with shaking at 150 rpm (24 h) at 30°C.
  2. The spent medium is filtered through sterile miracloth. The collected mycelium is washed with $CaCl_2$ buffer, resuspended in $CaCl_2$ buffer (5 g mycelium in 50 mL buffer) containing Glucanex (500 mg)[a] and incubated at 30°C for 2 h in an orbital shaker (80 rpm) (Li, Tang, Lin, & Cai, 2017).
  3. The protoplast suspension is then filtered through sterile miracloth, gently resuspended in the same volume of cold STC buffer and centrifuged at $1026 \times g$ at 4°C for 15 min. Protoplasts should be visualized as a small white strip at the bottom of the tube.
  4. The supernatant is carefully decanted and the protoplasts are resuspended in 25 mL of cold STC buffer. The suspension is then centrifuged at $1478 \times g$ at 4°C for 10 min.
  5. After removing the supernatant the protoplasts are resuspended in 1 mL of cold STC buffer, and kept on ice. At this stage, the number of protoplasts (with an appropriate dilution) is estimated by counting them with a Malassez cell counting chamber under a microscope. The count should be between 10×10^7 and 10×10^8 protoplast mL^{-1}.

- Transformation of protoplasts

  6. In a cell culture tube, 1 μL ATA, 10 μg of pAN-52-lac, 2 μg of pAB4-1 and 150 μL of protoplasts are mixed well and incubated for 30 min at RT.

[a] The enzyme mix used for cell wall digestion originally used in the reference publication is no longer available. Satisfactory results have been obtained with the Glucanex preparation from Sigma that is therefore used as replacement.

7. PEG buffer (850 μL) is added, mixed well, and the suspension is incubated at RT for 20 min.
8. After adding 10 mL STC buffer, the suspension is centrifuged at 1864 × *g* at 4°C for 10 min and the supernatant is removed.
9. The protoplasts are resuspended in 1 mL of STC buffer.

- Plating and selection

10. About 50 μL of solution from step 9 are plated on solid selective minimum medium (∅ 90 mm Petri dish plates) and incubated for 4 days at 30°C. Note that it is possible (and easy) to screen for laccase-secreting transformants already at this stage by plating transformants on ABTS- containing plates. See Section 2.3.2 for further details.

- Mass production of spores

It is necessary to obtain spores (from the parental strain or from transformants) for conservation (storage) as well as for the initiation of further steps of screening and enzyme production.

Equipment

1. Stationary incubator set at 30°C.
2. Large (∅ 150 mm) Petri dish plates.
3. Miracloth filter.
4. Malassez cell counting chamber.
5. Microscope.

Buffers and reagents

1. Solid minimum medium: the same composition as above is used for transformants selection.
2. 500 mL physiological water supplemented with 10 drops of tween 80.
3. 1 L of sterile water in a 1 L Erlenmeyer flask.

Procedure

1. Minimum medium agar plates are prepared. About 15 large plates should be necessary for generating enough spores to inoculate 8 L of culture in a bioreactor.
2. Plating is carried out, using about 50 μL of spores stock solution per plate, and these are incubated for 10 days at 30°C to allow vegetative growth and a subsequent sporulation. A 1-L Erlenmeyer flask full of sterile water is placed in a top bench incubator chamber to maintain a humid atmosphere and prevent plates from drying out over time.
3. After 10 days, the dark spores are collected with 10 mL of saline/tween 80 solution per plate and filtered through a sterile film (Miracloth) to eliminate traces of agar and mycelium debris.

4. A few microliters of a known dilution of the sample is introduced to a Malassez cell counting chamber, and the number of spores is estimated under a microscope.
5. The spores can be stored in 15% of glycerol saline solution and kept at −80°C until further use.

2.3 Screening for laccase secreting transformants in *S. cerevisiae* and *A. niger*

Clone-to-clone variations in expression and/or secretion levels are high both in *S. cerevisiae* and in *A. niger*. It is therefore important to reach transformation frequencies allowing screening on a sufficient number of clones (typically a few tens to several hundred in the case of libraries) in order to select the best secretors. Laccase secretors are selected either for their ability to oxidize a laccase substrate (e.g., using an activity test with ABTS or guaiacol) or for their specific reaction with an antilaccase antibody (immunodetection is especially useful in the case of production of nonactive enzyme, such as a T1-depleted enzyme). Screening tests can be performed on activity plates (primary screening) and thereafter on liquid cultures of isolated clones for both organisms. Deep-well plates and microtiter plates are adapted to the cultivation of *A. niger* and *S. cerevisiae* transformants, respectively.

2.3.1 Screening of clones from deep-well or microtiter plate cultures

This is a convenient method for rapid screening of a large number of parallel liquid cultures of transformants from both *S. cerevisiae* and *A. niger*.

a. Cultivation of *S. cerevisiae* transformants in microtiter plate

Equipment

1. Floor centrifuge (equipped with microplate holder).
2. Rotary incubator.

Buffers and reagents

1. Sterile SD medium (see Section 2.3.1).

Procedure

1. A 96-well microtiter plate is filled with 300 μL of SD medium.
2. Each yeast colony is inoculated with a sterile toothpick.
3. Cells are grown for 3 days at 30°C.
4. An aliquot of 10 μL of this cell suspension is taken and inoculated 300 μL of fresh SD medium.
5. The culture is grown for 24 h at 30°C.

6. Cells are pelleted at 3000 rpm for 2 min, and the supernatant (SN) is retained in a clean microtiter for measuring activity and protein concentration (see Section 2.3.3).

b. Cultivation of *A. niger* transformants in deep-well plates

Equipment

1. Polypropylene 24 deep-well plates (Whatman INC, Piscataway, USA).
2. Breathable sterile sealing film (Breathseal, Greiner Bio-One).
3. Shaking incubator.
4. Malassez cell counting chamber.
5. Microscope.

Buffers and reagents

1. Trace elements stock solution (see Section 2.2.2, part b).
2. Minimum medium: 70 mM $NaNO_3$, 7 mM KCl, 200 mM Na_2HPO_4, 2 mM $MgSO_4$, 50 g L^{-1} glucose, 50 mg L^{-1} $CuSO_4$, $5H_2O$. Then add 1‰ of trace element stock solution adjusted to pH 5 with a 1 M citric acid solution.
3. Tween 80–water: 10 drops of tween 80 in 10 mL of water.
4. Sterile 1 M citric acid.

Procedure

1. Samples of stored spores, prepared according to the protocol of Section 2.2.2 (mass production of spores), are thawed.
2. Each well of a 24 deep-well plate (containing 3 mL of sterile minimum medium) is inoculated at final concentration of 1×10^6 spores mL^{-1}, and 100 μL of sterile tween 80–water is added to each well to avoid adhesion of spores.
3. The deep-well plate is covered with a breathable sterile sealing film.
4. Incubation is carried out for 6 days at 30°C with orbital shaking set at 200 rpm. The pH is adjusted daily to 5.0 with 1 M citric acid. White balls of mycelium start to be visible after 3 days of culture (Fig. 6).
5. At the end of the culture, 1 mL of culture medium is filtered and analyzed for activity of production.
6. All cultures and analysis are carried out in triplicate.

2.3.2 Screening of laccase producing transformants on petri dish plates

Activity plate screening is based on the oxidation of chromogenic phenolic or nonphenolic laccase substrate. A colored halo (brownish and greenish for guaiacol and ABTS oxidation, respectively) developing with time around isolated colony attests for the secretion of active laccases (Fig. 7).

Fig. 6 Deep-well plate culture of laccase producing *A. niger* transformants. (A) View of the breathable film covering the plate during cultivation. (B) Aspect of a 3 mL culture of *A. niger* in minimum medium supplemented with glucose and oligoelements after 6 days of growth at 30°C. Each well corresponds to the growth of a single transformant. The presence of *white balls* attests for the development of the fungus.

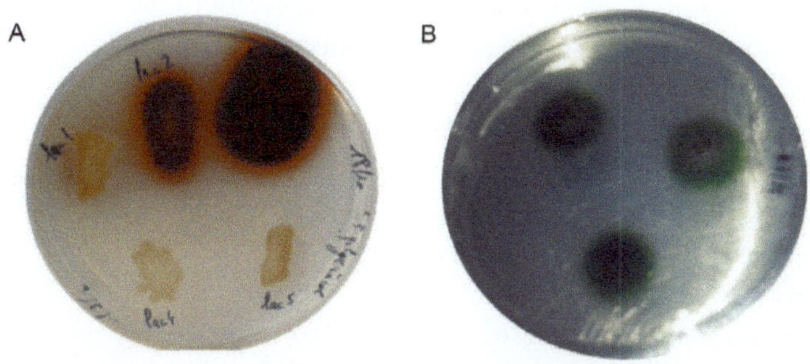

Fig. 7 Activity plates: guaiacol oxidation by yeast laccase producing variants (A) and ABTS oxidation by *A. niger* laccase producing variants (B).

This procedure is a particularly powerful primary screen applicable to a large number of clones at the same time directly on transformation plates, as the chromogenic substrates included in the selective media (ABTS or guaiacol) is nontoxic to transformants. Moreover, for colonies of similar size, both the extent of the halo and the color depth are indicative of differences in secretion levels and/or in specific activities (library screening).

Equipment

1. Stationary incubator set at 30°C.
2. Regular Petri dish plates.

Buffers and reagents

1. Solid minimum medium for *A. niger*. Same composition as above for transformants selection. Make the solution 50 mg L^{-1} $CuSO_4$, $5H_2O$, and 0.2 m*M* ABTS final from sterile stock solutions prior to gelification (hardening).
2. Solid SD medium for *S. cerevisiae*: YNB without AA, 0.67%; galactose, 2%; adenine, 30 mg L^{-1}; casein hydrolysate, 0.5%; tryptophan, 40 mg L^{-1}; $CuSO_4$, 100 μ*M*; 15 g L^{-1} agar; 50 m*M* succinate buffer pH 5.3. Add 500 μL L^{-1} guaiacol prior to gelification (hardening).

Procedure

1. Activity ABTS minimum medium plate is used for the selection of *A. niger* transformants. About 50 μL of transformants solution are applied per plate, followed by incubation for 4 days at 30°C. Development of a green halo around a colony, visible to the eye, attests for active laccase secretion.
2. SD medium plates are used for the selection of *S. cerevisiae* transformants. 200 μL of transformants are spread per plate, which are incubated for 2–3 days at 30°C until colonies form. Development of a green halo around a colony attests for active laccase secretion.

2.3.3 Screening of activity on 96-well plate

96-Well ABTS plate screening is an easy way to follow laccase activity.

Equipment

1. Three 96-well microtiter plates.
2. Microtiter plate reader.

Buffers and reagents

1. Buffer A: ABTS solution (5 m*M* final) freshly prepared in any buffer compatible with laccase activity and ABTS stability. We routinely use 100 m*M* acetate buffer pH 5.7 or Britton and Robinson buffer for extended pH studies.

Procedure

1. The ABTS activity assay is based on the rate of production of the chromophoricradical cation $ABTS^{\bullet +}$ at 414 nm ($\varepsilon = 3.5 \times 10^4\ L\ mol^{-1}\ cm^{-1}$ at pH 6.0).
2. The enzyme concentration is determined by the method of Bradford colorimetric assay on microtiter plates (protocol from furnisher).

Generally, Bradford assays are performed on 150 μL of nondiluted SN. The laccase protein concentration is estimated from a standard curve obtained from BSA solution (bovine serum albumin from 0 to 80 μg mL^{-1}).

3. Supernatants are diluted with buffer A (dilution factors are SN/20 and SN/40).
4. 20 μL of SN at an appropriate dilution is added to 180 μL of buffer A solution to initiate the reaction (final volume 200 μL). The specific activity of the laccase enzyme is calculated from the initial rate of the absorbance increase at 414 nm within 5 s of shaking, at 30°C. Kinetics are followed for 1 h within 30 s intervals with a microtiter plate reader (Fig. 8).
5. The specific activity of laccase toward ABTS is described as the number of micromoles of ABTS oxidized per minute per milligram of enzyme.

Note

1. This test is perfectly amenable for determining the enzymatic parameters of pure laccase as previously exemplified (Cusano et al., 2009).

2.3.4 ELISA screening

Sandwich ELISA, a variant of ELISA, is a highly sensitive technique for sample antigen detection. It allows quantification of the antigen between two layers of antibody molecules: the capture layer and the detection layer. An immobilized polyclonal anti-LAC3 antibody (multiepitope recognition) is used to capture the antigen. The same polyclonal biotinylated anti-LAC3

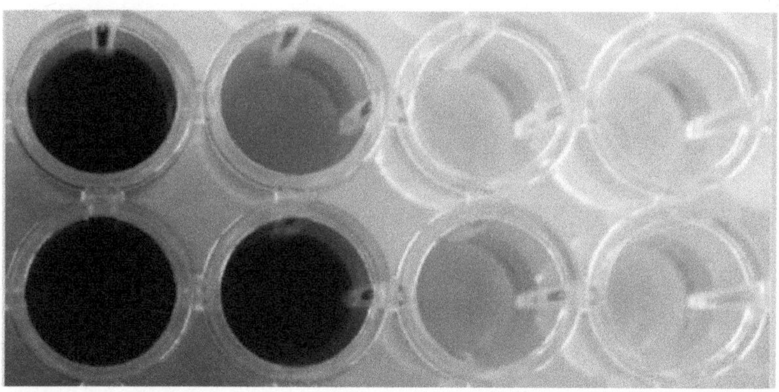

Fig. 8 Microtiter plate assay for laccase-mediated ABTS oxidation. The *green color* attests for laccase activity. *Adapted from Cusano, A. M., Mekmouche, Y., Meglecz, E., & Tron, T. (2009). Plasticity of laccase generated by homeologous recombination in yeast.* FEBS Journal, 276, *5471–5480.*

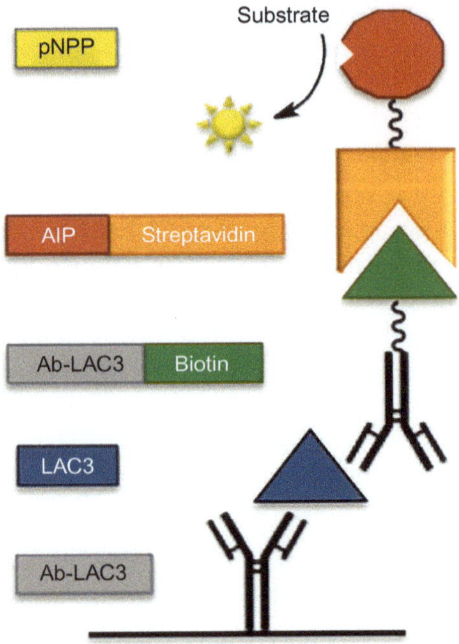

Fig. 9 Scheme of the "Sandwich ELISA" technique. The coated polyclonal antibody anti-LAC3 (Ab-LAC3) captures the antigen (LAC3). The secondary antibody, the same polyclonal biotinylated anti-LAC3 antibody (Ab-LAC3-biotin) serves for detection. The detection is made possible with the enzyme conjugate AlP-Streptavidin. The substrate pNPP (4-nitrophenyl phosphate disodium salt hexahydrate) is hydrolyzed into a colorimetric product detected at 405 nm.

antibody serves for the detection (Fig. 9). This method is particularly suited to the detection of low-activity or inactive laccase variants. The advantage of this method is that the antigen does not need to be purified before analysis.

Equipment

1. 96-Well microplates coated for ELISA.
2. Any microplate reader.
3. Multichannel precision pipettes.

Buffers and reagents

1. Coating buffer: 0.2 *M* sodium carbonate/bicarbonate, pH 9.4.
2. Capture antibody: antilaccase antibody Ab-LAC (Thermo Fisher) diluted in coating buffer.
3. PBST washing buffer: 0.1 *M* phosphate, 0.15 *M* sodium chloride, pH 7.2 containing 0.05% tween 20.
4. Blocking buffer: 2% (w/v) BSA in PBST washing buffer.
5. Standard dilution buffer: purified laccase diluted in blocking buffer.

6. Detection antibody: biotinylated polyclonal Ac-LAC antibody.
7. Biotin-secondary antibody.
8. Enzyme conjugate: Streptavidin Alkaline Phosphatase (Jackson Immuno Research).
9. Substrate: pNPP (Sigma S0942).
10. Glycine buffer: 0.1 *M* glycine, pH 10.4, with 1 m*M* $MgCl_2$ and 1 m*M* $ZnCl_2$.

Procedures

The steps are as follows:

- Antibodies preparation (capture and detection)
  1. Purification of Ab-LAC: following procedure from manufacturer (Thermo Scientific 89952).
  2. Biotinylation of Ab-LAC: following the protocol of a Biotin Protein Labeling Kit (Sigma Aldrich).
  3. Dilute to $2\,\mathrm{mg\,mL^{-1}}$, adding an equal volume of glycerol (final concentration: 50%), and store at −20°C.
- Surface preparation
  4. The capture antibody is diluted (20,000 ×) in coating buffer.
  5. 100 μL of diluted capture antibody is added to each well; the microplates are then covered and incubated overnight at 4°C.
  6. Discarding the solution, the plates are washed three times with 200 μL per well of PBST washing buffer.
  7. 200 μL of blocking buffer is added per well, the plates are then covered and incubated for 1 h at RT.
- Antigen capture
  8. The blocking buffer is discarded and 100 μL of sample or standards (100–2000 pg) is added. The plates are incubated for 1 h at RT.
  9. The solution is discarded and the plate washed three times with 200 μL per well of PBST washing buffer.
  10. The biotinylated detection antibody is diluted 5000-fold in blocking buffer.
  11. To each well is added 100 μL of diluted biotinylated detection antibody, the plate is then covered and incubated for 1 h at RT.
  12. The solution is discarded and the plate is washed five times with 200 μL per well of PBST washing buffer.
- Detection and quantification
  13. The enzyme conjugate AlP-Streptavidin is diluted 1000-fold.
  14. 100 μL of the diluted AlP-Streptavidin is added to the plate, which is then covered and incubated for 1 h at RT.

15. After removing the solution, the plate is washed seven times with 200 μL per well of PBST washing buffer.
16. The substrate solution is prepared: one tablet of pNPP (5 mg) is used in 10 mL of glycine buffer.
17. The plate is incubated at RT until the desired color intensity is reached. Ideally, a clear gradient is obtained with the variable concentrations of standard used.
18. The absorbance at 405 nm is recorded with a plate reader to estimate the amount of bound antigen from the standard curve established from different concentrations of pure LAC3 antigen.

2.4 Fermentation in *A. niger*

As stated earlier, the production of laccase and variants is carried out in 5 or 10 L bioreactors following production procedures already described (Mekmouche et al., 2014; Record et al., 2002). Prior to the fermentation, it is necessary to produce a sufficient stock of spore allowing an inoculum of 2×10^6 spores mL^{-1} for 10 L culture: this is described below.

2.4.1 Preparation of spore inoculum and storage

See Section 2.2.2.

A stock of stored spores is thawed to prepare the inoculum, and as many plates as necessary for a 10 L culture are prepared. Usually, 15 plates are necessary to inoculate a 10 L fermentor.

2.4.2 Production in bioreactor and growth parameters

The initial growth parameters are presented in Table 2. Parameters such as agitation or addition of antifoam are adjusted daily. Specific activity, O_2 consumption, and protease activity are measured routinely (Mekmouche et al., 2014).

Growth conditions in a 10 L bioreactor have been described previously (Mekmouche et al., 2014). The recombinant enzyme is the major secretion

Table 2 Typical growth parameters for heterologous expression of fungal laccase in *A. niger*

Parameters	Aeration (VVM)	Inoculum Size Spore (mL^{-1})	pH	Glucose ($g L^{-1}$)	Rotation (rpm)
	1.2	2×10^6	5.5	50	200–600

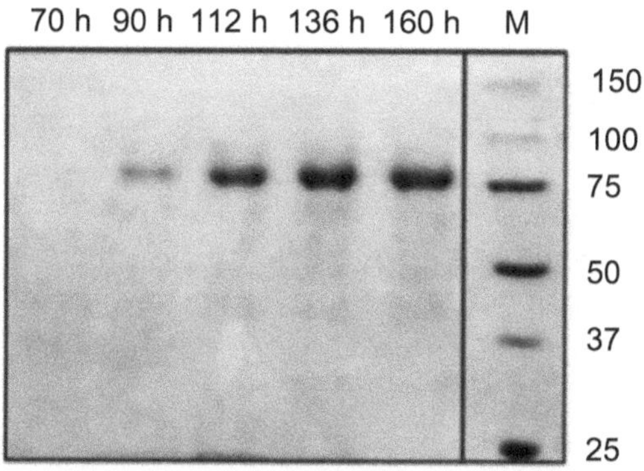

Fig. 10 Time course of laccase production during up-scaled fermentation. SDS-PAGE (7.5%) is stained with Coomassie blue. 10 mL of culture supernatant were loaded per well as a function of cultivation time (from 0 to 184 h); *M*: molecular weight markers in kDa. *Adapted from Mekmouche, Y., Zhou, S., Cusano, A. M., Record, E., Lomascolo, A., Robert, V., et al. (2014). Gram-scale production of a basidiomycetous laccase in Aspergillus niger.* Journal of Bioscience and Bioengineering, 117, *25–27 with permission.*

product: it is therefore possible to follow the time course of production of the enzyme simply by loading an SDS-PAGE with extracellular culture medium (Fig. 10).

3. General purification procedure and characterization

Purification of laccase variants can be obtained easily following a protocol adapted from Cusano et al. (2009) based on four steps: (1) filtration to eliminate the mycelium, (2) concentration, (3) capture (anionic exchange), and (4) polishing (gel filtration), altogether allowing for a satisfactory level of homogeneity (Fig. 11).

3.1 Purification procedure

3.1.1 Assay procedures

Although several substrates are regularly used for laccase activity determination, we will here describe only the syringaldazine (SGZ) colorimetric assay for volumetric activity determination and an electrophoretic colorimetric assay based on phenylenediamine (PPD) oxidation (zymogram).

a. SGZ assay procedure

Equipments

1. UV–vis spectrophotometer.
2. Disposable cuvettes of 1.5 mL.

Fig. 11 Initial steps for the purification of recombinant laccases produced in *A. niger*. (A) Setting for the filtration of the cultivation medium through a Separ (SEFAR MEDIFAB 30-170/54) filter allowing for the elimination of the mycellium; (B) tangential ultrafiltration and concentration of 10 L of cultivation medium (UF cartridge regenerated cellulose cutoff 30 kDa from Millipore); and (C) NaCl gradient elution of laccase on a DEAE column ($2.5 \times 25\ cm^2$) loaded with the ultrafiltrate.

Buffers and reagents

1. Syringaldazine stock solution: $8\,\text{mg}\,\text{mL}^{-1}$ in MeOH.
2. Buffer A: 100 m*M* sodium acetate pH 5.7.
3. Bradford reagent (Millipore).

Buffers and reagents

1. The SGZ activity assay is based on the rate of production of the corresponding quinone at 525 nm ($\varepsilon = 65{,}000\,\text{L}\,\text{mol}^{-1}\,\text{L}\,\text{cm}^{-1}$) (Bauer & Rupe, 1971).
2. The laccase concentration is determined by two colorimetric methods. The first method is the Bradford colorimetric assay (Bradford, 1976) and the second one is based on oxidized T1 absorption (when enzyme is purified) at 610 nm ($\varepsilon = 5600\,\text{L}\,\text{mol}^{-1}\,\text{L}\,\text{cm}^{-1}$) (Simaan et al., 2011).
3. For routine assay a working concentration of 88 μ*M* of syringaldazine is used. Laccase is added at an appropriate concentration to buffer A solution in the disposable cuvette to a final volume of 970 μL.
4. The addition of 30 μL of SGZ initiates the reaction. The volumetric activity of the laccase enzyme is calculated from the initial rate measurement of the absorbance increase at 525 nm, followed over the time course of 2 min.
5. The specific activity of laccase toward SGZ is described as the number of micromoles of SGZ oxidized per minute per milligram of the laccase.

b. Zymogram assay procedure

It may be convenient to visualize laccase distribution in protein samples throughout purification. Beside (or in addition to) protein staining and immunodetection, laccase conversion of chromogenic substances to colored compounds is sensitive and quick. Laccase activity is easily detectable *in gel* after protein separation on a regular PAGE-SDS gel, providing any denaturation of the protein sample is avoided prior to loading the gel (i.e., no reducing agent is present in the loading buffer and no heat treatment has been used).

Equipment

1. SDS-PAGE gel (e.g., 7.5%).
2. Electrophoresis apparatus.

Buffers and reagents

1. Modified Laemmli sample buffer: 4% SDS, 20% glycerol, 0.004% bromophenol blue, and 0.125 *M* Tris–HCl, pH approx. 6.8.
2. Buffer A: 20 m*M* acetate buffer pH 3.6.
3. *p*-Phenylenediamine.
4. Buffer B: 100 m*M* phosphate buffer pH 7.5.

Procedure

1. Protein samples are diluted in modified Laemmli sample buffer (no reducing agent) and then loaded onto a PAGE SDS: it is important not to heat the protein sample.
2. The PAGE-SDS gel is run under appropriate conditions.
3. The gel is soaked in buffer A for 10 min.
4. The gel is rinsed three times with buffer A.
5. The gel is soaked in a freshly prepared solution of PPD (2 mg mL^{-1} in buffer A) until color is developed. The reaction is quenched by soaking the gel in buffer B.
6. Rinse the gel three times with buffer B.
7. Extensively wash with water prior to (short-term) storage.

3.1.2 Laccase purification procedure

Equipment

1. Vacuum filtration setting (e.g., 110 mm Büchner funnel + 5 L flask + vacuum pump).
2. Filters (SEFAR MEDIFAB 30-170/54).
3. High speed floor centrifuge, centrifuge tube.
4. Ultrafiltration apparatus (e.g., Prep/Scale Spiral-Wound Ultrafiltration Modules: TFF-2. MWCO 30 kDa 150 mL working volume).
5. High flow rate peristaltic pump (e.g., Cole Palmer Masterflex I/P Easy-Load Pump).
6. Empty glass gravity chromatographic column $2.5 \times 25\,cm^2$ (e.g., Econo-Column® Bio-Rad).
7. Size-exclusion chromatographic column (e.g., HiPrep™ 26/60 Sephacryl® S-300 HR, GE Healthcare).
8. Centrifugal concentrator 30,000 MWCO PES (e.g., Vivaspin 20 Sartorius).
9. Stirred-cell ultrafiltration apparatus (e.g., Amicon) with 30,000 MWCO PES membranes.
10. UV–vis spectrophotometer.
11. FPLC apparatus equipped with double wavelength detection 280 and 610 nm (e.g., Äkta purifier 10, GE Healthcare).

Buffers and reagents

1. Buffer C: 20 m*M* phosphate buffer pH 6.0.
2. Buffer D: 20 m*M* phosphate buffer pH 6.0, 300 m*M* NaCl.
3. Buffer E: 20 m*M* phosphate buffer pH 6.0, 200 m*M* NaCl.
4. Syringaldazine stock solution: 8 mg mL^{-1} in MeOH.

5. ABTS solution: buffer containing 5 m*M* ABTS, freshly prepared. It is possible to use any buffer compatible with laccase activity and ABTS stability. We generally use 100 m*M* acetate buffer pH 5.7 or Britton and Robinson buffer for pH studies.

Procedure

1. Filter the *A. niger* dark brown fermentation medium (typically 10 L) on a Büchner funnel equipped with a low porosity filter (SEFAR MEDIFAB 30-170/54) under vacuum. Wash the mycelium several times (three to five times) with a minimum volume of buffer A (for a total volume representing 10% of the initial cultivation volume).
2. Concentrate the filtered medium using an ultrafiltration apparatus equipped with a MWCO 30 kDa membrane down to the minimal volume (depends on apparatus and membrane specifications, usually ≈200 mL for a 2.3 sq. ft. membrane) at a flow rate compatible with the maximum operated pressure recommended by the membrane manufacturer (flow rate of ≈3 L min^{-1}). At this stage, most of the dark brown material goes to the permeate tank and the intensity of the blue-green color of the retentate visually translates laccase production.
3. Add 50 mL of DEAE-Sepharose resin previously equilibrated with buffer C to the retentate (batch process). Allow binding under slow magnetic stirring for 1 h at 4°C.
4. Pour the DEAE slurry into an empty glass chromatographic column $2.5 \times 25\ cm^2$ and pack it by gravity.
5. Wash the resin with 250 mL buffer C. Depending on the laccase variant the resin should be more or less intensely blue colored.
6. Connect the column to a low flow peristaltic pump and elute laccase by sharply raising the ionic strength of the buffer. Laccases derived from LAC3 produced in our laboratory elute usually with 50–75 mL of buffer D (containing 300 m*M* NaCl) at a flow rate of 2 mL min^{-1}. The presence of colorless laccase variants (e.g., delta T1) may be confirmed by immunodetection.
7. Protein concentration is evaluated by Bradford assay and laccase concentration is determined by UV–vis spectroscopy ($\varepsilon_{610nm} = 5600$ L mol^{-1} cm^{-1}).
8. If necessary the protein concentration of the sample is adjusted to maximize loading of size-exclusion chromatography.
9. An appropriate volume of sample is applied to the top of a preequilibrated (buffer E) $2.6 \times 60\ cm^2$ Sephacryl S-300 HR column (or equivalent) connected to a FPLC system. Gel filtration is carried

out at a flow rate of 1 mL min^{-1} (buffer E) collecting 5 mL fractions. Progress is followed with detectors set at 280 and 610 nm. Repeated injections are necessary given the size of the sample obtained after DEAE. Fractions with a ratio between $14 < \lambda_{280nm}/\lambda_{610nm} < 16$ and an activity within 20% of the peak fraction are pooled (Fig. 12).

10. Pooled fractions are exchanged with buffer A to remove NaCl using a 30-kDa cutoff ultrafiltration membrane in a stirred-cell filtration unit (e.g., Amicon). The sample is concentrated to 10 mg mL^{-1} and 10% of glycerol (v/v) is added to the solution.
11. The sample is filtered under a sterile atmosphere using a 0.2 μm filter; 500 μL is then dispensed to 1 mL aliquots in cryotubes which are then stored at −20°C for long-term storage.

Representative yields for several laccase variants are presented in Table 3. Yields and catalytic properties for LAC3 and UNIK variants (unique lysine variants) are twofold higher than those of the K_0 variant (no lysine). Eventually, yields are markedly higher than those obtained in *S. cerevisiae* (about 40 times for LAC3 heterologous production) (Table 3).

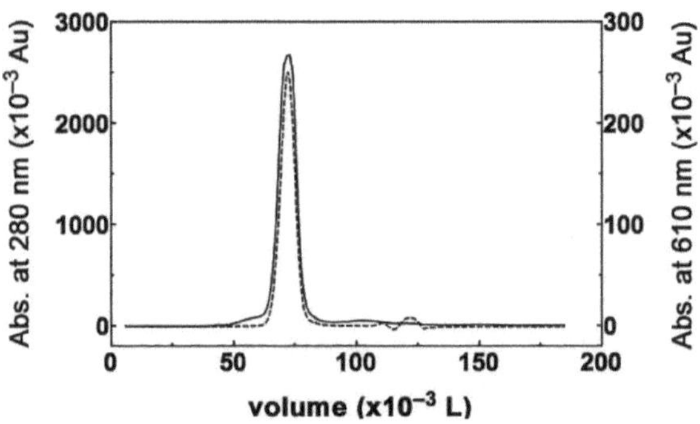

Fig. 12 Typical S300 chromatogram obtained for LAC3. Two wavelengths (280 nm *solid line*, 610 nm *dashed line*) are selected to follow the elution.

Table 3 Typical purification yields obtained for various LAC variants

	LAC3	$UNIK_{161}$	$UNIK_{157}$	K_0
Production	8 L	8 L	8 L	8 L
	40,000 U L^{-1}	28,000 U L^{-1}	25,000 U L^{-1}	8900 U L^{-1}
Purification	900 mg	700 mg	600 mg	400 mg
	170 U mg^{-1}	160 U mg^{-1}	160 U mg^{-1}	90 U mg^{-1}

Notes

1. Filtration of the cultivation medium may lead to a rapid browning of the solution. This is probably a manifestation of laccase activity consecutive to a reoxygenation of the solution during the elimination of the O_2-consuming *A. niger* mycelium.
2. After ultrafiltration (step 2), the volume of the retentate is the minimal one that can be obtained, as it is constrained by the volume of the cartridge and tubings. It is usually necessary to rinse the system one or two times with a minimal volume of 200 mL. This variation of the volume does not affect the capture of the enzyme on an anionic exchanger.
3. At each step of the purification process, a small aliquot is taken out to load on SDS-PAGE, measure the protein concentration (Bradford assay and T1 absorbance for the last step) and the specific activity.

4. Applications

Over the past few years, efforts have been made to investigate alternative routes for the oxidation of organic compounds and to perform selective catalytic chemical transformations. To date, these reactions are highly important at an industrial level and the current processes are among the most polluting ones. Appearing as highly promising sustainable alternatives to traditional catalysts, artificial enzymes often combine a transition metal catalyst moiety with a protein environment (Diéguez, Bäckvall, & Pàmies, 2018; Schwizer et al., 2018). Inspired from this new field, we recently engaged ourselves in studying interactions between laccase and devices or materials. We study three types of interactions: (i) bimolecular; (ii) oriented covalent; and (iii) oriented supramolecular (Fig. 13).

We have recently described the first examples of catalytic systems combining the O_2-reducing power of a laccase with the catalytic power of inorganic moieties represented either by a palladium-based water-soluble catalyst (aerobic oxidation of veratryl alcohol) or a ruthenium-based complex (photo-induced oxidation of styrene compounds) (Fig. 14). In both cases, substrate is oxidized under mild conditions by the inorganic moiety probably redox cycling, via the enzyme, with the four-electron reduction of dioxygen into water. In these hybrid systems, the enzyme and the inorganic catalysts form functional bimolecular systems and O_2 acts as a renewable, cheap, and clean oxidant.

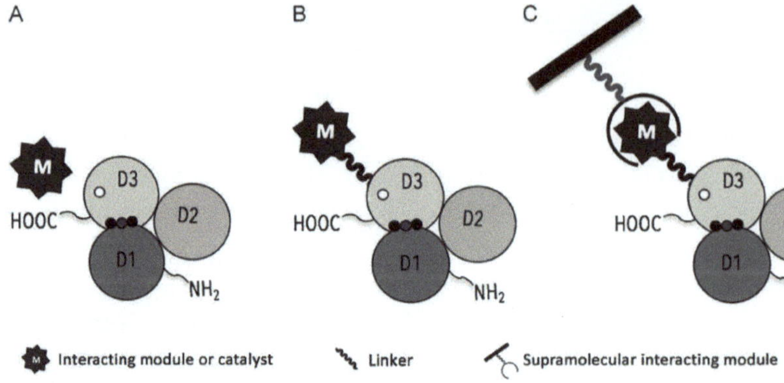

Fig. 13 General strategy for producing hybrid catalysts. Three cupredoxin domains (D1, D2, and D3) represent the laccase structure. *White circle*, T1 copper atom; *gray circle*, T2 copper atom; and *black circle*, T3 copper atoms. (A) Bimolecular interaction; (B) oriented covalent interaction; and (C) oriented supramolecular interaction between laccase and heterogeneous material.

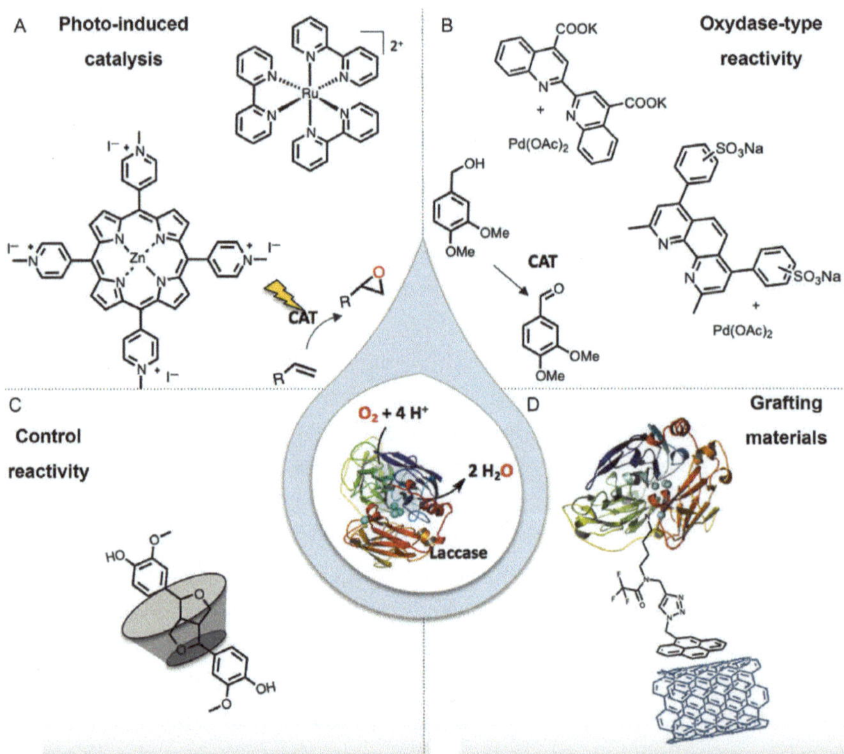

Fig. 14 Hybrid catalysts combining inorganic or organic moieties with the powerful oxidoreductase laccase. (A) Interaction of small photosensitizers with laccase to promote photo-induced oxidation of alkenes to epoxides; (B) laccase as a palladium oxidase. Primary alcohol oxidation mediated by the Pd/laccase/O_2 system under mild conditions; (C) controlled polymerization of coniferyl alcohol mediated by β-cyclodextrin with fungal laccase; and (D) supramolecular interaction of a pyrene-grafted laccase and MWCNT electrode to promote oriented direct electron transfer.

4.1 General procedure for an oriented covalent grafting of the laccase surface

4.1.1 Preparation of laccase variants for single-site anchoring

The well-known topography of laccases makes them easy material for surface-targeted modification. The design of variants for surface-targeted AA functionalization is addressed in Section 2.2.1. Briefly, a single lysine (K) residue is precisely positioned at the surface of a lysine-less variant through site-directed mutagenesis. The variants created this way are designed UNIK.

4.1.2 Reductive amination of lysine

Highly selective modifications of primary amino groups present at the surface of proteins can be obtained under mild conditions. The chemical modification of UNIK variants is adapted from the procedure described by McFarland and Francis (2005). The primary amino group of lysine present at the surface of UNIK is reacted with an aldehyde-containing molecule under reducing conditions. Two recent publications from our group describe a strategy for achieving a controlled orientation in the grafting of the surface of a fungal laccase (Lalaoui et al., 2016; Robert et al., 2017). The grafting procedure of two molecules is detailed hereafter. The first molecule is an original 2,2,2-trifluoro-*N*-(4-oxo-butyl)-*N*-(1-pyren-4-ylmethyl-1*H*-[1,2,3]triazol-4-ylmethyl)-acetamide reactant (compound **1**, Fig. 15) synthesized to favor the anchoring of laccase variants on carbon nanotubes containing electrodes through supramolecular interaction. This molecule features: (i) an aldehyde moiety for reductive amination and (ii) a pyrene moiety favoring interaction with the electrode material. Fig. 14A summarizes a schematic view of this multifunctional platform.

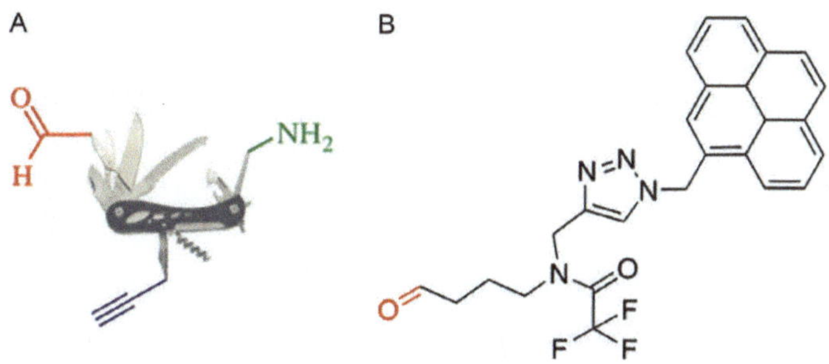

Fig. 15 (A) Artistic view of a multifunctional molecule as a graft for laccase. (B) Structure of the 2,2,2-trifluoro-*N*-(4-oxo-butyl)-*N*-(1-pyren-4-ylmethyl-1*H*-[1,2,3]triazol-4-ylmethyl)-acetamide reactant (compound **1**).

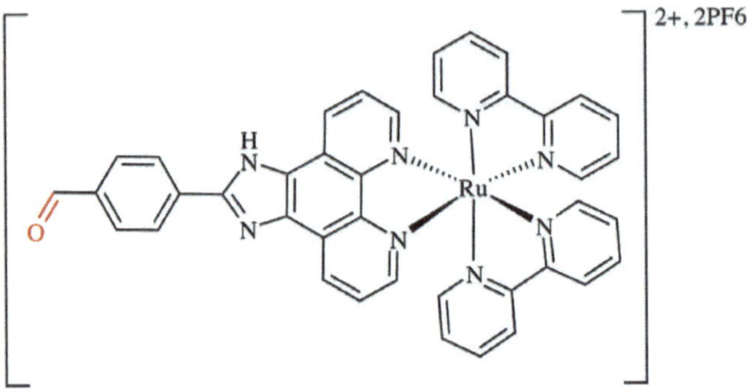

Fig. 16 Structure of compound **2** [(bpy)$_2$Ru(fimbzl)][PF$_6$]$_2$·2H$_2$O (fimbzl=4 ([1,10] phenanthroline[5,6-*d*]imidazol-2-yl)benzaldehyde).

The second molecule is an original heteroleptic ruthenium(II)–poly-pyridine-type complex with one of the bipyridine ligands containing a fused imidazole motif and a benzaldehyde (compound **2**, Fig. 16).

This complex has been selected for the following reasons: (i) its ability to photoreduce laccases upon visible light excitation (Schneider et al., 2015; Simaan et al., 2011); (ii) its intense absorption band in the visible spectrum at $\lambda_{max}=450\,nm$ ($\varepsilon_{450nm}=14{,}500\,L\,cm^{-1}\,mol^{-1}$) (Kalyanasundaram, 1982) which does not overlap the $\lambda_{max}=605\,nm$ of the T1 Cu(II) site of laccases; and (iii) its fluorescence properties ($\lambda_{ex}=450\,nm$, $\lambda_{em}=600\,nm$).

Equipment

1. $1.5\times150\,cm^2$ glass chromatography column.
2. Sephadex® G-25 fine (GE Healthcare).
3. Peristaltic pump (e.g., Gilson).
4. Centrifugeable ultrafiltration devices, 30,000 MWCO PES (e.g., Vivaspin 20, Sartorius).
5. Floor centrifuge and centrifuge tubes.
6. UV–vis spectrometer.

Buffers and reagents

1. Buffer A: 20 m*M* sodium phosphate buffer pH 6.0 containing 10% glycerol.
2. Buffer B: 25 m*M* sodium phosphate buffer pH 7.4 containing 100 m*M* sodium formate.
3. Buffer C: 50 m*M* sodium acetate buffer pH 5.5, 100 m*M* NaCl.
4. Buffer D: 50 m*M* sodium acetate buffer pH 5.5.
5. Custom aldehyde-donor molecule (e.g., compound **1** or compound **2**). Prepare a 3 m*M* stock solution in buffer A. This solution can contain up to 30% DMSO to solubilize the molecule of interest.

6. Water-stable $[Cp^{*}Ir\text{-}(bipy)(H_2O)]SO_4$ complex (Ir catalyst) synthesized as previously described (McFarland & Francis, 2005). Prepare a 14 m*M* stock solution in deionized water, aliquot, and store at −20°C.

Procedure

1. A 200 μ*M* stock solution of laccase is prepared in buffer B, the concentration being based on the oxidized T1 absorption at 610 nm.
2. In a 5-mL round bottom flask equipped with a Teflon magnetic stirring bar, 625 μL of laccase, 417 μL of the custom aldehyde-donor molecule, and 90 μL of Ir catalyst are combined, adjusting to a final volume of 2.5 mL with buffer A. The reaction is carried out for 72 h with permanent stirring at RT. The solution is then concentrated to 1 mL with a centrifugeable ultrafiltration device.
3. The 1 mL mixture is loaded onto a G-25 gel filtration column preequilibrated with buffer C (5 CV) for desalting and buffer exchange. Elution is carried out with a linear flow of 1 mL min^{-1} of buffer C, collecting 500 μL fractions. Fractions containing laccase are identified with a colorimetric assay (ABTS or SGZ oxidation), pooled, and concentrated to 250 μL with an ultracentrifugation device (Vivaspin 20, 4500 rpm).
4. The concentrated protein solution is exchanged with buffer D (to remove NaCl) prior to analysis.

Notes

1. During the grafting procedure, laccase is solubilized at neutral to basic pH. In these conditions, laccase is reversibly inactivated and easily reduced by the formate-containing solution as testified by the loss of the characteristic blue color of the oxidized T1 copper site. The G25 column allows for the elimination of the excess of unreacted aldehyde and reduction catalyst and thus a progressive reoxidation of laccase (reapparition of the blue color) can be observed as elution progresses.
2. For photosensitive molecules, the desalting column (G25) and the 5-mL balloon are covered with a piece of black plastic to avoid any light-induced processes.

4.1.3 Evaluation of the grafting

The efficiency of grafting can be evaluated simply by use of UV–vis spectrophotometry, providing the graft absorbs at a wavelength clearly distinct from the transition at λ_{max} = 605 nm characteristic of the T1 Cu(II) site of laccases. For example, compounds **1** and **2** have each an intense absorption band at, respectively, λ_{max} = 345 nm (ε_{350nm} = 37,500 L mol^{-1} cm^{-1}) and

$\lambda_{max}=450$ nm ($\varepsilon_{450nm}=14{,}500$ L mol^{-1} cm^{-1}). Therefore, the $A_{\lambda max}$ graft/$A_{\lambda max}$ laccase ratio reflects well the grafting efficiency. With transition metal-containing complexes (like compound **2**) the metal/Cu ratio in the grafted enzymes (therefore the efficiency of the grafting) can be deduced from inductively coupled plasma mass spectrometry (ICP–MS) analysis. Compounds **1** and **2** being also fluorescent molecules, their grafting to the laccase surface can be visualized after electrophoresis on an SDS-PAGE gel under appropriate irradiation. However, this method is not quantitative. Ultimately, confirmation of the grafting location can be obtained by mass spectrometry analysis of peptides produced by trypsin digestion of the grafted laccase and separated by HPLC (UV–vis or fluorescence detection). A potential grafting at the amino terminus can be evaluated by AA sequencing (Edman degradation).

Equipment

1. RP-C18 HPLC column (150/4) (e.g., Nucleosil Macherey Nagel).
2. Desalting columns (e.g., PD-10 Amersham).
3. HPLC equipped with UV/fluorescence detector.
4. UV–vis spectrometer.
5. 7.5% SDS-PAGE gel.
6. Electrophoresis apparatus.
7. Thermomixer (e.g., Eppendorf).
8. Digital gel imaging system (e.g., UVITEC Cambridge).

Buffers and reagents

1. Buffer A: Britton and Robinson buffer pH 5.0.
2. Buffer B: 250 m*M* Tris–HCl, 4 m*M* EDTA, 6 *M* guanidine, pH 8.5.
3. Dithiothreitol (60-fold molar excess over disulfide bridges) in buffer F.
4. 2-Iodoacetamide.
5. Buffer C: 125 m*M* of NH_4HCO_3 pH 8.0.
6. Trypsin.
7. HPLC solvent A: water 0.1% TFA.
8. HPLC solvent B: acetonitrile 0.1% TFA.

Procedure

- UV–vis spectrometry analysis
  1. Solutions of both unreacted and functionalized laccases are prepared in buffer A at a convenient concentration (≈50 μ*M*). The spectra of both laccase solutions (Fig. 17) are recorded. The number of grafted molecules per laccase can be deduced from the A_{345nm}/A_{605nm} and A_{450nm}/A_{605nm} ratios for compounds **1** (Fig. 15) and **2** (Fig. 16), respectively.

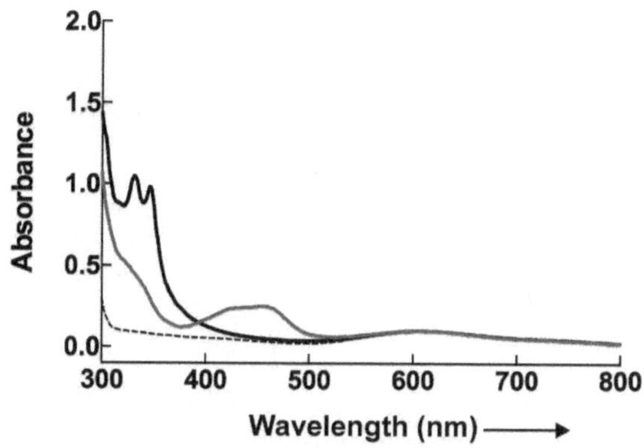

Fig. 17 UV–vis spectra of unreacted and grafted laccases. *Dashed line*, unreacted LAC3; *solid dark line*, UNIK grafted with compound **1**; and *solid gray line*, UNIK grafted with compound **2**. *Adapted from Lalaoui, N., Rousselot-Pailley, P., Robert, V., Mekmouche, Y., Villalonga, R., Holzinger, M., et al. (2016). Direct electron transfer between a site-specific pyrene-modified laccase and carbon nanotube/gold nanoparticle supramolecular assemblies for bioelectrocatalytic dioxygen reduction.* ACS Catalysis, 6, *1894–1900; Robert, V., Monza, E., Tarrago, L., Sancho, F., De Falco, A., Schneider, L., et al. (2017). Probing the surface of a laccase for clues towards the design of chemo-enzymatic catalysts.* ChemPlusChem, 82, *607–614.*

- SDS-PAGE analysis

  This method is qualitative and highlights the grafting of a fluorescent probe at the surface of the enzyme. Under the conditions of electrophoretic migration, unbound molecules will not follow the same profile as laccase because compound **1** is uncharged and compound **2** is positively charged.

  1. A 7% SDS-PAGE gel is prepared.
  2. Samples are heat denatured according to standard SDS-PAGE electrophoresis protocol.
  3. The gel is loaded with unreacted enzyme and grafted laccase samples (5 μg/well).
  4. After electrophoresis, the gel is examined under irradiation at the appropriate wavelength to excite the fluorophore. Thereafter, the gel is stained with Coomassie Blue and examined under white light irradiation. An example of analysis of samples subjected to reductive alkylation with compound **2** is given in Fig. 18.

- Mass spectrometry analysis

  In case the graft is a transition metal complex the exogenous metal (graft)-to-laccase ratios can be confirmed by analyzing the metal content of the grafted enzymes by ICP–MS. Eventually, the precise location of the graft at the surface of enzymes can be confirmed by MALDI-TOF-MS

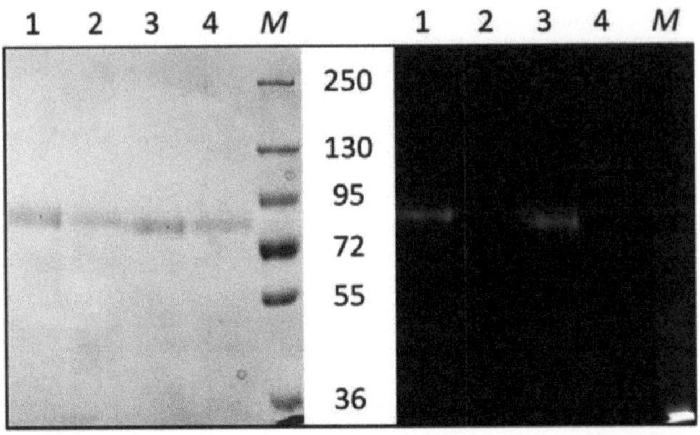

Fig. 18 Representative separation of grafted enzymes on a 7% SDS-PAGE gel. *Left panel*: white light irradiation (Coomassie blue staining); *right panel*: UV irradiation (360 nm). *M*: molecular weight markers. Apparent masses are expressed in kDa. Lane 1: 1–UNIK161 (5 mg); lane 2: UNIK161 (5 mg); lane 3: 1–LAC3 (5 mg); and lane 4: LAC3 (5 mg). *Adapted from Robert, V., Monza, E., Tarrago, L., Sancho, F., De Falco, A., Schneider, L., et al. (2017). Probing the surface of a laccase for clues towards the design of chemo-enzymatic catalysts.* ChemPlusChem, 82, *607–614.*

analysis of trypsin peptides. Samples are prepared according to a procedure adapted from Kharrat et al. (1998).

1. The modified enzyme solution at a desired concentration is placed in a 1-mL microtube.
2. Prior to digestion, the protein is reduced with an excess of dithiothreitol (DTT) (60-fold molar excess over disulfide bridges) in buffer B for 20 h at 40°C, under nitrogen and in the dark.
3. *S*-alkylation of the protein is carried out by adding an excess of 2-iodoacetamide (threefold molar excess over DTT). This mixture is maintained under constant gentle stirring at room temperature for 20 min using a thermomixer.
4. The *S*-alkylated protein is desalted on a PD-10 column preequilibrated with buffer C.
5. The alkylated protein is digested with trypsin (5%, w/w) for 20 h in buffer A at 37°C.
6. The resulting peptides are purified by HPLC equipped with a RP-C18 column.
7. Example of conditions: injection of 50–100 μL of sample; elution is monitored both by UV–vis (λ = 280 and 450 nm) and by fluorescence (λ_{ex} = 450 nm) with the following elution program (Table 4). Peaks displaying fluorescence are collected and submitted to MALDI-TOF analysis (Microflex II (2008) Bruker).

Table 4 Elution conditions for the separation of peptides digested from modified laccase

Time (min)	Solvent A	Solvent B	Flow Rate ($mL\,min^{-1}$)
0	100	0	1
45	50	50	1
50	50	50	1
55	0	100	1
60	0	100	1

4.2 Hybridization with materials

The development of new processes for a sustainable chemistry lies at the heart of the great challenges for tomorrow. In heterogeneous catalysis or supported catalysis, major breakthroughs are expected in terms of functionalization and surface modification as well as in the development of nanocatalysts. Along with these developments, immobilization of enzymes on solid supports for various applications (biosensors, medical fields, biocatalyst carriers, biofuel cells, etc.) has gained a spectacular growing interest as judged by the number of reviews published in this field (Bernal, Rodríguez, & Martínez, 2018; Bezerra, de Farias Lemos, de Sousa, & Gonçalves, 2015; Górecka & Jastrzębska, 2011; Hartmann & Dirk Jung, 2010; Milton & Minteer, 2017; Singh, Tiwari, Singh, & Lee, 2013; Zhou & Hartmann, 2013). As for synthetic catalysts, many advantages arise from immobilization, such as improved stability, reusability, and localization. Additionally, the architecture, the chemical nature, and mechanical properties of the solid support can be tuned to improve enzyme efficiency. Indeed, there is growing interest in achieving a comprehensive view of processes that control all mechanistic aspects related to stability, efficiency, and selectivity of reactions catalyzed by these systems that are often much less controlled than in homogeneous catalysis. Molecular biology has also facilitated strategies of enzyme immobilization (Bernal et al., 2018), techniques that include entrapment, encapsulation, physical adsorption, ionic binding, affinity and, most recently, covalent attachment. Covalent binding of an enzyme to appropriately functionalized surfaces can prevent leaching. The ultimate aim is to understand fundamental synthesis and processing concepts in order to tailor materials fully controlling all involved mechanistic aspects.

We aim at developing such materials. Enzyme electrodes for example represent promising electrocatalytic biomaterials for bioenergy conversion.

By combining methodologies described above (i.e., large production of a site-directed variant of laccase followed by an oriented postfunctionalization of the enzyme surface), we have obtained a homogeneous population of pyrene-functionalized laccase π-stacked at a MWCNT material. The resulting biocathode is robust and operates at high catalytic current densities (up to 3 mA cm^{-2}). Compared to our previous attempt based on His-tagged laccases (Balland et al., 2008), the precise control of the orientation of our fungal laccase during its immobilization on an electroactive surface has helped to maximize the rate of direct electron transfer between the electrode and the enzyme (Lalaoui et al., 2016). We are also studying the effect of the orientation of the enzyme grafted at the surface of superparamagnetic particles on particles activity (Lu Ren, unpublished data).

5. Conclusions

Fungal laccases are known to be particularly robust catalysts. A significant number of reports have focused on the application of this eco-friendly enzyme in many applications. Improving the productivity of heterologous systems is a challenging task from fundamental to practical points of views. In this chapter, an overview of the large-scale production of laccase variants, their purification, and their use in different applications has been given. Exploring the plasticity of laccase through mutagenesis and expression in the yeast *S. cerevisiae*, it is efficient to use the GRAS organism *A. niger* for mass production of chosen variants. Precisely controlled chemical postfunctionalizations of the enzyme surface are leading to new biohybrid catalysts or materials with unprecedented performances regarding laccase function.

Acknowledgment

We thank the Plateforme Analyse et Valorisation de la Biodiversité (AVB-AMU) for help in producing the recombinant enzyme. This study was supported by grants from the Agence Nationale de la Recherche (ANR-09-BLANC-0176 and ANR-15-CE07-0021-01).

References

Alcalde, M. (2015). Engineering the ligninolytic enzyme consortium. *Trends in Biotechnology*, *33*, 155–162.

Alcalde, M., Ferrer, M., Plou, F. J., & Ballesteros, A. (2006). Environmental biocatalysis: From remediation with enzymes to novel green processes. *Trends in Biotechnology*, *24*, 281–287.

Alves, A. M. C. R., Record, E., Lomascolo, A., Scholtmeijer, K., Asther, M., Wessels, J. G. H., et al. (2004). Highly efficient production of laccase by the basidiomycete & Pycnoporus cinnabarinus. *Applied and Environmental Microbiology*, *70*, 6379–6384.

Anastas, P. T., & Warner, J. C. (1998). *Green chemistry: Theory and practice*. UK: Oxford University Press.

Armstrong, F., & Blundell, K. (2007). *Energy...beyond oil*. Oxford, UK: Oxford University Press.

Balland, V., Hureau, C., Cusano, A. M., Liu, Y., Tron, T., & Limoges, B. (2008). Oriented immobilization of a fully active monolayer of histidine-tagged recombinant laccase on modified gold electrodes. *Chemistry—A European Journal*, *14*, 7186–7192.

Bauer, R., & Rupe, C. O. (1971). Use of syringaldazine in a photometric method for estimating "free" chlorine in water. *Analytical Chemistry*, *43*, 421–425.

Becker, D. M., & Guarente, L. (1991). High-efficiency transformation of yeast by electroporation. *Methods in Enzymology*, *194*, 182–187.

Bernal, C., Rodríguez, K., & Martínez, R. (2018). Integrating enzyme immobilization and protein engineering: An alternative path for the development of novel and improved industrial biocatalysts. *Biotechnology Advances*, *36*, 1470–1480.

Bezerra, C. S., de Farias Lemos, C. M. G., de Sousa, M., & Gonçalves, L. R. B. (2015). Enzyme immobilization onto renewable polymeric matrixes: Past, present, and future trends. *Journal of Applied Polymer Science*, *132*, 42125.

Bradford, M. M. (1976). A rapid and sensitive method for the quantitation of microgram quantities of protein utilizing the principle of protein-dye binding. *Analytical Biochemistry*, *72*(1), 248–254.

Bulter, T., Alcalde, M., Sieber, V., Meinhold, P., Schlachtbauer, C., & Arnold, F. H. (2003). Functional expression of a fungal laccase in Saccharomyces cerevisiae by directed evolution. *Applied and Environmental Microbiology*, *69*, 987–995.

Camarero, S., Pardo, I., Cañas, A. I., Molina, P., Record, E., Martínez, A. T., et al. (2012). Engineering platforms for directed evolution of laccase from *Pycnoporus cinnabarinus*. *Applied and Environmental Microbiology*, *78*, 1370–1384.

Cannatelli, M. D., & Ragauskas, A. J. (2017). Two decades of laccases: Advancing sustainability in the chemical industry. *The Chemical Record*, *17*, 122–140.

Couto, S. R., & Toca-Herrera, J. L. (2006). Industrial and biotechnological applications of laccases: A review. *Biotechnology Advances*, *24*, 500–513.

Couto, S. R., & Toca-Herrera, J. L. (2007). Laccase production at reactor scale by filamentous fungi. *Biotechnology Advances*, *25*, 558–569.

Cracknell, J. A., Vincent, K. A., & Armstrong, F. A. (2008). Enzymes as working or inspirational electrocatalysts for fuel cells and electrolysis. *Chemical Reviews*, *108*, 2439–2461.

Cusano, A. M., Mekmouche, Y., Meglecz, E., & Tron, T. (2009). Plasticity of laccase generated by homeologous recombination in yeast. *FEBS Journal*, *276*, 5471–5480.

Diéguez, M., Bäckvall, J.-E., & Pàmies, O. (2018). *Artificial Metalloenzymes and MetalloDNAzymes in catalysis: From design to applications*. Wiley-VCH Verlag GmbH & Co. KGaA.

Fernández-Fernández, M., Sanromán, M. A., & Moldes, D. (2013). Recent developments and applications of immobilized laccase. *Biotechnology Advances*, *31*, 1808–1825.

Gietz, R. D., & Woods, R. A. (2002). Transformation of yeast by lithium acetate/single-stranded carrier DNA/polyethylene glycol method. *Methods in Enzymology*, *350*, 87–96.

Gordon, L., Khalaj, V., Ram, A. F. J., Archer, D. B., Brookman, J. L., Trinci, A. P. J., et al. (2000). Glucoamylase: Green fluorescent protein fusions to monitor protein secretion in Aspergillus niger. *Mutagenesis*, *146*, 415–426.

Górecka, E., & Jastrzębska, M. (2011). Immobilization techniques and biopolymer carriers. *Biotechnology and Food Sciences*, *75*, 65–86.

Gutiérrez-Sánchez, C., Pita, M., Vaz-Domínguez, C., Shleev, S., & De Lacey, A. L. (2012). Gold nanoparticles as electronic bridges for laccase-based biocathodes. *Journal of the American Chemical Society*, *134*, 17212–17220.

Hartmann, M., & Dirk Jung, D. (2010). Biocatalysis with enzymes immobilized on mesoporous hosts: The status quo and future trends. *Journal of Materials Chemistry*, *20*, 844–857.

Jones, S. M., & Solomon, E. I. (2015). Electron transfer and reaction mechanism of laccases. *Cellular and Molecular Life Sciences*, *72*, 869–883.

Kalyanasundaram, K. (1982). Photophysics, photochemistry and solar energy conversion with tris(bipyridyl)ruthenium(II) and its analogues. *Coordination Chemistry Reviews*, *46*, 159–244.

Kates, R. W. (2017). Sustainability science. In D. Richardson, N. Castree, M. F. Goodchild, A. Kobayashi, W. Liu, & R. A. Marston (Eds.), *The international encyclopedia of geography*. John Wiley & Sons, Ltd.

Kharrat, R., Mansuelle, P., Sampieri, F., Crest, M., Oughideni, R., Van Rietschoten, J., et al. (1998). Maurotoxin, a four disulfide bridge toxin from Scorpio maurus venom: Purification, structure and action on potassium channels. *FEBS Letters*, *406*, 284–290.

Kiiskinen, L. L., Kruus, K., Bailey, M., Ylösmäki, E., Siika-Aho, M., & Saloheimo, M. (2004). Expression of Melanocarpus albomyces laccase in Trichoderma reesei and characterization of the purified enzyme. *Microbiology*, *150*, 3065–3074.

Kirk, O., Borchert, T. V., & Fuglsang, C. C. (2002). Industrial enzyme applications. *Current Opinion in Biotechnology*, *13*, 345–351.

Klonowska, A., Gaudin, C., Asso, M., Fournel, A., Réglier, M., & Tron, T. (2005). LAC3, a new low redox potential laccase from Trametes sp. strain C30 obtained as a recombinant protein in yeast. *Enzyme and Microbial Technology*, *36*, 34–41.

Kunamneni, A., Camarero, S., García-Burgos, C., Plou, F. J., Ballesteros, A., & Alcalde, M. (2008). Engineering and applications of fungal laccases for organic synthesis. *Microbial Cell Factories*, *7*, 32.

Lalaoui, N., Rousselot-Pailley, P., Robert, V., Mekmouche, Y., Villalonga, R., Holzinger, M., et al. (2016). Direct electron transfer between a site-specific pyrene-modified laccase and carbon nanotube/gold nanoparticle supramolecular assemblies for bioelectrocatalytic dioxygen reduction. *ACS Catalysis*, *6*, 1894–1900.

Lazarides, T., Sazanovich, I. V., Simaan, A. J., Kafentzi, M. C., Delor, M., Mekmouche, Y., et al. (2013). Visible light-driven O_2 reduction by a porphyrin-laccase system. *Journal of the American Chemical Society*, *135*, 3095–3103.

Le Goff, A., Holzinger, M., & Cosnier, S. (2015). Recent progress in oxygen-reducing laccase biocathodes for enzymatic biofuel cells. *Cellular and Molecular Life Sciences*, *72*, 941–952.

Li, D., Tang, Y., Lin, J., & Cai, W. (2017). Methods for genetic transformation of filamentous fungi. *Microbial Cell Factories*, *16*(1), 168.

Liu, Y., Cusano, A. M., Wallace, E. C., Mekmouche, Y., Ullah, S., Robert, V., et al. (2014). Characterization of C-terminally engineered laccases. *International Journal of Biological Macromolecules*, *69*, 435–441.

Mate, D. M., & Alcade, M. (2015). Laccase engineering: From rational design to directed evolution. *Biotechnology Advances*, *33*, 25–40.

Mate, D. M., & Alcalde, M. (2017). Laccase: A multi-purpose biocatalyst at the forefront of biotechnology. *Microbial Biotechnology*, *10*, 1457–1467.

Mate, D., García-Burgos, C., García-Ruiz, E., Ballesteros, A. O., Camarero, S., & Alcalde, M. (2010). Laboratory evolution of high-redox potential laccases. *Chemical Biology*, *17*, 1030–1041.

McFarland, J. M., & Francis, M. B. (2005). Reductive alkylation of proteins using iridium catalyzed transfer hydrogenation. *Journal of the American Chemical Society, 127*, 13490–13491.

Mekmouche, Y., Schneider, L., Rousselot-Pailley, P., Faure, B., Simaan, A. J., Bochot, C., et al. (2015). Laccases as palladium oxidases. *Chemical Science, 6*, 1247–1251.

Mekmouche, Y., Zhou, S., Cusano, A. M., Record, E., Lomascolo, A., Robert, V., et al. (2014). Gram-scale production of a basidiomycetous laccase in Aspergillus niger. *Journal of Bioscience and Bioengineering, 117*, 25–27.

Meyer, H.-P., Eichhorn, E., Hanlon, S., Lütz, S., Schürmann, M., Wohlgemuth, R., et al. (2013). The use of enzymes in organic synthesis and the life sciences: Perspectives from the Swiss Industrial Biocatalysis Consortium (SIBC). *Catalysis Science & Technology, 3*, 29–40.

Mikolasch, A., & Schauer, F. (2009). Fungal laccases as tools for the synthesis of new hybrid molecules and biomaterials. *Applied Microbiological and Biotechnology, 82*, 605–624.

Milton, R. D., & Minteer, S. D. (2017). Direct enzymatic bioelectrocatalysis: Differentiating between myth and reality. *Journal of the Royal Society Interface, 14*(131), 20170253.

Nestl, B. M., Hammer, S. C., Nebel, B. A., & Hauer, B. (2014). New generation of biocatalysts for organic synthesis. *Angewandte Chemie International Edition, 53*, 3070–3095.

Oldenburg, K. R., Vo, K. T., Michaelis, S., & Paddon, C. (1997). Recombination-mediated PCR-directed plasmid construction in vivo in yeast. *Nucleic Acids Research, 25*, 451–452.

Orr-Weaver, T. L., Szostak, J. W., & Rothstein, R. J. (1983). Genetic applications of yeast transformation with linear & gapped plasmids. *Methods in Enzymology, 101*, 228–245.

Pardo, I., & Camarero, S. (2015). Laccase engineering by rational and evolutionary design. *Cellular and Molecular Life Sciences, 72*, 897–910.

Pistone, L., Ottolina, G., De, S., Romero, A. A., Martins, L. O., & Luque, R. (2016). Encapsulated laccases for the room-temperature oxidation of aromatics: Towards synthetic low-molecular-weight lignins. *ChemSusChem, 9*, 756–762.

Pita, M., Gutierrez-Sanchez, C., Olea, D., Velez, M., Garcia-Diego, C., Shleev, S., et al. (2011). High redox potential cathode based on laccase covalently attached to gold electrode. *Journal of Physical Chemistry C, 115*, 13420–13428.

Pita, M., Mate, D. M., Gonzalez-Perez, D., Shleev, S., Fernandez, V. M., Alcalde, M., et al. (2014). Bioelectrochemical oxidation of water. *Journal of the American Chemical Society, 136*, 5892–5895.

Punt, P. J., & van den Hondel, C. A. (1992). Transformation of filamentous fungi based on hygromycin B and phleomycin resistance markers. *Methods in Enzymology, 216*, 447–457.

Qiu, H., Xu, C., Huang, X., Ding, Y., Qu, Y., & Gao, P. (2009). Immobilization of laccase on nanoporous gold: Comparative studies on the immobilization strategies and the particle size effects. *Journal of Physical Chemistry C, 113*, 2521–2525.

Record, E., Punt, P. J., Chamkha, M., Labat, M., van den Hondel, C. A. M. J. J., & Asther, M. (2002). Expression of the Pycnoporus cinnabarinus laccase gene in Aspergillus niger and characterization of the recombinant enzyme. *European Journal of Biochemistry, 269*, 602–609.

Riva, S. (2006). Laccases: Blue enzymes for green chemistry. *Trends in Biotechnology, 24*, 219–226.

Robert, V., Mekmouche, Y., Rousselot-Pailley, P., & Tron, T. (2011). Engineering laccases: In search for novel catalysts. *Current Genomics, 12*, 123–129.

Robert, V., Monza, E., Tarrago, L., Sancho, F., De Falco, A., Schneider, L., et al. (2017). Probing the surface of a laccase for clues towards the design of chemo-enzymatic catalysts. *ChemPlusChem, 82*, 607–614.

Rodgers, C. J., Blanford, C. F., Giddens, S. R., Skamnioti, P., Armstrong, F. A., & Gurr, S. J. (2010). Designer laccases: A vogue for high-potential fungal enzymes? *Trends in Biotechnology*, *28*, 63–72.

Sambrook, J., Fritsch, E. F., & Maniatis, T. (1989). *Molecular cloning: A laboratory manual* (2nd ed.). . Cold Spring Harbor, NY: Cold Spring Harbor Laboratory.

Schneider, L., Mekmouche, Y., Rousselot-Pailley, P., Simaan, A. J., Robert, V., Réglier, M., et al. (2015). Visible-light-driven oxidation of organic substrates with dioxygen mediated by a [Ru(bpy)3](2+)/laccase system. *ChemSusChem*, *8*, 3048–3051.

Schwizer, F., Okamoto, Y., Heinisch, T., Gu, Y., Pellizzoni, M. M., Lebrun, V., et al. (2018). Artificial metalloenzymes: Reaction scope and optimization strategies. *Chemical Reviews*, *118*, 142–231.

Shleev, S., Christenson, A., Serezhenkov, V., Burbaev, D., Yaropolov, A., Gorton, L., et al. (2005). Electrochemical redox transformations of T1 and T2 copper sites in native Trametes hirsuta laccase at gold electrode. *Biochemistry Journal*, *385*, 745–754.

Shleev, S., Pita, M., Yaropolov, A. I., Ruzgas, T., & Gorton, L. (2006). Direct heterogeneous electron transfer reactions of Trametes hirsuta laccase at bare and thiol-modified gold electrodes. *Electroanalysis*, *18*, 1901–1908.

Simaan, A. J., Mekmouche, Y., Herrero, C., Moreno, P., Aukauloo, A., Delaire, J. A., et al. (2011). Photoinduced multielectron transfer to a multicopper oxidase resulting in dioxygen reduction into water. *Chemistry—A European Journal*, *17*, 11743–11746.

Singh, R. K., Tiwari, M. K., Singh, R., & Lee, J. K. (2013). From protein engineering to immobilization: Promising strategies for the upgrade of industrial enzymes. *International Journal of Molecular Sciences*, *14*, 1232–1277.

Solomon, E. I., Heppner, D. E., Johnston, E. M., Ginsbach, J. W., Cirera, J., Qayyum, M., et al. (2014). Copper active sites in biology. *Chemical Reviews*, *114*, 3659–3853.

Solomon, E. I., Szilagyi, R. K., DeBeer George, S., & Basumallick, L. (2004). Electronic structures of metal sites in proteins and models: Contributions to function in blue copper proteins. *Chemical Reviews*, *104*, 419–458.

Tarrago, L., Modolo, C., Yemloul, M., Robert, V., Rousselot-Pailley, P., & Tron, T. (2018). Controlling the polymerization of coniferyl alcohol with cyclodextrins. *New Journal of Chemistry*, *42*, 11770–11775.

Tron, T. (2013). Laccases. In R. H. Kretsinger, V. N. Uversky, & E. A. Permyakov (Eds.), *Encyclopedia of metalloproteins* (pp. 1066–1070). New York: Springer.

van Hartingsveldt, W., Mattern, I. E., van Zeijl, C. M., Pouwels, P. H., & van den Hondel, C. A. (1987). Development of a homologous transformation system for Aspergillus niger based on the pyrG gene. *Molecular Genetics & Genomics*, *206*, 71–75.

Witayakran, S., & Ragauskas, A. J. (2009). Synthetic applications of laccase in green chemistry. *Advanced Synthesis & Catalysis*, *351*, 1187–1209.

Zhou, Z., & Hartmann, M. (2013). Progress in enzyme immobilization in ordered mesoporous materials and related applications. *Chemical Society Reviews*, *42*, 3894–3912.

Zucca, P., Cocco, G., Sollai, F., & Sanjust, E. (2016). Fungal laccases as tools for biodegradation of industrial dyes. *Biocatalysis*, *1*, 82–108.

Further reading

Bessa, D., Pereira, F., Moreira, R., Johansson, B., & Queirós, O. (2012). Improved gap repair cloning in yeast: Treatment of the gapped vector with Taq DNA polymerase avoids vector self-ligation. *Yeast*, *29*, 419–423.

Lippard, S. J. (2006). The inorganic side of chemical biology. *Nature Chemical Biology*, *2*, 504–507.

Mezard, C., & Nicolas, A. (1994). Homologous, homeologous, and illegitimate repair of double-strand breaks during transformation of a wild-type strain and a rad52 mutant strain of Saccharomyces cerevisiae. *Molecular and Cellular Biology*, *14*, 1278–1292.

Mezard, C., Pompon, D., & Nicolas, A. (1992). Recombination between similar but not identical DNA sequences during yeast transformation occurs within short stretches of identity. *Cell*, *70*, 659–670.

Solomon, E. I., Baldwin, M. J., & Lowery, M. D. (1992). Electronic structures of active sites in copper proteins: Contributions to reactivity. *Chemical Reviews*, *92*, 521–542.

Yang, J., Li, W., Bun, N. T., Deng, X., Lin, J., & Ye, X. (2017). Laccases: Production, expression regulation, and applications in pharmaceutical biodegradation. *Frontiers in Microbiology*, *8*, 832.

Zumarraga, M., Camarero, S., Shleev, S., Martinez-Arias, A., Ballesteros, A., Plou, F. J., et al. (2008). Altering the laccase functionality by in vivo assembly of mutant libraries with different mutational spectra. *Proteins*, *71*, 250–260.

CHAPTER THREE

Production and spectroscopic characterization of lytic polysaccharide monooxygenases

Glyn R. Hemsworth[a,b,†], Luisa Ciano[c,†], Gideon J. Davies[c], Paul H. Walton[c,*]

[a]School of Molecular and Cellular Biology, Faculty of Biological Sciences, University of Leeds, Leeds, United Kingdom
[b]Astbury Centre for Structural Molecular Biology, University of Leeds, Leeds, United Kingdom
[c]Department of Chemistry, University of York, York, United Kingdom
*Corresponding author: e-mail address: paul.walton@york.ac.uk

Contents

Abstract

Lytic polysaccharide monooxygenases (LPMOs, also known as PMOs) are a recently discovered family of enzymes that play a key role in the breakdown of polysaccharide substrates. The ability of LPMOs to introduce chain breaks, using an oxidative mechanism, has particularly attracted attention as the world seeks more cost-effective and environmentally friendly ways of producing second-generation biofuels for the future. LPMOs are copper-dependent enzymes and have an unusual active site which includes the N-terminal residue of the protein in the copper coordination sphere. This N-terminal histidine side chain is also methylated in fungal enzymes, the molecular reason for

† These authors contributed equally to this chapter.

Methods in Enzymology, Volume 613
ISSN 0076-6879
https://doi.org/10.1016/bs.mie.2018.10.014

which is still a debated topic. The production of these enzymes poses several challenges if we are to understand their chemical mechanisms. Here, we describe the methods that have been used in the field to produce LPMOs and provide information on the workflows that we use for our electron paramagnetic resonance (EPR) spectroscopy experiments. EPR has been a particularly powerful tool in the study of these enzymes and our objective with this chapter is to provide some helpful information for researchers for whom this technique might be daunting or theoretically difficult to access.

1. Introduction

The production of second-generation biofuels from lignocellulosic feedstocks has been hotly pursued for the last decade as a more sustainable fuel source in comparison to oil and first-generation biofuels (several reviews are available, e.g., Merino & Cherry, 2007; Naik, Goud, Rout, & Dalai, 2010; Saini, Saini, & Tewari, 2015). One of the major challenges toward the realization of the cost-effective production of these fuels, however, has been the nature of the substrate, cellulose, and its resistance to degradation (Himmel et al., 2007). The study of enzymes involved in the breakdown of this substrate was pioneered by Reese and Mandels through the 1950s (Mandels & Reese, 1957; Reese, Siu, & Levinson, 1950), 1960s (Mandels, Parrish, & Reese, 1962; Mandels & Reese, 1960), and 1970s (Mandels, Hontz, Nystrom, Lee, & Lynd, 1974; Mandels, Weber, & Parizek, 1971), and we continue to make new discoveries in this area to this day (see recent reviews by Bischof, Ramoni, & Seiboth, 2016; Payne et al., 2015). What is now viewed as the "classical" model of enzymatic cellulose deconstruction requires the combined action of essentially three distinct enzymatic activities to allow the liberation of single glucose molecules from this challenging substrate (reviewed in Payne et al., 2015). The glycoside hydrolases involved include endoglucanases, cellobiohydrolases, and β-glucosidases, which synergize with one another to bring about this molecular conversion for metabolism in nature. These naturally occurring enzymes, however, despite their efficiency in degrading their substrates, have arguably been too slow to make the enzymatic conversion of biomass cost-effective in the biorefinery. As such, new enzymatic activities have been sought after (as reviewed by Harris, Xu, Kreel, Kang, & Fukuyama, 2014; Horn, Vaaje-Kolstad, Westereng, & Eijsink, 2012). In this context, Reese, in his original 1950 publication, suggested that there was likely an additional activity required to disrupt the crystalline structure of cellulose to allow glycoside hydrolases access to the substrate (Reese et al., 1950). This activity is now thought to be provided by a family of enzymes known as lytic polysaccharide monooxygenases (LPMOs) (see Beeson, Vu, Span, Phillips, & Marletta, 2015; Ciano, Davies, Tolman, & Walton, 2018; Hemsworth, Johnston, Davies, & Walton, 2015 for recent reviews).

Prior to the demonstration of LPMO activity in 2010 (Vaaje-Kolstad et al., 2010), LPMOs were "by sequence" defined in two families in the carbohydrate active enzymes database (http://www.cazy.org) (Lombard, Golaconda Ramulu, Drula, Coutinho, & Henrissat, 2014)—glycoside hydrolase-61 (GH61) and carbohydrate binding module-33 (CBM33) (see Morgenstern, Powlowski, & Tsang, 2014 for a review). In 2010, Harris et al. (2010) had demonstrated that proteins classified as GH61s from *Thielavia terrestris* and *Thermoascus aurantiacus* were able to reduce the enzyme load required for the depolymerization of cellulose twofold in a metal-dependent manner. The biochemical basis for this effect became clear when Vaaje-Kolstaad et al. showed that the structurally related protein, CBP21 from *Serratia marcescens* (then classified as a CBM33 in CAZy), was able to oxidatively induce chain breaks into chitin in a manner dependent on reducing agent and O_2, thereby establishing CBM33s as LPMOs for the first time (Vaaje-Kolstad et al., 2010). Subsequent work demonstrated that GH61s were also oxidative enzymes and that they required a copper cofactor to function (Phillips, Beeson, Cate, & Marletta, 2011; Quinlan et al., 2011). LPMOs have now been discovered from diverse organisms across the tree of life and have been implicated in the deconstruction of a range of polysaccharides beyond cellulose and chitin (Agger et al., 2014; Bennati-Granier et al., 2015; Borisova et al., 2015; Couturier et al., 2018; Frommhagen et al., 2015; Lo Leggio et al., 2015; Simmons et al., 2017; Vu, Beeson, Span, Farquhar, & Marletta, 2014). Due to their redox function, LPMOs are now defined in CAZy as auxiliary activities (AA) (Levasseur, Drula, Lombard, Coutinho, & Henrissat, 2013) with six families currently classified—AA9, AA10, AA11, AA13, AA14 and AA15.

LPMOs are unusual in being monooxygenases which utilize only a single copper ion in their active site (see Solomon et al., 2014 for a useful review on copper-dependent enzymes). Indeed, there have been recent suggestions that LPMOs may, under certain conditions, also function as peroxygenases, but further work is still required to determine the true physiological cosubstrate for LPMOs (Bissaro et al., 2017; Kuusk et al., 2018). LPMOs bind their single copper cofactor in a motif that has been dubbed the "histidine brace" (Quinlan et al., 2011), an arrangement of two histidine residues, one of which

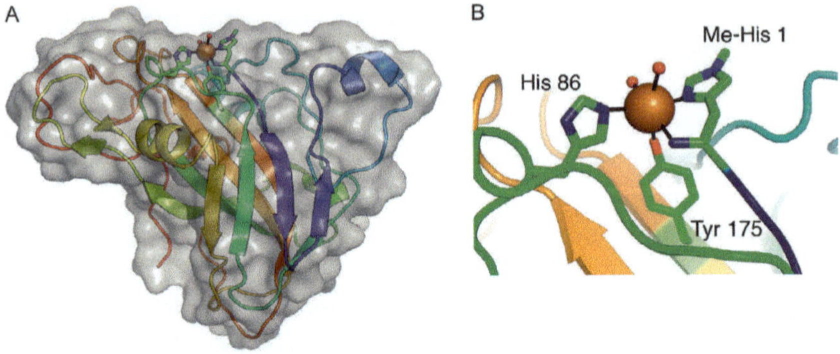

Fig. 1 Example structure of an AA9 LPMO. (A) Overall structure of *Thermoascus aurantiacus* AA9 LPMO shown in cartoon colored from N- (*blue*) to C-terminus (*red*). (B) Close-up view of the active site copper ion (*orange sphere*) coordinated by the histidine brace which includes the N-terminal methylated histidine of the protein.

is the N-terminal residue, which coordinate the copper ion in a T-shaped geometry (Fig. 1) (see reviews by Ciano et al., 2018; Hemsworth, Davies, & Walton, 2013; Vaaje-Kolstad, Forsberg, Loose, Bissaro, & Eijsink, 2017; Vu & Ngo, 2018). In addition, for fungal enzymes at least, there is a modification to the N-terminal histidine side chain in the form of τ-*N*-methylation which is only detected when proteins are expressed in fungal hosts (Frandsen et al., 2016; Li, Beeson, Phillips, Marletta, & Cate, 2012; Lo Leggio et al., 2015; Quinlan et al., 2011; Vu, Beeson, Phillips, Cate, & Marletta, 2014; Vu, Beeson, Span, et al., 2014). The importance of this modification is still debated, but it is clear that LPMOs remain active in its absence when expressed in, e.g., *Pichia pastoris* (Bennati-Granier et al., 2015; Bey et al., 2013; Couturier et al., 2018; Wu et al., 2013). The requirement for the N-terminal histidine in the mature protein has dictated the strategies that have been used to produce these enzymes which we will discuss in this chapter. We will also describe our workflows for conducting electron paramagnetic resonance (EPR) spectroscopy on these enzymes. EPR has been a powerful tool in contributing toward unraveling the steps along the reaction coordinate catalyzed by LPMOs. EPR is a conceptually challenging technique, however, which requires expertize to both successfully simulate the data and to interpret the resulting spin Hamiltonian parameters in a meaningful way to gain insights into the electronic aspects of the copper active site. We hope that the topics covered in this chapter will, therefore, prove useful not only for those working on LPMOs, but more broadly for researchers interested in studying diverse copper-dependent enzymes.

2. Expression and purification of LPMOs

LPMOs have been identified from diverse organisms including bacteria (Vaaje-Kolstad, Horn, van Aalten, Synstad, & Eijsink, 2005; Vaaje-Kolstad et al., 2010), viruses (Chiu et al., 2015), fungi (Karkehabadi et al., 2008; Langston et al., 2011; Phillips et al., 2011; Quinlan et al., 2011; Saloheimo, Nakari-Setälä, Tenkanen, & Penttilä, 1997), and more recently higher eukaryotes (Sabbadin et al., 2018). This diversity can pose challenges toward the choice of expression strategy in order to produce pure proteins to be studied. The fact that these proteins are typically secreted in their native hosts, leading to the generation of the N-terminal histidine integral to the active site, must also be considered when expressing these proteins for study (for reviews, please see Hemsworth et al., 2013; Vaaje-Kolstad et al., 2017). Endogenously, LPMOs will often also contain disulfide bonds, and eukaryotic family members can be further modified by N- and O-glycosylation, as well as by methylation of the N-terminal histidine. The choice of expression system must therefore be tailored toward the target enzyme and can have a significant impact on the protein yield, and potentially on the activity of the enzyme though this has not been investigated in great detail.

2.1 Recombinant expression and purification from the *E. coli* periplasm

Escherichia coli is the workhorse of the academic lab when it comes to protein production. Bacterial LPMOs of the AA10 class are therefore most often purified from this organism. The most common approach has been to secrete the resultant protein to the periplasm (Courtade, Le, Sætrom, Brautaset, & Aachmann, 2017; Forsberg, Mackenzie, et al., 2014; Hemsworth, Taylor, et al., 2013; Vaaje-Kolstad, Horn, et al., 2005; Vaaje-Kolstad et al., 2010). It is also possible, however, to express these enzymes in the cytoplasm using specialized *E. coli* strains (Forsberg, Røhr, et al., 2014; Forsberg et al., 2011; Gregory et al., 2016) and these have largely been our methods of choice for LPMO production. Although fungal enzymes are rarely produced using these strategies, we were able to express and purify an AA11 from *Aspergillus oryzae* using a periplasmic secretion system in *E. coli* (Hemsworth, Henrissat, Davies, & Walton, 2014).

In order to target the protein to the periplasm, we carefully design our constructs for cloning into a pelB-containing vector and typically use HiFi

DNA Assembly (New England Biolabs), In-Fusion (Takara), or PIPE cloning (Klock & Lesley, 2009). This strategy allows the target gene to be placed directly after the signal sequence without the addition of any residues from use of restriction enzymes, and so negates the need to use site-directed mutagenesis later to remove any undesired residues. Others have found success using the CBP21 leader peptide as the signal sequence (Crouch, Labourel, Walton, Davies, & Gilbert, 2016; Gardner et al., 2014), and one should always consider screening other potential leader peptides in order to improve the chances of obtaining good yields of soluble protein in the periplasm. In addition, careful thought needs to be put toward the use of affinity tags using the periplasmic secretion strategy. An affinity tag would typically be placed at the C-terminus in this case. We have tended to avoid using His-tags for fear that they might interfere with copper binding and produce unwanted copper-binding sites during our EPR analyses and also, potentially, reactivity studies. If a His-tag is used to purify the protein and not removed prior to such an analysis, then careful controls must be performed as Cu binding to the tag cannot be excluded if the metal ion is added in excess. EPR can offer a quick and powerful way to check the sample for the presence of any additional/adventitious binding sites (see later). Other affinity tags can also be used of course, with a strep-tag recently used for the purification of the latest LPMO discovered in family AA15 providing a prominent example (Sabbadin et al., 2018).

Once the desired constructs are obtained and confirmed to be correct by Sanger sequencing, plasmids are transformed typically into BL21(DE3) *E. coli* cells as the first option, but other *E. coli* strains should also be considered. Transformants are plated out on selective LB agar plates and allowed to grow at 37°C overnight. A 5 mL LB liquid culture is then inoculated using a single colony the next day and the cells are allowed to grow at 37°C in a shaker at 180 rpm throughout the day. Once the culture is visibly turbid, a 500 μL aliquot is used to inoculate a 50 mL culture which is left to grow overnight in the same conditions. The next day, 2 L baffled flasks, each containing 500 mL of LB, are then inoculated using 5 mL each from the overnight culture and these are incubated at 37°C, shaking at 180 rpm. The OD_{600} is monitored closely, and once this has reached 0.6 the temperature is typically cooled to 16°C for about an hour. Protein expression can then be induced by the addition of isopropyl β-D-1-thiogalactopyranoside (IPTG) to a final concentration of 1 m*M* and the cultures are left overnight. The next day, cells are harvested by centrifugation at 8000 x *g*, and the spent media discarded: the cell pellets can be frozen at −80°C until ready for subsequent use.

We use a modified periplasmic preparation protocol rather than the cold osmotic shock protocol (Neu & Heppel, 1965) that is often used as we have found that we recover better yields of protein using this approach. Typically, cell paste is thoroughly resuspended in 3 × volumes of 20% sucrose (20 m*M* Tris can also be included to stabilize pH) and 40 μL of 10 mg/mL lysozyme is added for every gram of cell paste being used before leaving the suspension on ice for 1 h. A 60 μL aliquot of 1 *M* $MgSO_4$ is then added for every gram of cell paste being used and the cells are left on ice for an additional 20 min. Cell debris is then harvested by centrifugation at 10,000 × *g* for 20 min and the supernatant is removed as Fraction I to a fresh container and kept on ice. The pellet is resuspended in ice-cold water and this is also left on ice for a further 1 h (some DNAse can be added at this point to try and reduce the viscosity of the solution produced at the end of the process). The cell debris is again removed by centrifugation at 10,000 × *g* for 20 min and the supernatant is removed as Fraction II. Fractions I and II can be checked at this point to see whether the LPMO is abundantly present in one fraction over the other, however, usually both fractions are combined for subsequent purification.

If using an affinity tag, standard chromatographic approaches can be used for the purification of LPMO. We typically perform purifications by FPLC using an affinity chromatography step for the initial purification and gel filtration for a final "polishing" step. In the absence of a tag we would typically aim to purify via ion-exchange chromatography followed by gel filtration (see Fig. 2A and B, for example, gels). We would recommend using Good's buffers throughout any purification, these represent biologically compatible buffers which should not interact significantly with biologically relevant cations such as copper (Good et al., 1966). If attempting to measure copper binding to LPMO it is imperative that all buffers are thoroughly demetallated prior to the final step of the purification: this is typically achieved by cycling the buffer through a 10 mL chelex column (Bio-Rad) for 24 h before it is used in gel filtration. The protein should also be thoroughly treated with EDTA (typically at 10 m*M* concentration or higher for at least an hour) to remove any metal that might already be bound, since LPMOs have very high affinity for copper. Indeed, it is probably this high affinity for copper that resulted in low-level activity, which could have been mistaken for hydrolytic activity, during early studies on GH61s (Karlsson et al., 2001). The high affinity of the enzyme for copper allows it to scavenge its cofactor even when present at very low concentration and could also have contributed to the initial confusion over the identity of the correct metal cofactor following the initial

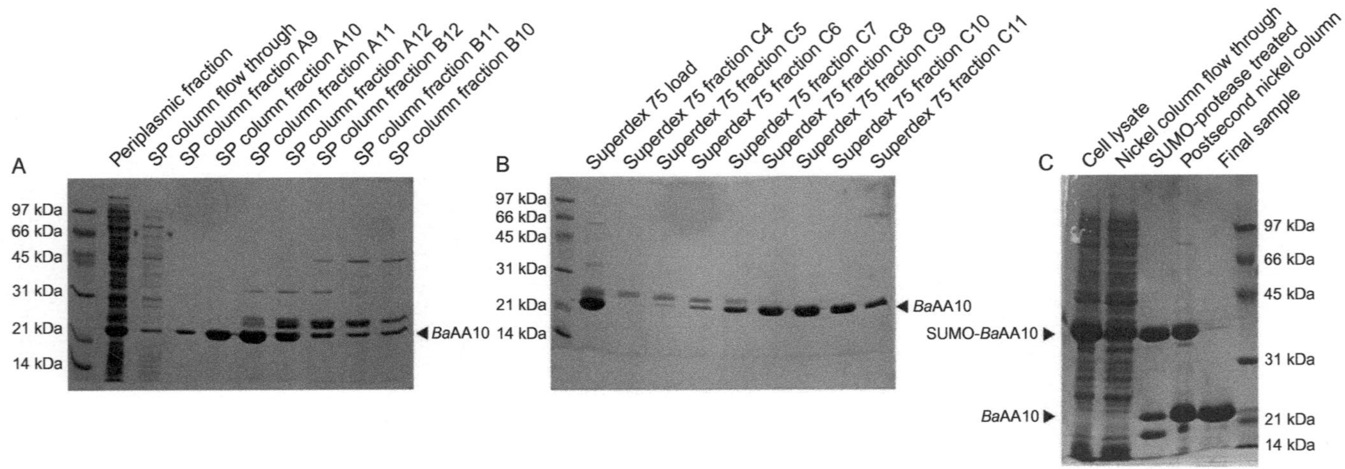

Fig. 2 Example gels for LPMO purification strategies. (A) Gel showing initial purification of *Ba*AA10 from the periplasm using an SP ion-exchange column. (B) Subsequent purification of BaAA10 using a Superdex 75 size-exclusion column to yield the final sample (Hemsworth, Taylor, et al., 2013). (C) SDS-PAGE gel showing stages from the purification of *Ba*AA10 using a SUMO-tagged construct (Gregory et al., 2016).

structure determinations for these enzymes (Karkehabadi et al., 2008; Vaaje-Kolstad, Houston, Riemen, Eijsink, & van Aalten, 2005; Vaaje-Kolstad et al., 2010).

If an apo-enzyme is not desired, but rather a copper-loaded sample is to be used for study, prior to the final gel-filtration step, $CuCl_2$ can be added in a slight excess (typically 1.5 ×) in order to load the protein with copper. The sample is then typically passed through a 16/600 Superdex 75 column (GE Healthcare) to remove any excess copper, giving a highly purified, copper-loaded sample ready for study. In order to store protein, we typically concentrate to 10 mg/mL or more using a 3, 5, or 10 kDa Vivaspin column depending on the size of the LPMO (avoiding cellulose-based membranes concentrators). The sample is then flash frozen in liquid nitrogen for storage at −80°C.

Before discussing other strategies for LPMO production in *E. coli* and other organisms, there is another option for protein secretion to obtain mature LPMOs—to secrete the protein from Gram-positive bacteria into the culture medium (Nakagawa et al., 2015). We have not used this approach ourselves, but it could be an excellent strategy for large scale LPMO production with industry often using Gram-positive expression systems for the production of many stable and industrially useful enzymes.

2.2 Recombinant expression and purification from the *E. coli* cytoplasm

Given that LPMOs are typically extracellular enzymes that contain disulfide bonds, the periplasmic expression protocol described earlier often represents the first option for protein production (Courtade et al., 2017; Forsberg, Mackenzie, et al., 2014; Hemsworth, Taylor, et al., 2013; Vaaje-Kolstad, Horn, et al., 2005; Vaaje-Kolstad et al., 2010). This protocol is not the only option, however, and there are reports of LPMOs being expressed intracellularly using cleavable tags at the N-terminus to aid purification (Forsberg, Røhr, et al., 2014; Forsberg et al., 2011; Gregory et al., 2016). For a successful outcome, the protease to be used to remove that tag must be carefully selected in order to leave the native N-terminal histidine available for copper binding. Factor Xa (Forsberg, Røhr, et al., 2014) and enterokinase are two such proteases. We prefer the use of a SUMO-tag (Champion-pET-SUMO, Invitrogen) which can be removed using SUMO-protease (Invitrogen) (Gregory et al., 2016). This method has the advantage that the SUMO-tag can act as a solubility-enhancing tag in addition to providing an N-terminal histidine tag to aid in the purification of the protein.

SUMO-protease recognizes the tertiary structure of the SUMO domain and cleaves directly after a diglycine motif which can be placed immediately upstream of the N-terminal His-coding region, so that cleavage results in the release of the native protein. Constructs can be carefully designed using the cloning strategies discussed earlier in order to achieve this result without the use of restriction enzymes. This strategy was used to improve the yield of the AA10 from *Bacillus amyloliquefaciens* as much as threefold (Gregory et al., 2016). While this protein lacked disulfide bonds which would have aided in the correct folding of the protein intracellularly, we have since used this strategy to express disulfide-containing LPMOs intracellularly using specialized cells such as the NEB Express T7 cells (New England Biolabs) (Unpublished data).

The specific protocol for protein expression in *E. coli* is much the same as that discussed earlier with cells grown in shake flasks and the temperature typically being lowered to 16°C or 20°C upon IPTG induction to ensure correct protein folding. Cells are harvested the same way by centrifugation at 8000 × *g*, with spent media being discarded and cell pellets frozen at −80°C until ready for lysis.

Since the protein is now present in the cytoplasm, there is no need to perform the laborious periplasmic prep protocol in order to release the protein from the cells. Cells are simply resuspended in Buffer A (typically 50 m*M* HEPES pH 7, 250 m*M* NaCl, 30 m*M* imidazole) and can be lysed by one's preferred method (typically, sonication or French press is used in our laboratories). The cell lysate can then be loaded directly onto a nickel resin to which the His_6-SUMO-tagged LPMO will bind (Fig. 2C). Protein is released by the application of a gradient from 0% to 100% Buffer B (Buffer A + 300 m*M* imidazole) and 1.8 mL fractions are collected across the gradient. Peak fractions containing His_6-SUMO-LPMO are combined ready for tag removal. In order to prepare for tag cleavage, it is important to remove excess imidazole and NaCl from the sample which can be detrimental to SUMO-protease function: this can be achieved by buffer exchange on a 10 kDa MW cutoff concentrator, using a desalting column or by dialysis (being careful to avoid cellulose-based dialysis membranes). The protease buffer should also contain a reducing agent to ensure optimal SUMO-protease activity since it is a cysteine protease. We, therefore, typically buffer-exchange our samples into 20 m*M* Tris pH 8, 5 m*M* β-mercaptoethanol. Small-scale tag-cleavage reactions should be set up the first time that this strategy is used to find an optimal ratio of protease to SUMO-LPMO, which can then be scaled up to the whole sample.

We typically perform tag removal at a SUMO-LPMO concentration of 1–2 mg/mL overnight at 4°C.

Once the cleavage has been performed, the released tag, any uncleaved protein, and the SUMO-protease are removed from the sample by passing it back down a nickel column ensuring that the eluate, which will contain the mature LPMO, is collected (Fig. 2C). The differences in the LPMO's affinity for copper over nickel ensure that there is no transmetallation of the enzyme. The enzyme sample can then be concentrated, copper loaded, or thoroughly EDTA treated as described earlier, before a final polishing step using a 16/600 Superdex 75 column (GE Healthcare) to yield the final sample (Fig. 2C).

2.3 Recombinant expression and purification from yeast

For many LPMOs, expression in *E. coli* has not been possible and so laboratories have turned to other expression systems for challenging targets. *P. pastoris* is generally the most common yeast expression system used for protein production in academic laboratories and has been used for the production of several LPMOs (Bennati-Granier et al., 2015; Kittl, Kracher, Burgstaller, Haltrich, & Ludwig, 2012; Tanghe et al., 2015). While *P. pastoris* lacks the machinery required to perform the methylation of the N-terminal histidine found in fungal LPMOs, it maintains the ability to glycosylate proteins which may be important to ensure the correct folding of LPMOs from eukaryotic organisms.

The most common vector used for LPMO production in *P. pastoris* is the commercially available pICZα (Invitrogen) (Bennati-Granier et al., 2015; Couturier et al., 2018; Kittl et al., 2012), although other vectors such as the pPpT4 vector can also be used (Tanghe et al., 2015). These are shuttle vectors which allow all of the cloning to be performed in *E. coli* for speed and ease before the vector is introduced into the *P. pastoris* host for gene expression. As for expression in *E. coli*, similar factors need to be considered during construct design in *P. pastoris*. The vectors contain signal sequences to direct the protein for secretion and so the LPMO sequence must be carefully cloned such that the codon for the N-terminal histidine is immediately placed after the cleavage site for the signal peptide. Alternatively, if the protein is from a fungal source the native leader sequence of the LPMO gene can also be used rather than the signal peptide available in the vector (Kittl et al., 2012; Tanghe et al., 2015).

The target gene is typically placed under the control of the AOX1 promoter in these vectors, allowing induction of protein expression using methanol (Bennati-Granier et al., 2015; Couturier et al., 2018; Kittl et al., 2012). Cells are typically grown in defined media based on the protocol provided by Invitrogen so we will not redescribe this here. Thanks to the use of the signal sequence the protein is directed out of the cell into the media and so the protein needs to be directly purified from here. Affinity tags can be incorporated at the C-terminus of the protein to assist in the purification, but in many cases the protein has been purified without the use of an affinity tag, for example, Kittl et al. (2012) used a multistep purification process involving hydrophobic chromatography, ion-exchange chromatography, and finally size-exclusion chromatography.

Copper-loading procedures are much the same as those described earlier for LPMOs expressed in *E. coli*, with the addition of a small excess of copper prior to the final gel-filtration step in the purification. If the protein is to be used for crystallography, it is often necessary to deglycosylate it. In our laboratories this step is typically achieved using Endoglycosidase H (New England Biolabs) following the manufacturer's protocols. If, however, the protein is to be used in assays, we typically do not deglycosylate the protein as this can affect both protein stability and, potentially, protein–substrate interactions.

2.4 Recombinant expression and purification from fungi

The majority of LPMOs that have been studied to date are found in fungi and so there are clear advantages to expressing these proteins in fungal hosts. These advantages include correct methylation on the N-terminal histidine side chain as well as protein secretion into the extracellular milieu following correct glycosylation and protein folding. Fungal recombinant expression systems, however, are far less well established in academic laboratories. We have, therefore, benefited greatly from our long-term collaboration with Novozymes who use a highly specialized *A. oryzae* expression system for LPMO production on a large scale (Frandsen et al., 2016; Lo Leggio et al., 2015; Quinlan et al., 2011). The high yields offered by this system are able to meet the high sample demands for EPR (see later). These systems are not readily available for use by others, however, and so examples of recombinant LPMO expression from fungal sources are harder to find in the literature. Phillips et al. (2011) early on were able to purify *Neurospora crassa* LPMOs directly from secretomes when the fungus was grown on cellulose. Since then, the Marletta lab and their colleagues have been able to use a *N. crassa* expression system to produce both *N. crassa* LPMOs (Vu, Beeson, Phillips, et al., 2014) and LPMOs from other fungi (Span, Suess, Deller, Britt, & Marletta, 2017). In this approach, a knock-in system is used to insert the LPMO gene downstream of a glyceraldehyde-3-phosphate dehydrogenase promoter (GAP) (Bardiya & Shiu, 2007). The DNA is then incorporated into the *N. crassa* genome by recombination, and colonies in which the gene has been successfully incorporated need to be screened for LPMO production. As for *P. pastoris*, the LPMO should be secreted into the growth media allowing direct purification from the cleared media. Affinity tags can be incorporated at the C-terminus of the protein to assist in this respect if they do not interfere with subsequent planned experiments.

3. EPR spectroscopy

Once pure protein is obtained, a key technique that has been used in the field to gain insight into the copper environment in LPMOs is EPR spectroscopy. The fact that copper in its oxidized (+2) state ($S = ½$, Kramers' doublet) has a single unpaired electron makes it EPR active. Therefore, by collecting EPR spectra from pure LPMOs it has been possible to gain additional insight into the electronic state of the copper center in the resting protein. In addition, upon the inclusion of polysaccharide substrates in the samples, it is sometimes possible to observe marked changes in the EPR spectra (Fig. 3) (Bissaro, Isaksen, Vaaje-Kolstad, Eijsink, & Røhr, 2018; Borisova et al., 2015; Frandsen et al., 2016; Simmons et al., 2017), suggesting that significant changes occur at the copper active site in the presence of substrate, which might be of key importance to activating oxygen and hence catalysis. In order to interpret these spectra and understand what the changes in the EPR mean, it is necessary to simulate the data. While there are algorithms that will attempt to simulate the data automatically, these often fall short of a truly meaningful simulation. Optimal EPR simulation is, therefore, still best performed by an expert EPR spectroscopist and requires considerable patience and knowhow in order to obtain the best possible outcome. Here, we describe the steps taken to prepare samples for EPR analysis and provide information on how we interpret our EPR spectra.

3.1 EPR sample preparation for LPMOs

Most EPR spectroscopy on LPMOs has been performed in frozen solution at X-band frequencies (ca. 9.3 GHz). At these frequencies, sample holders

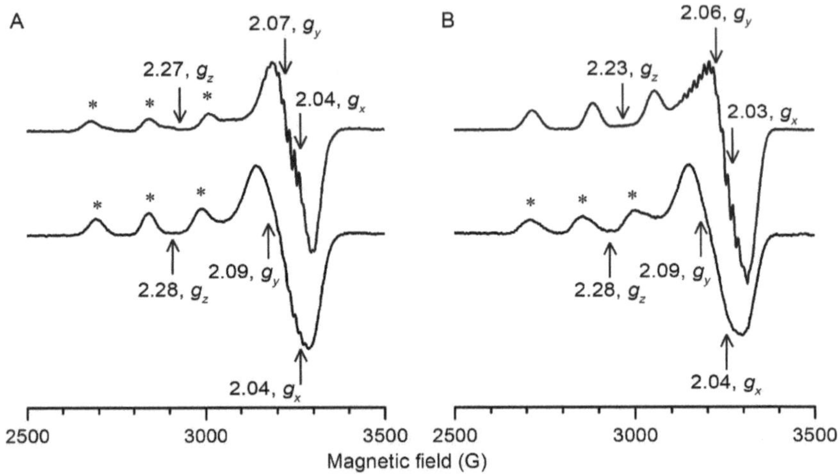

Fig. 3 Examples of changes observed in the X-band CW EPR spectra of an AA9 upon addition of substrate. (A) *Ls*AA9A–G6 (*top*) and *Ls*AA9A with no cellohexaose (*bottom*) in low chloride conditions. (B) *Ls*AA9A–G6 (*top*) and *Ls*AA9A with no cellohexaose (*bottom*) in high chloride conditions. * Indicates signals from copper site where water is the exogenous ligand, other signals are from species with chloride as exogenous ligand. *Figure adapted from Frandsen, K. E. H., Simmons, T. J., Dupree, P., Poulsen, J.-C. N., Hemsworth, G. R., Ciano, L., et al. (2016). The molecular basis of polysaccharide cleavage by lytic polysaccharide monooxygenases.* Nature Chemical Biology, 12, *298–303.*

are typically quartz tubes with roughly 3–5 mm internal diameters. The height of the resonator cavity will be of the order of 10–15 mm, meaning that ideally a sample volume of ca. 300–600 μL is required. In reality, by optimizing the collection parameters, good EPR spectra can be obtained with much smaller volumes, in the order of 80–120 μL with a concentration of at least 0.15 m*M* Cu. Albeit often seen by biochemists as a highly demanding technique in terms of protein consumption, EPR is nondestructive, and often samples can be fully recovered and reused after the experiments. In our hands, the repeated cooling/warming cycles that a sample typically experiences during collection of EPR spectra do not lead to protein degradation. Obtaining a signal with good signal-to-noise ratio depends on several experimental variables including the capabilities of the EPR spectrometer, the temperature at which the spectra are collected, the number of scans and the experimental settings. However, good spectra with excellent signal-to-noise ratios can be obtained with the sample concentrations and volumes described above at 100–170 K with acquisition settings reported in Table 1. Of particular note is the power setting which must be at a level that does not

Table 1 Standard acquisition parameters for frozen solution X-band EPR spectra of LPMOs

Centre field	3000 G
Sweep	2300–3700 G
Modulation amplitude	4 G
Modulation frequency	100 kHz
Time constant	163.8 ms
Sweep time	92 s
Conversion time	90 ms
Power	5–10 mW
Scans	1–8
Temperature	150–170 K

cause saturation, especially in concentrated samples. This level should be such that relaxation rates are greater than absorption rates. For copper spectroscopy, relatively high power settings can be used because of the high relaxation rates associated with a metal center. For detailed spectra, higher sample concentrations may be needed, possibly up to 1–2 mM concentration. Great caution is needed at these concentrations, however, since face-to-face dimerization of LPMOs may occur with the result that two nearby Cu centers will change their coordination geometries or will lead to dipolar relaxation effects that perturb spin Hamiltonian values. Our experience in this regard is that concentrations of ca. 2 mM lead to spectral changes that could be associated with protein dimerization.

EPR spectra collected in aqueous solution often require the use of a glassing agent such as glycerol or sucrose to minimize the formation of ice crystals. Such crystals can increase microwave scattering, reduce the microwave cavity quality factor, cause changes in protein structure, pH, or ionic strength, or even shatter the microwave tube! For smaller tubes which are used with high frequency EPR this can be a particular problem. Tube breakages for X-band EPR tubes are less common, especially if thick wall tubes are used. For work with LPMO solutions, glycerol is typically the glassing agent of choice (10%–50% v/v), although sucrose can also be used (2 mg/μL). When combined with very rapid freezing the resulting sample usually forms a good glass. The cooling method needs to be carefully considered, however, in that very rapid freezing is required to prevent undesirable processes from occurring at the metal center. For instance, at room temperature the copper coordination may include the three atoms of the ligating histidine brace along with a fourth exogenous ligand. Cooling slowly to typical EPR collection temperatures may drive the formation of a five coordinate Cu site in which a further water molecule is recruited into the coordination sphere. The result would be that the EPR spectrum collected at low temperature is not representative of the solution phase structure. Therefore, very rapid freezing is required to "trap" the room temperature species. Rapid freezing is best achieved by plunging the EPR sample tube into a cold liquid (e.g., methanol in dry ice Chaplin et al., 2016) rather than into liquid nitrogen where—despite the very low temperature of the liquid—the resulting bubbles of N_2 insulate the tube from the liquid to prevent very rapid freezing. Notwithstanding the advantages of rapid cooling of samples, the experimenter needs to be aware that any heterogeneity in protein structure, where the interconversion is slower than the cooling timescale, may be trapped by the cooling process, such that sample heterogeneity is increased, the result of which is a severe loss of spectral resolution. It is not unusual for rapidly cooled samples to have broader spectral peaks than samples cooled less rapidly (Guzzi, Stirpe, Verbeet, & Sportelli, 2001). Necessarily, therefore, different cooling techniques need to be explored. In our experience, a happy compromise can be found for X-band LPMO samples when a sample at room temperature is plunged into the stream of cooling N_2 gas which cools the resonator. Also, on occasion for high concentration samples, it has been possible to collect spectra of LPMO solutions at room temperature. The slow tumbling rate of the protein in solution means that the spectrum is still anisotropic, from which it is possible to link the spin Hamiltonian parameters at room temperature with those at low temperature (unpublished data).

The type of glassing agent that is employed in these studies is an important consideration in experimental design. The principal consideration is that glassing agents have a "poly-ol" structure which can mimic that of the natural polysaccharide substrate of the enzyme and possibly interfere with the Cu active site by binding close to it. Accordingly, the dependency of the spectrum on the concentration of glassing agent should be established in any experiment. Unfortunately, an examination of the LPMO literature shows that such dependency experiments are relatively rare and only appear to have been done in a handful of cases (Sabbadin et al., 2018). When there is evidence to suggest that the glassing agent is interfering with the Cu center, then it is recommended that the agent be changed to another (e.g., DMSO, Ficoll) or that the spectrum is collected without any glassing agent at all. The latter method appears to work surprisingly well for LPMO samples (Bissaro et al., 2018; Borisova et al., 2015; Chaplin et al., 2016; Forsberg, Mackenzie, et al., 2014; Forsberg et al., 2016; Forsberg, Røhr, et al., 2014).

One of the great advantages of EPR is that spectra can be obtained on a sample in any state of matter: liquid, solid, slurry, frozen glass, or even a gas. For LPMOs, of course, since the substrates are usually insoluble polysaccharides, the interaction of the LPMO with its natural substrate can be studied directly. Moreover, introducing a solid phase sample into a solution to create a slurry essentially removes the need for a glassing agent. Necessarily, the nature of the solid phase means that protein binding may be slow or incomplete and it is advisable to leave the slurry sample for some time before then acquiring the EPR spectra. In the majority of cases, at least part of the protein binds very rapidly (Frandsen et al., 2016; Simmons et al., 2017), but for some samples we have noticed that full binding can take up to several hours and/or cannot be achieved, so we would recommend that samples are left incubating overnight in the absence of oxygen and/or a reducing agent. It is also important that the introduced sample is isotropically distributed within the EPR tube, otherwise alignment effects could complicate the spectrum. Beyond these considerations, it appears as if solid state slurries of LPMO solutions with their insoluble substrates are perfectly good samples for both CW and pulsed EPR studies (Bissaro et al., 2018; Borisova et al., 2015; Frandsen et al., 2016; Simmons et al., 2017).

3.2 Multifrequency collection

EPR spectra of LPMOs in the Cu(II) states give spin Hamiltonian parameters in which there is either a typical "type 2" Cu center with large $|A_z|$ values (ca. 400–600 MHz) (Borisova et al., 2015; Chaplin et al., 2016; Couturier et al., 2018; Forsberg, Mackenzie, et al., 2014; Forsberg et al., 2016; Frandsen et al., 2016; Hansson et al., 2017; Hemsworth et al., 2014; Quinlan et al., 2011; Sabbadin et al., 2018; Simmons et al., 2017; Span et al., 2017) or a more rhombic Cu(II) center that has a reduced $|A_z|$ value but large $|A_x|$ and $|A_y|$ values (Bissaro et al., 2018; Forsberg, Røhr, et al., 2014; Gregory et al., 2016; Hemsworth, Taylor, et al., 2013). In both cases the perpendicular region of the spectrum cannot be unambiguously simulated from the X-band data alone. For this reason, if an in-depth analysis is required, collecting spectra at different frequencies is paramount. Depending on the kind of information desired, one can decide to use lower (e.g., S-band) or higher (e.g., Q-band) frequency than X-band, with the latter being usually the preferred method. A couple of examples of this approach are present in the LPMO literature (Couturier et al., 2018; Hansson et al., 2017), but it is easy to imagine that more and more will appear as finer details of the electronic structure at the Cu active site are investigated. For Q-band spectra, quartz tubes of ca. 2 mm in diameter are employed, while the samples are typically 15–30 μL in volume at a concentration of copper of 1 m*M* or higher. The same, if not greater, caution in optimizing the acquisition settings as for X-band should be exercised when collecting Q-band spectra. In our experience, obtaining good quality Q-band spectra is a nontrivial problem, especially if protein concentration is an issue due to protein aggregation at high concentration or a limited amount of protein is available, but it can sometimes be overcome with patience and trials of different settings. Typical settings used for our experiments include center field at 1150 mT, sweep width 150 mT, modulation width 0.6–1 mT, time constant 0.3 s, scan time 1–2 min, power 0.5–1 mW.

Collecting CW Q-band spectra has a further benefit in that EPR spectra of LPMOs can sometimes contain high-field features (Bissaro et al., 2018). These features under simulation (see later) are very hard to assign unambiguously as they could arise either from large hyperfine-coupling constants or from low principal *g* values. There are indeed cases in the literature where such features have been simulated with low *g* values. It is, however, well known in the wider EPR community that the high-field signals may arise from a spectral artifact known as an overshoot feature, associated with large $|A_z|$ hyperfine-coupling values (see Fig. 4). By collecting CW EPR spectra at different frequencies, the normal choice being Q-band, this misinterpretation can be avoided.

3.3 Simulation procedures

Once the CW spectra have been recorded, simulations have to be performed in order to gain information about the copper active site. Several simulation packages are available to users, some of them developed by EPR instrument manufacturers (e.g., the WinEPR and SimFonia suites Bruker, n.d.), others made available by researchers in the field (e.g., SpinCount The Hendrich Metalloprotein Group, n.d. or EasySpin Stoll & Schweiger, 2006). Regardless of the program used to carry out the simulations, some general considerations are needed on how to approach such a task, in particular for

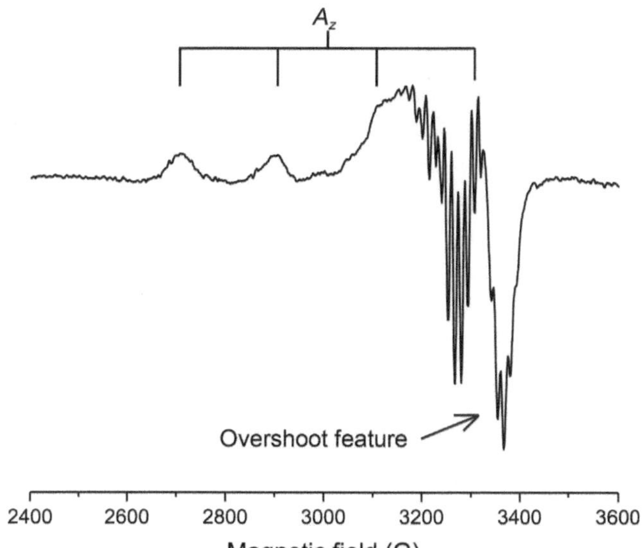

Fig. 4 X-band CW EPR spectrum of an AA10 from *Bacillus amyloliquefaciens*, *Ba*AA10, after addition of squid pen chitin, showing the overshoot feature associated with the large $|A_z|$ value (unpublished data, courtesy of Peter Lindley, York).

those who might not have encountered EPR before. Determining the spin Hamiltonian parameters for the parallel region of the spectrum is in general straightforward, as at least three of the four hyperfine peaks to the Cu can be clearly observed between 2500 and 3100 G in the X-band spectra of LPMOs. Careful analysis of this region of the spectrum can also reveal the presence of multiple species or adventitious binding sites, if a second set of hyperfine peaks is present either as individual peaks or as shoulders of the main component of the spectrum. Specifically, we would recommend particular care if the LPMO carries an affinity tag, as highlighted in Section 2.1. In these circumstances, EPR provides a very powerful tool for the identification of second binding sites in amounts as low as 10% of the main species. As mentioned earlier, most LPMOs show a type 2 copper active site according to the Peisach–Blumberg (P–B) classification (Peisach & Blumberg, 1974), with the SOMO mostly $d(x^2-y^2)$ in character, and a close to square planar geometry. While the P–B classification is a useful "first-step" guide to the general structure of the copper ion in the active sites of LPMOs, we would caution against its in-depth use for the assignment of coordinating atoms (e.g., determining between CuN_2O_2 and

CuN_3O coordination spheres), due to the multiple contributing factors to both g_z and $|A_z|$ values, the most problematic of which is the unknown degree of covalency in the metal–ligand bonds. In this regard, it is indeed possible to notice the high degree of variability in the published EPR parameters of LPMOs in a P–B plot (Fig. 5 and Table 2) even for proteins that, from a structural point of view, show the same coordination environment (recently reviewed by Ciano et al., 2018; Vu & Ngo, 2018). Finally, in the context of the broad interpretation of spin Hamiltonian parameters of LPMOs, there are reports of compressed trigonal bipyramidal structures with $d(z^2)$ SOMOs, but this assignment is not in accord with the spectral envelopes observed in all LPMO EPR spectra (Vu & Ngo, 2018).

The perpendicular region of the spectrum, in contrast, is not as straightforward to interpret and simulate, especially from X-band data alone. Here, the partial overlap of the g_x/g_y and $|A_x|/|A_y|$ values and the usually broad appearance of the spectrum do not allow unambiguous determination of the spin Hamiltonian parameters of the copper ion. Great care should therefore be used in these circumstances. The problem can be overcome by simultaneously simulating both X- and Q-band data, when multifrequency collection of the spectra is a possibility. Simultaneous fits of the data from two different frequencies significantly narrows the set of possible spin Hamiltonian values, with the further advantage that artifacts arising from overshoot features, whether present, can be all but eliminated.

Analysis of the superhyperfine (SHF) coupling, when visible, can provide insights into the coordination environment of the copper active site and possible changes triggered by the binding of substrate (Frandsen et al., 2016). Simulation of the SHF coupling can be a very time-consuming and, at times, frustrating challenge, but it can provide key information about the active site. In our experience, a reliable simulation of SHF coupling patterns can take many days to perform accurately. Moreover, it is important to point out that the most direct and reliable method to obtain coupling values to coupled nuclei is the use of pulsed EPR techniques such as ENDOR or HYSCORE (not discussed herein), in which the orientation-selective feature of the techniques allows some separation of the individual coupling components of each coordinating atom. Pulsed techniques are not easily accessible and can require high amounts of protein, but careful analysis of CW spectra can give surprisingly accurate information about the type and number of nuclei coupled to the unpaired electron and their coupling values. The risk of overparameterizing the problem should be taken into account, and we would recommend proceeding in steps and adding further

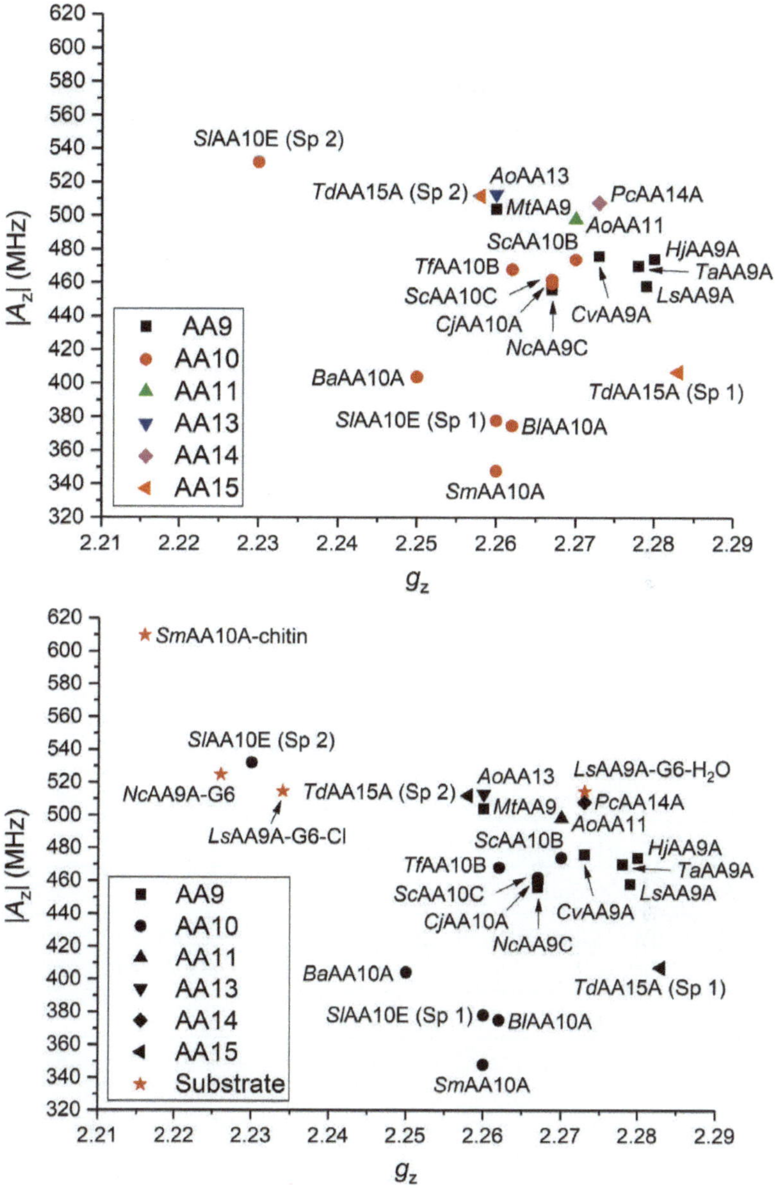

Fig. 5 Peisach–Blumberg plots of published EPR data for LPMOs (data and references reported in Table 2). For clarity and consistency, all enzymes have been named using the CAZy abbreviation, where two letters (*in italics*) define the organism the protein originates from, followed by the Auxiliary Activity family and the specific enzyme, if reported and more than one are present in the same organism (e.g., *Ls*AA9A). *Top*: P–B plot of LPMOs divided by family, according to the legend on the plot. *Bottom*: P–B plot as above, with the parameters for substrate-bound enzymes included; "G6" is used for cellohexaose.

Table 2 Table of g_z and $|A_z|$ LPMO EPR values published to date, used for the plots in Fig. 5

LPMO	References	g_z	$\|A_z\|$ (MHz)
*Nc*AA9C	Borisova et al. (2015)	2.267	456
*Ta*AA9A	Quinlan et al. (2011)	2.278	470
*Hj*AA9A	Hansson et al. (2017)	2.280	474
*Mt*AA9	Span et al. (2017)	2.260	504
*Cv*AA9A	Simmons et al. (2017)	2.273	476
*Ls*AA9A	Frandsen et al. (2016)	2.279	458
*Ba*AA10A	Hemsworth, Taylor, et al. (2013)	2.25	404
*Cj*AA10A	Forsberg et al. (2016)	2.267	462
*Sm*AA10A	Forsberg, Røhr, et al. (2014)	2.260	348
*Sc*AA10C		2.267	459
*Bl*AA10		2.262	375
*Tf*AA10B		2.262	468
*Sc*AA10B	Forsberg, Mackenzie, et al. (2014)	2.270	474
*Sl*AA10E	Chaplin et al. (2016)	2.260 (Sp 1) 2.230 (Sp 2)	378 (Sp 1) 532 (Sp 2)
*Ao*AA11	Hemsworth et al. (2014)	2.27	498
*Ao*AA13	Lo Leggio et al. (2015)	2.26	513
*Pc*AA14A	Couturier et al. (2018)	2.273	508
*Td*AA15A	Sabbadin et al. (2018)	2.283 (Sp 1) 2.258 (Sp 2)	407 (Sp 1) 512 (Sp 2)
*Nc*AA9C-G6	Borisova et al. (2015)	2.226	525
*Ls*AA9A-G6-H_2O	Frandsen et al. (2016)	2.273	515
*Ls*AA9A-G6-Cl		2.234	515
*Sm*AA10A-chitin	Bissaro et al. (2018)	2.216	610

coupled nuclei only if the experimental data cannot be satisfactory simulated otherwise. Simultaneous simulations of both first- and second-derivative spectra can be of great help in establishing the coupling values, although usually only the largest component of the coupling, when anisotropic, can be determined. If possible, isotopic labeling (e.g., ^{15}N) should be performed in order to restrict the set of possible values which would give a good fit to the experimental data.

Finally, some authors have used DFT calculations to augment their interpretations of EPR data (Bissaro et al., 2018). While useful in a broad context, there are well-known problems in using DFT to give accurate representations of spin Hamiltonian parameters (Neese, 2009, 2017). Indeed, these problems appear to be evident in DFT calculations performed on LPMOs. The key difficulties stem from the issues associated with the inaccurate modeling of the Fermi contact term (which, in turn, arises from configuration interaction of the SOMO with s-orbitals and polarization of core electrons at the copper ion), the relatively poor modeling of medium-distance electron–electron interactions, the important role of dispersion forces, and the tendency of DFT calculations to overly apply covalency to metal–ligand bonding. In partial mitigation of these problems, we would recommend using a basis set in which core functions are enhanced (e.g., CP(PPP), Neese, 2002). For the latter issue, a functional should be selected in which the degree of Hartree–Fock (HF) exchange has been manually increased, e.g., the B3LYP-D functional has 20% HF exchange set as default, but this should be increased to ~38% with the final value being determined by trial and error. However, as is evident in the preceding sentence, this is necessarily an empirical procedure and our experience in this regard is that the spin Hamiltonian parameters obtained from DFT calculations for LPMOs can be significantly different from the experimental values. As such, great caution must be exercised in using DFT-determined spin Hamiltonian parameters for in-depth interpretation of the active site electronics, even when there appears to be good agreement with the experimental values. Indeed, whether the results of DFT calculations on spin Hamiltonian parameters can give useful absolute results at all in such a complicated system is debatable. Our advice in this regard therefore is that trends in calculated parameters may be meaningful (when the same basis sets, functionals, and methods are used between systems), as might the absolute signs of calculated hyperfine-coupling values, but that further values should *not* be used as a basis for in-depth analysis of the electronic structure of the SOMO without a critical evaluation of the underlying issues. In this context, our further advice is that a ligand field analysis in which many of the "problematic" values are parameterized offers a more reliable and suitably caveated analysis of LPMO EPR spectra, from which appropriately framed conclusions can be drawn.

Acknowledgments

G.R.H. gratefully acknowledges support from a BBSRC (Biotechnology and Biological Sciences Research Council) David Phillips Fellowship (grant # BB/N019970/1). L.C., G.J.D., and P.H.W. also gratefully acknowledge the support of the BBSRC under grant BB/L001926/1.

References

Agger, J. W., Isaksen, T., Várnai, A., Vidal-Melgosa, S., Willats, W. G. T., Ludwig, R., et al. (2014). Discovery of LPMO activity on hemicelluloses shows the importance of oxidative processes in plant cell wall degradation. *Proceedings of the National Academy of Sciences of the United States of America, 111*, 6287–6292.

Bardiya, N., & Shiu, P. K. T. (2007). Cyclosporin A-resistance based gene placement system for *Neurospora crassa. Fungal Genetics and Biology, 44*, 307–314.

Beeson, W. T., Vu, V. V., Span, E. A., Phillips, C. M., & Marletta, M. A. (2015). Cellulose degradation by polysaccharide monooxygenases. *Annual Review of Biochemistry, 84*, 923–946.

Bennati-Granier, C., Garajova, S., Champion, C., Grisel, S., Haon, M., Zhou, S., et al. (2015). Substrate specificity and regioselectivity of fungal AA9 lytic polysaccharide monooxygenases secreted by *Podospora anserina. Biotechnology for Biofuels, 8*, 90.

Bey, M., Zhou, S., Poidevin, L., Henrissat, B., Coutinho, P. M., Berrin, J.-G., et al. (2013). Cello-oligosaccharide oxidation reveals differences between two lytic polysaccharide monooxygenases (family GH61) from *Podospora anserina. Applied and Environmental Microbiology, 79*, 488–496.

Bischof, R. H., Ramoni, J., & Seiboth, B. (2016). Cellulases and beyond: The first 70 years of the enzyme producer *Trichoderma reesei. Microbial Cell Factories, 15*, 106.

Bissaro, B., Isaksen, I., Vaaje-Kolstad, G., Eijsink, V. G. H., & Røhr, Å. K. (2018). How a lytic polysaccharide monooxygenase binds crystalline chitin. *Biochemistry, 57*, 1893–1906.

Bissaro, B., Røhr, A. K., Müller, G., Chylenski, P., Skaugen, M., Forsberg, Z., et al. (2017). Oxidative cleavage of polysaccharides by monocopper enzymes depends on H_2O_2. *Nature Chemical Biology, 851*, 1.

Borisova, A. S., Isaksen, T., Dimarogona, M., Kognole, A. A., Mathiesen, G., Várnai, A., et al. (2015). Structural and functional characterization of a lytic polysaccharide monooxygenase with broad substrate specificity. *Journal of Biological Chemistry, 290*, 22955–22969.

Bruker, n.d. EPR Simulation Suites. https://www.bruker.com/products/mr/epr/epr-software/simulation-suites/overview.html.

Chaplin, A. K., Wilson, M. T., Hough, M. A., Svistunenko, D. A., Hemsworth, G. R., Walton, P. H., et al. (2016). Heterogeneity in the histidine-brace copper coordination sphere in auxiliary activity family 10 (AA10) lytic polysaccharide monooxygenases. *Journal of Biological Chemistry, 291*, 12838–12850.

Chiu, E., Hijnen, M., Bunker, R. D., Boudes, M., Rajendran, C., Aizel, K., et al. (2015). Structural basis for the enhancement of virulence by viral spindles and their in vivo crystallization. *Proceedings of the National Academy of Sciences of the United States of America, 112*, 3973–3978.

Ciano, L., Davies, G. J., Tolman, W. B., & Walton, P. H. (2018). Bracing copper for the catalytic oxidation of C–H bonds. *Nature Catalysis, 1*, 571–577.

Courtade, G., Le, S. B., Sætrom, G. I., Brautaset, T., & Aachmann, F. L. (2017). A novel expression system for lytic polysaccharide monooxygenases. *Carbohydrate Research, 448*, 212–219.

Couturier, M., Ladevèze, S., Sulzenbacher, G., Ciano, L., Fanuel, M., Moreau, C., et al. (2018). Lytic xylan oxidases from wood-decay fungi unlock biomass degradation. *Nature Chemical Biology*, *14*, 306–310.

Crouch, L. I., Labourel, A., Walton, P. H., Davies, G. J., & Gilbert, H. J. (2016). The contribution of non-catalytic carbohydrate binding modules to the activity of lytic polysaccharide monooxygenases. *Journal of Biological Chemistry*, *291*, 7439–7449.

Forsberg, Z., Mackenzie, A. K., Sørlie, M., Røhr, Å. K., Helland, R., Arvai, A. S., et al. (2014). Structural and functional characterization of a conserved pair of bacterial cellulose-oxidizing lytic polysaccharide monooxygenases. *Proceedings of the National Academy of Sciences of the United States of America*, *111*, 8446–8451.

Forsberg, Z., Nelson, C. E., Dalhus, B., Mekasha, S., Loose, J. S. M., Crouch, L. I., et al. (2016). Structural and functional analysis of a lytic polysaccharide monooxygenase important for efficient utilization of chitin in *Cellvibrio japonicus*. *Journal of Biological Chemistry*, *291*, 7300–7312.

Forsberg, Z., Røhr, Å. K., Mekasha, S., Andersson, K. K., Eijsink, V. G. H., Vaaje-Kolstad, G., et al. (2014). Comparative study of two chitin-active and two cellulose-active AA10-type lytic polysaccharide monooxygenases. *Biochemistry*, *53*, 1647–1656.

Forsberg, Z., Vaaje-Kolstad, G., Westereng, B., Bunæs, A. C., Stenstrøm, Y., MacKenzie, A., et al. (2011). Cleavage of cellulose by a CBM33 protein. *Protein Science*, *20*, 1479–1483.

Frandsen, K. E. H., Simmons, T. J., Dupree, P., Poulsen, J.-C. N., Hemsworth, G. R., Ciano, L., et al. (2016). The molecular basis of polysaccharide cleavage by lytic polysaccharide monooxygenases. *Nature Chemical Biology*, *12*, 298–303.

Frommhagen, M., Sforza, S., Westphal, A. H., Visser, J., Hinz, S. W. A., Koetsier, M. J., et al. (2015). Discovery of the combined oxidative cleavage of plant xylan and cellulose by a new fungal polysaccharide monooxygenase. *Biotechnology for Biofuels*, *8*, 101.

Gardner, J. G., Crouch, L., Labourel, A., Forsberg, Z., Bukhman, Y. V., Vaaje-Kolstad, G., et al. (2014). Systems biology defines the biological significance of redox-active proteins during cellulose degradation in an aerobic bacterium. *Molecular Microbiology*, *94*, 1121–1133.

Good, N. E., Winget, G. D., Winter, W., Connolly, T. N., Izawa, S., & Singh, R. M. (1966). Hydrogen ion buffers for biological research. *Biochemistry*, *5*, 467–477.

Gregory, R. C., Hemsworth, G. R., Turkenburg, J. P., Hart, S. J., Walton, P. H., & Davies, G. J. (2016). Activity, stability and 3-D structure of the Cu(ii) form of a chitin-active lytic polysaccharide monooxygenase from *Bacillus amyloliquefaciens*. *Dalton Transactions*, *45*, 16904–16912.

Guzzi, R., Stirpe, A., Verbeet, M., & Sportelli, L. (2001). Structural heterogeneity of blue copper proteins: An EPR study of amicyanin and of wild-type and Cys3Ala/Cys26Ala mutant azurin. *European Biophysics Journal*, *30*, 171–178.

Hansson, H., Karkehabadi, S., Mikkelsen, N., Douglas, N. R., Kim, S., Lam, A., et al. (2017). High-resolution structure of a lytic polysaccharide monooxygenase from Hypocrea jecorina reveals a predicted linker as an integral part of the catalytic domain. *Journal of Biological Chemistry*, *292*, 19099–19109.

Harris, P. V., Welner, D., McFarland, K. C., Re, E., Navarro Poulsen, J.-C., Brown, K., et al. (2010). Stimulation of lignocellulosic biomass hydrolysis by proteins of glycoside hydrolase family 61: Structure and function of a large, enigmatic family. *Biochemistry*, *49*, 3305–3316.

Harris, P. V., Xu, F., Kreel, N. E., Kang, C., & Fukuyama, S. (2014). New enzyme insights drive advances in commercial ethanol production. *Current Opinion in Chemical Biology*, *19*, 162–170.

Hemsworth, G. R., Davies, G. J., & Walton, P. H. (2013). Recent insights into copper-containing lytic polysaccharide mono-oxygenases. *Current Opinion in Structural Biology*, *23*, 660–668.

Hemsworth, G. R., Henrissat, B., Davies, G. J., & Walton, P. H. (2014). Discovery and characterization of a new family of lytic polysaccharide monooxygenases. *Nature Chemical Biology*, *10*, 122–126.
Hemsworth, G. R., Johnston, E. M., Davies, G. J., & Walton, P. H. (2015). Lytic polysaccharide monooxygenases in biomass conversion. *Trends in Biotechnology*, *33*, 747–761.
Hemsworth, G. R., Taylor, E. J., Kim, R. Q., Gregory, R. C., Lewis, S. J., Turkenburg, J. P., et al. (2013). The copper active site of CBM33 polysaccharide oxygenases. *Journal of the American Chemical Society*, *135*, 6069–6077.
Himmel, M. E., Ding, S.-Y., Johnson, D. K., Adney, W. S., Nimlos, M. R., Brady, J. W., et al. (2007). Biomass recalcitrance: Engineering plants and enzymes for biofuels production. *Science (New York, NY)*, *315*, 804–807.
Horn, S. J., Vaaje-Kolstad, G., Westereng, B., & Eijsink, V. G. (2012). Novel enzymes for the degradation of cellulose. *Biotechnology for Biofuels*, *5*, 45.
Karkehabadi, S., Hansson, H., Kim, S., Piens, K., Mitchinson, C., & Sandgren, M. (2008). The first structure of a glycoside hydrolase family 61 member, Cel61B from Hypocrea jecorina, at 1.6 A resolution. *Journal of Molecular Biology*, *383*, 144–154.
Karlsson, J., Saloheimo, M., Siika-Aho, M., Tenkanen, M., Penttilä, M., & Tjerneld, F. (2001). Homologous expression and characterization of Cel61A (EG IV) of *Trichoderma reesei*. *European Journal of Biochemistry*, *268*, 6498–6507.
Kittl, R., Kracher, D., Burgstaller, D., Haltrich, D., & Ludwig, R. (2012). Production of four *Neurospora crassa* lytic polysaccharide monooxygenases in *Pichia pastoris* monitored by a fluorimetric assay. *Biotechnology for Biofuels*, *5*, 79.
Klock, H. E., & Lesley, S. A. (2009). The polymerase incomplete primer extension (PIPE) method applied to high-throughput cloning and site-directed mutagenesis. *Methods in Molecular Biology*, *498*, 91–103.
Kuusk, S., Bissaro, B., Kuusk, P., Forsberg, Z., Eijsink, V. G. H., Sørlie, M., et al. (2018). Kinetics of H2O2-driven degradation of chitin by a bacterial lytic polysaccharide monooxygenase. *Journal of Biological Chemistry*, *293*, 523–531.
Langston, J. A., Shaghasi, T., Abbate, E., Xu, F., Vlasenko, E., & Sweeney, M. D. (2011). Oxidoreductive cellulose depolymerization by the enzymes cellobiose dehydrogenase and glycoside hydrolase 61. *Applied and Environmental Microbiology*, *77*, 7007–7015.
Levasseur, A., Drula, E., Lombard, V., Coutinho, P. M., & Henrissat, B. (2013). Expansion of the enzymatic repertoire of the CAZy database to integrate auxiliary redox enzymes. *Biotechnology for Biofuels*, *6*, 41.
Li, X., Beeson, W. T., Phillips, C. M., Marletta, M. A., & Cate, J. H. D. (2012). Structural basis for substrate targeting and catalysis by fungal polysaccharide monooxygenases. *Structure*, *20*, 1051–1061.
Lo Leggio, L., Simmons, T. J., Poulsen, J.-C. N., Frandsen, K. E. H., Hemsworth, G. R., Stringer, M. A., et al. (2015). Structure and boosting activity of a starch-degrading lytic polysaccharide monooxygenase. *Nature Communications*, *6*, 5961.
Lombard, V., Golaconda Ramulu, H., Drula, E., Coutinho, P. M., & Henrissat, B. (2014). The carbohydrate-active enzymes database (CAZy) in 2013. *Nucleic Acids Research*, *42*, D490–D495.
Mandels, M., Hontz, L., Nystrom, J., Lee, R., & Lynd, I. B. (1974). Enzymatic hydrolysis of waste cellulose. *Biotechnology and Bioengineering*, *16*, 1471–1493.
Mandels, M., Parrish, F. W., & Reese, E. T. (1962). Sophorose as an inducer of cellulase in *Trichoderma viride*. *Journal of Bacteriology*, *83*, 400–408.
Mandels, M., & Reese, E. T. (1957). Induction of cellulase in *Trichoderma viride* as influenced by carbon sources and metals. *Journal of Bacteriology*, *73*, 269–278.
Mandels, M., & Reese, E. T. (1960). Induction of cellulase in fungi by cellobiose. *Journal of Bacteriology*, *79*, 816–826.

Mandels, M., Weber, J., & Parizek, R. (1971). Enhanced cellulase production by a mutant of *Trichoderma viride*. *Applied Microbiology, 21*, 152–154.

Merino, S. T., & Cherry, J. (2007). Progress and challenges in enzyme development for biomass utilization. *Advances in Biochemical Engineering/Biotechnology, 108*, 95–120.

Morgenstern, I., Powlowski, J., & Tsang, A. (2014). Fungal cellulose degradation by oxidative enzymes: From dysfunctional GH61 family to powerful lytic polysaccharide monooxygenase family. *Briefings in Functional Genomics, 13*, 471–481.

Naik, S. N., Goud, V. V., Rout, P. K., & Dalai, A. K. (2010). Production of first and second generation biofuels: A comprehensive review. *Renewable and Sustainable Energy Reviews, 14*, 578–597.

Nakagawa, Y. S., Kudo, M., Loose, J. S. M., Ishikawa, T., Totani, K., Eijsink, V. G. H., et al. (2015). A small lytic polysaccharide monooxygenase from *Streptomyces griseus* targeting α- and β-chitin. *The FEBS Journal, 282*, 1065–1079.

Neese, F. (2002). Prediction and interpretation of the 57Fe isomer shift in mössbauer spectra by density functional theory. *Inorganica Chimica Acta, 337*, 181–192.

Neese, F. (2009). Prediction of molecular properties and molecular spectroscopy with density functional theory: From fundamental theory to exchange-coupling. *Coordination Chemistry Reviews, 253*, 526–563.

Neese, F. (2017). Quantum chemistry and EPR parameters. In *EMagRes* (pp. 1–22). Chichester, UK: John Wiley & Sons, Ltd.

Neu, H. C., & Heppel, L. A. (1965). The release of enzymes from *Escherichia coli* by osmotic shock and during the formation of spheroplasts. *Journal of Biological Chemistry, 240*, 3685–3692.

Payne, C. M., Knott, B. C., Mayes, H. B., Hansson, H., Himmel, M. E., Sandgren, M., et al. (2015). Fungal cellulases. *Chemical Reviews, 115*, 1308–1448.

Peisach, J., & Blumberg, W. E. (1974). Structural implications derived from the analysis of electron paramagnetic resonance spectra of natural and artificial copper proteins. *Archives of Biochemistry and Biophysics, 165*, 691–708.

Phillips, C. M., Beeson, W. T., Cate, J. H., & Marletta, M. A. (2011). Cellobiose dehydrogenase and a copper-dependent polysaccharide monooxygenase potentiate cellulose degradation by *Neurospora crassa*. *ACS Chemical Biology, 6*, 1399–1406.

Quinlan, R. J., Sweeney, M. D., Lo Leggio, L., Otten, H., Poulsen, J.-C. N., Johansen, K. S., et al. (2011). Insights into the oxidative degradation of cellulose by a copper metalloenzyme that exploits biomass components. *Proceedings of the National Academy of Sciences of the United States of America, 108*, 15079–15084.

Reese, E. T., Siu, R. G. H., & Levinson, H. S. (1950). The biological degradation of soluble cellulose derivatives and its relationship to the mechanism of cellulose hydrolysis. *Journal of Bacteriology, 59*, 485–497.

Sabbadin, F., Hemsworth, G. R., Ciano, L., Henrissat, B., Dupree, P., Tryfona, T., et al. (2018). An ancient family of lytic polysaccharide monooxygenases with roles in arthropod development and biomass digestion. *Nature Communications, 9*, 756.

Saini, J. K., Saini, R., & Tewari, L. (2015). Lignocellulosic agriculture wastes as biomass feedstocks for second-generation bioethanol production: Concepts and recent developments. *3 Biotech, 5*, 337–353.

Saloheimo, M., Nakari-Setälä, T., Tenkanen, M., & Penttilä, M. (1997). cDNA cloning of a *Trichoderma reesei* cellulase and demonstration of endoglucanase activity by expression in yeast. *European Journal of Biochemistry, 249*, 584–591.

Simmons, T. J., Frandsen, K. E. H., Ciano, L., Tryfona, T., Lenfant, N., Poulsen, J. C., et al. (2017). Structural and electronic determinants of lytic polysaccharide monooxygenase reactivity on polysaccharide substrates. *Nature Communications, 8*, 1064.

Solomon, E. I., Heppner, D. E., Johnston, E. M., Ginsbach, J. W., Cirera, J., Qayyum, M., et al. (2014). Copper active sites in biology. *Chemical Reviews, 114*, 3659–3853.

Span, E. A., Suess, D. L. M., Deller, M. C., Britt, R. D., & Marletta, M. A. (2017). The role of the secondary coordination sphere in a fungal polysaccharide monooxygenase. *ACS Chemical Biology*, *12*, 1095–1103.

Stoll, S., & Schweiger, A. (2006). EasySpin, a comprehensive software package for spectral simulation and analysis in EPR. *Journal of Magnetic Resonance*, *178*, 42–55.

Tanghe, M., Danneels, B., Camattari, A., Glieder, A., Vandenberghe, I., Devreese, B., et al. (2015). Recombinant expression of *Trichoderma reesei* Cel61A in *Pichia pastoris*: Optimizing yield and N-terminal processing. *Molecular Biotechnology*, *57*, 1010–1017.

The Hendrich Metalloprotein Group n.d. Spincount. http://www.chem.cmu.edu/groups/hendrich/facilities/index.html.

Vaaje-Kolstad, G., Forsberg, Z., Loose, J. S., Bissaro, B., & Eijsink, V. G. (2017). Structural diversity of lytic polysaccharide monooxygenases. *Current Opinion in Structural Biology*, *44*, 67–76.

Vaaje-Kolstad, G., Horn, S. J., van Aalten, D. M. F., Synstad, B., & Eijsink, V. G. H. (2005). The non-catalytic chitin-binding protein CBP21 from *Serratia marcescens* is essential for chitin degradation. *Journal of Biological Chemistry*, *280*, 28492–28497.

Vaaje-Kolstad, G., Houston, D. R., Riemen, A. H. K., Eijsink, V. G. H., & van Aalten, D. M. F. (2005). Crystal structure and binding properties of the *Serratia marcescens* chitin-binding protein CBP21. *Journal of Biological Chemistry*, *280*, 11313–11319.

Vaaje-Kolstad, G., Westereng, B., Horn, S. J., Liu, Z., Zhai, H., Sørlie, M., et al. (2010). An oxidative enzyme boosting the enzymatic conversion of recalcitrant polysaccharides. *Science (New York, NY)*, *330*, 219–222.

Vu, V. V., Beeson, W. T., Phillips, C. M., Cate, J. H. D., & Marletta, M. A. (2014). Determinants of regioselective hydroxylation in the fungal polysaccharide monooxygenases. *Journal of the American Chemical Society*, *136*, 562–565.

Vu, V. V., Beeson, W. T., Span, E. A., Farquhar, E. R., & Marletta, M. A. (2014). A family of starch-active polysaccharide monooxygenases. *Proceedings of the National Academy of Sciences of the United States of America*, *111*, 13822–13827.

Vu, V. V., & Ngo, S. T. (2018). Copper active site in polysaccharide monooxygenases. *Coordination Chemistry Reviews*, *368*, 134–157.

Wu, M., Beckham, G. T., Larsson, A. M., Ishida, T., Kim, S., Payne, C. M., et al. (2013). Crystal structure and computational characterization of the lytic polysaccharide monooxygenase GH61D from the Basidiomycota fungus *Phanerochaete chrysosporium*. *Journal of Biological Chemistry*, *288*, 12828–12839.

Further reading

Courtade, G., Wimmer, R., Røhr, A. K., Preims, M., Felice, A. K. G., Dimarogona, M., et al. (2016). Interactions of a fungal lytic polysaccharide monooxygenase with β-glucan substrates and cellobiose dehydrogenase. *Proceedings of the National Academy of Sciences of the United States of America*, *113*, 5922–5927.

CHAPTER FOUR

A novel overproduction system for the structural determination of a proton-pumping hydrogen-producing [NiFe]-hydrogenase

Rhiannon M. Evans*, Stephen E. Beaton
Inorganic Chemistry Laboratory, University of Oxford, Oxford, United Kingdom
*Corresponding author: e-mail address: rhiannon.evans@chem.ox.ac.uk

Contents

Abstract

The recent novel overproduction system for the membrane-bound oxygen-sensitive [NiFe]-hydrogenase-2 (Hyd-2) from *Escherichia coli* is detailed. Hyd-2 is an efficient and reversible catalyst for the interconversion of H_2 and $2H^+$. Produced at low levels during anaerobic respiration, Hyd-2 is instrumental in the generation of proton-motive force (PMF), and likewise uses PMF to generate H_2. The structure of the Hyd-2 complex could yield immense information on the mechanism of proton pumping in this group of enzymes, which are of energetic and pathogenic importance. The overproduction of the soluble "catalytic core" where H_2 oxidation/H^+ reduction takes place, relies on gene deletions and use of a synthetic operon in the overproduction strain to ensure metal processing and redox center formation and incorporation are not limiting. Based on

Methods in Enzymology, Volume 613
ISSN 0076-6879
https://doi.org/10.1016/bs.mie.2018.10.006

previous evidence of a cytoplasmic excess of the active site-containing subunit (HybC), the Hyd-2 production bottleneck is relaxed by overproduction of the Fe—S cluster-containing subunit (HybO). The *hybO* gene is altered to prevent translocation of the HybOC catalytic core to the membrane. Protein yield is increased by an order of magnitude, allowing protein-intensive techniques such as X-ray crystallography to flourish. The structure of the entire Hyd-2 complex is inferred from the catalytic core and homology modeling of the ferredoxin and membrane integral partners, leading to the proposal that Hyd-2 is a dimer of tetramers.

Abbreviations

DmsC dimethylsulfoxide reductase, subunit C
Hyd-1 hydrogenase-1
Hyd-2 hydrogenase-2
Hyd-2-OP overproduced hyd-2
IPTG isopropyl β-D-1-thiogalactopyranoside
IscR iron–sulfur cluster regulator
MWCO molecular weight cutoff
NrfD formate-dependent nitrite reductase, subunit D
PES polyethersulfone
PMF proton-motive force
PsrC polysulfide reductase, subunit C
rcf relative centrifugal force
SH soluble hydrogenase
Tat twin arginine translocation
TM transmembrane
TMH transmembrane helix domain

1. Introduction

1.1 Hydrogen catalysis in nature

Molecular hydrogen (H_2) is utilized and produced by microorganisms as a fuel source and to dump excess reducing equivalents, mainly by hydrogenases (Lubitz, Ogata, Rüdiger, & Reijerse, 2014), of which there are several classification systems (Greening et al., 2015; Vignais & Billoud, 2007; Vignais, Billoud, & Meyer, 2001). Electrons are brought to the buried active site usually by a series of iron–sulfur clusters. At the NiFe, FeFe, or Fe-only active site, the overall reaction of $H_2 \rightleftharpoons 2H^+ + 2\ e^-$ is performed at rates comparable to platinum (Hambourger et al., 2008; Jones, Sillery, Albracht, & Armstrong, 2002; Woolerton & Vincent, 2009). The [NiFe]-hydrogenases are diverse in structure, function, and cellular compartmentalization (Greening et al., 2015) and have been implicated in the pathogenicity of several animal and plant pathogens, for example, *Salmonella enterica* sp. (Maier, Olczak, Maier, Soni, & Gunn, 2004), *Helicobacter pylori* (Olson & Maier, 2002), and *Pectobacterium atrospecticum* (Babujee et al., 2012).

Elucidation of the structural requirements and mechanism of action of hydrogenases will inspire and improve the design of synthetic H_2 catalysts (Ginovska-Pangovska, Dutta, Reback, Linehan, & Shaw, 2014) and our understanding of microbial cellular H_2 utilization for fuel, survival, and virulence. Molecular biological approaches to boost the production of hydrogenases to allow for more detailed structural characterization to be a reality are thus of paramount importance.

1.2 *Escherichia coli* hydrogenase-2 genetic and structural arrangement

Under anaerobic conditions, using glycerol as an electron donor and sodium fumarate as an electron acceptor, *E. coli* produces two membrane-bound [NiFe]-hydrogenases, termed Hyd-1 and Hyd-2 (Sargent, 2016). A third, formate hydrogen lyase (FHL) complex, "Hyd-3" can also be produced under anaerobic fermentative conditions (McDowall et al., 2014). Metabolic flexibility during H_2 metabolism is granted since electron transfer between Hyd-2 and the respiratory chain (via the quinone pool) is bidirectional: Hyd-2 operates as a redox pressure release valve, oxidizing H_2 at the periplasm while reducing electron acceptors at the cytoplasm (i.e., fumarate reduction to succinate by menaquinone-dependent fumarate reductase), or reducing protons and producing H_2 when electron acceptors are non-abundant and the quinone pool is overreduced, e.g., during glycerol fermentation (Pinske et al., 2015). Hydrogen oxidation is linked to the generation of proton-motive force (PMF), likewise, using the uncoupler carbonyl cyanide *m*-chlorophenylhydrazone it has been shown that the PMF can drive H_2 production (Pinske et al., 2015).

Hyd-2 is a special and interesting case as it lacks the heme-containing, membrane integral cytochrome-b subunit required for electron transfer to and from the quinone pool, which, for example, is present in the case of Hyd-1 (Volbeda et al., 2013). Instead, Hyd-2 comprises the "HybC" subunit, containing the [NiFe] active site, and "HybO" which contains three iron sulfur (Fe—S) clusters. This periplasmic HybOC "catalytic core" (Fig. 1) is anchored to the membrane via the C-terminal transmembrane (TM) helix (TMH) of HybO, which is separated from the catalytic core by several amino acids.

The catalytic core is translocated to the cytoplasmic membrane via a twin arginine translocation (Tat) system signal peptide at the N-terminus of

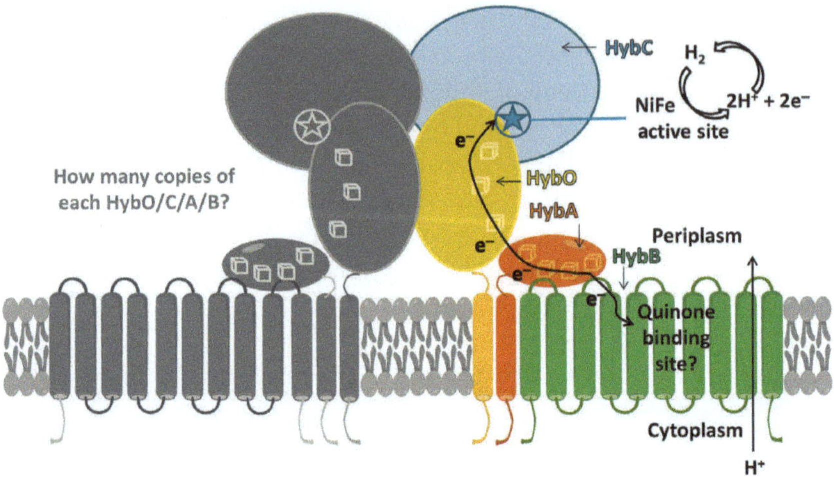

Fig. 1 The proposed arrangement of Hyd-2 comprising, minimally, at least one copy of each subunit: HybO contains a chain of three Fe—S clusters, HybC contains the [NiFe] active site, HybA is a ferredoxin partner thought to contain four Fe—S clusters, and HybB is the membrane integral cofactor-less protein. The catalytic core of Hyd-2 (i.e., HybOC) is anchored to the membrane via the TMH of HybO, which interacts with the TMH of HybA.

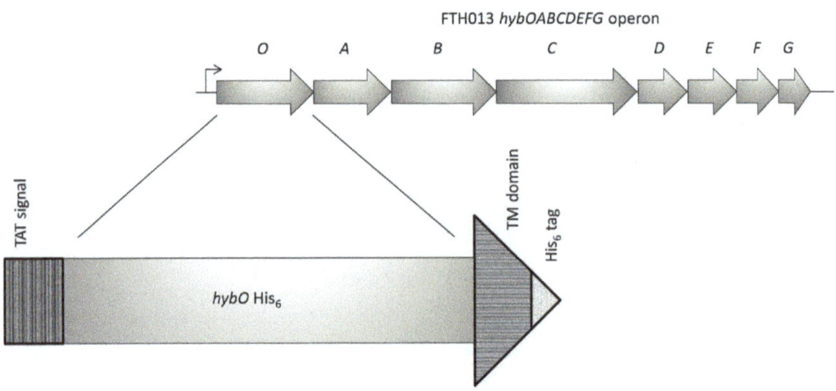

Fig. 2 The genetic arrangement of the Hyd-2 *hybOABCDEFG* operon based on strain FTH013 (Lukey et al., 2010) where the *hybO* gene is modified to include a C-terminal His_6 affinity tag. The HybO product includes an N-terminal Tat signal peptide, and the C-terminal TM domain is upstream of the His_6 tag.

HybO (Fig. 2) and is associated with the structurally unknown electron-transfer subunit "HybA," also processed by Tat. HybA is essential for in vivo electron transfer between the quinone pool and the catalytic core (Pinske et al., 2015), and is predicted to bind four Fe—S clusters. Anchored

Table 1 The strains of *E. coli* used in this study

Strain	**Genotype**	**References**
MC4100	F^-, Δ*argF-lacU169*, *araD139*, *rpsL150*, *relA1*, *ptsF*, *rbsR*, *flbB5301*	Casadaban and Cohen (1979)
FTH013	As MC4100 *hybO* Cterminal His_6 tag	Lukey et al. (2010)
IC010	As MC4100, Δ*hyaB*, Δ*hycE*	Deplanche, Caldelari, Mikheenko, Sargent, and Macaskie (2010)
IC011	As IC010, Δ*hybO*, Δ*hybA*	Lamont and Sargent (2016)
HJ001	As IC011, Δ*iscR*	
HJ001-hyp	As HJ001, *tatD::hypA1-X*	Beaton et al. (2018)

at the periplasmic side of the membrane via a single C-terminal TMH, HybA's interaction with HybO is thought to be via their respective TMHs (Fig. 1). The fourth subunit, "HybB," is a cofactor-less member of the NrfD/DmsC/PsrC family of proteins (Sargent, 2016), predicted to contain 10 TMH and be the site of proton translocation and menaquinone reduction (Fig. 1). HybB is essential for Hyd-2-dependent electron transfer from H_2 to the quinone pool (Dubini, Pye, Jack, Palmer, & Sargent, 2002) generating a PMF. Together, HybO, HybC, HybA, and HybB form the in vivo functional unit of Hyd-2, previously suggested to be of the physiological form $HybO_2C_2A_nB_n$ (Sawers & Boxer, 1986).

The *hyb* operon of Hyd-2 is regulated by IscR and nitrate/nitrite response regulator and its genetic arrangement is shown in Fig. 2. Expressed under anaerobic conditions, the *hybOAB* and *C* genes are responsible for encoding the HybOAB and C peptides of Hyd-2 (Fig. 1) whereas *hybDEFG* encode products required for maturation and assembly of Hyd-2 (reviewed by Böck, King, Blokesch, & Posewitz, 2006; Sargent, 2016).

Previous analyses of Hyd-2 by protein film electrochemistry (PFE, Lukey et al., 2010), infrared (Hexter, Chung, Vincent, & Armstrong, 2014), and electron paramagnetic resonance spectroscopies (Roessler, 2011) have utilized the gene products from strain FTH013 (Table 1), in which parent strain MC4100 has been modified at the *hybO* gene to include a hexahistidine (His_6) tag to enable Ni-affinity purification (Fig. 2). Production of Hyd-2 from this strain is poor (~0.015 mg of Hyd-2 per g of cells) making structural determination difficult (Lukey et al., 2010).

Assessment of the cytoplasmic concentrations of Hyd-2 subunits reveals an excess of HybC, suggesting that a bottleneck in the production of fully

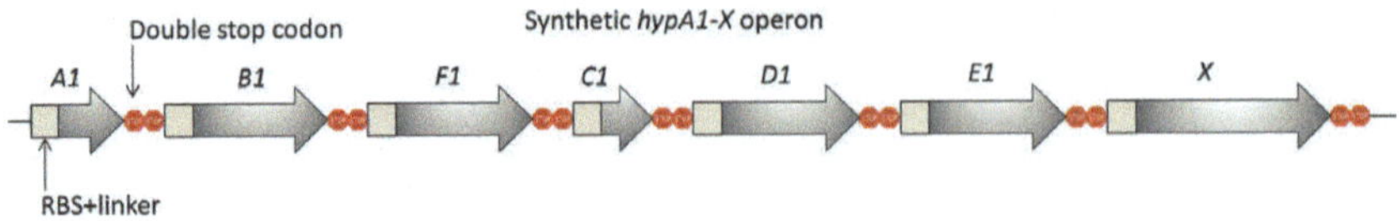

Fig. 3 The genetic arrangement of the synthetic *hypA1-X* operon encoding maturases from the bacterium *R. eutropha*. The synthetic operon was designed to include a ribosomal binding site (RBS) and linker molecule upstream of the start codon of each gene. The end of each gene includes a double stop codon. HypA and HypB gene products are involved in the processing of Ni while HypD–F are involved in the Fe-carbonyl/cyanide active site synthesis and HypC is a chaperone protein (Böck et al., 2006). HypX has a more elusive function (Buhrke et al., 2001; Burgdorf et al., 2005).

formed Hyd-2 lies in the production of HybO (Dubini et al., 2002). Increasing the production of HybO should relieve this bottleneck. In order for structural characterization to be a reality for the special case of Hyd-2, this theory was tested and a novel overproduction system developed (Beaton et al., 2018) in which *hybO* is expressed from a high copy number plasmid.

Here, we describe the practical aspects and considerations for the design of the overproduction system. We detail the preparation of overproduced Hyd-2, relative to its native counterpart (i.e., from FTH013), which has allowed for a systematic electrochemical comparison and the crystal structure of HybOC to be solved. The results provide important insight into the arrangement of the HybOCAB complex.

1.3 First approaches to overexpression of a [NiFe]-hydrogenase

The potential success of a Hyd-2 overproduction system was given credence by the work of Lamont and Sargent (2016), who demonstrated the ability to overproduce a nonnative hydrogenase in *E. coli*. Utilizing various synthetic operons, the soluble hydrogenase (SH) from *Ralstonia eutropha* was assembled in the *E. coli* host. The *R. eutropha* maturase *hypA1-X* synthetic operon (Fig. 3) could also substitute for the native *E. coli* genes: knockout variants incapable of forming native hydrogenases were shown to produce some H_2 evolution activity when the synthetic operon was expressed.

2. Overproduction of fully formed holo-Hyd-2

In order to make protein-intensive techniques such as crystallography a reality for Hyd-2, we have produced the catalytic core of Hyd-2 (HybOC, denoted Hyd-2-OP) from an overexpression plasmid (pQE-80-ΔTM ΔTat *hybO*, known simply as pO^{C}) in which the C-terminal TMH and the N-terminal Tat peptide of HybO are deleted (Figs. 1 and 2). The over-expression plasmid has been used in conjunction with strain "HJ001-hyp" (Table 1), based on work recently described (Lamont & Sargent, 2016).

2.1 Molecular biology strategies for the overproduction of Hyd-2

From previous investigations (Lukey et al., 2010), it has been shown that HybA copurifies with the HybOC dimer (to a low level) when purified from strain FTH013 (Table 1) via the his-tag on HybO. The TMH of HybO is thought to associate with the TMH of the ferredoxin partner HybA since there is no copurification from membrane isolates when the N-terminus of the HybO TMH undergoes trypsin cleavage (Ballantine & Boxer, 1986; Sargent, Ballantine, Rugman, Palmer, & Boxer, 1998). Therefore, in order to be able to purify just the catalytic core and avoid copurifying HybA, *hybA* is deleted in strain HJ001-hyp. Additionally, the TMH and Tat domains of *hybO* are deleted in the pO^{C} plasmid leading to the production of HybO that is incapable of associating with HybA via a TMH interaction and which will not be translocated to the membrane (by recognition of the Tat domain of either HybO or HybA). Importantly, studies of Tat pathway variants suggest that hydrogenases are fully formed prior to translocation (reviewed by Sargent, 2016). The HybC subunit does not have a Tat peptide, and as such, when the ΔTat *hybO* product associates with HybC this catalytic core will remain in the cell cytoplasm.

2.1.1 Design of overexpression plasmid pO^{C}

The *hybO* gene was amplified from strain FTH013 (Table 1) with primers flanking regions not including the Tat signal or TMH domains (Fig. 4). The forward primer used (P1 in Fig. 4) was of sequence 5′-*GAATTCGAGCTCGCGGCCGCACTAGTCACAGAGGAACAGG-TATG* GAGATGGCCGAATCGGTTACTAACC-3′ where the underlined region overlaps with codons corresponding to residues glutamic acid-1 through asparagine-8 (i.e., immediately after the Tat signal domain) and the italicized region overlaps with pQE-80L following amplification with Primers 3 and 4 (Fig. 4, i.e., overlapping with P4). The reverse primer used (P2 in Fig. 4) was of sequence 5′-*GTCGACGCGGCCGCCCTAGG* ttatta**GTGATGGTGATGGTGATG** GCCGCCCTCTTTAGCGTTAA CATCC-3′ where the underlined region overlaps with codons for aspartic

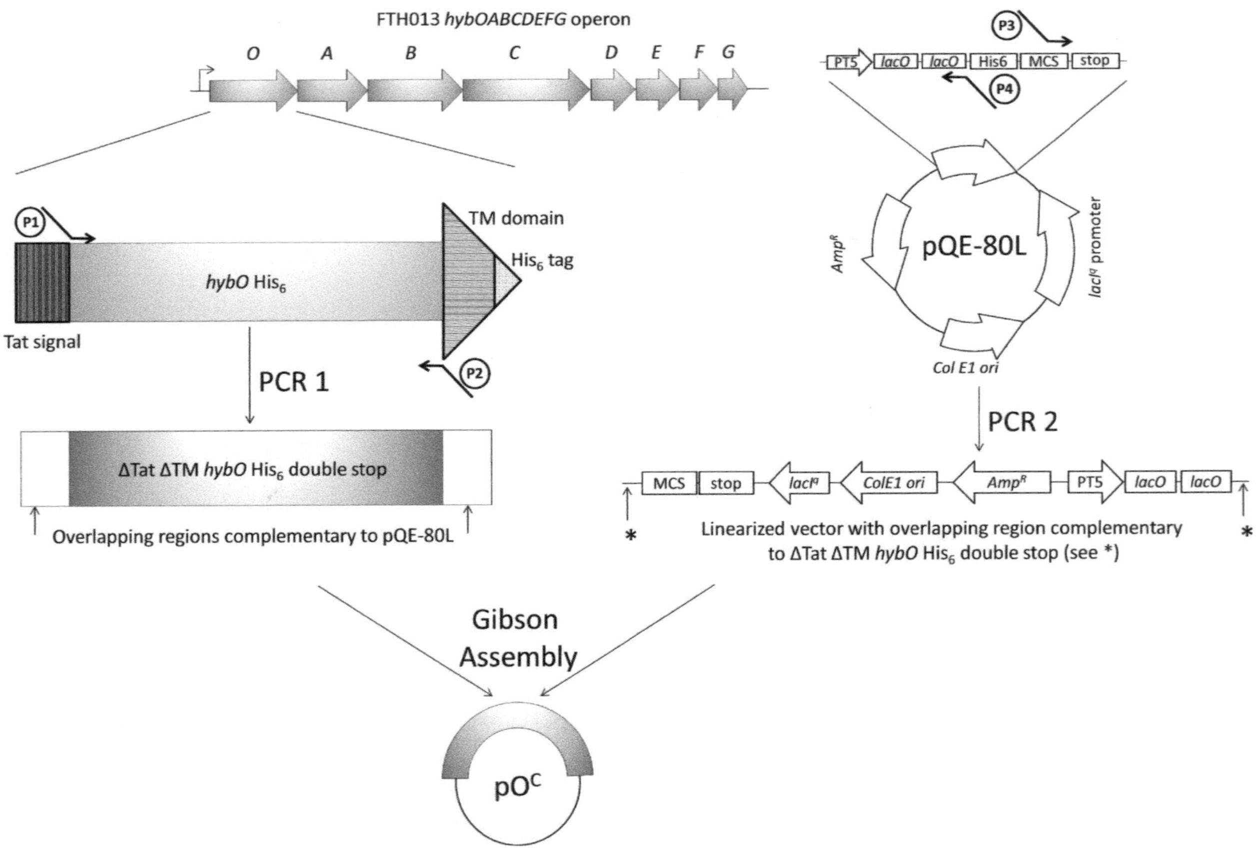

Fig. 4 The design and construction of the overproduction plasmid pO^C. The desired ΔTat ΔTM *hybO* His_6 double stop PCR fragment was amplified from strain FTH013 to create overlapping regions with the pQE-80L plasmid, itself amplified to remove the resident N-terminal His_6 tag and create overlapping regions to the amplified *hybO* gene.

acid-286 through glycine-293 (corresponding to asp-323 and gly-330 in native Hyd-2). The italicized region overlaps with pQE-80L following amplification with Primers 3 and 4 (Fig. 4, i.e., overlapping with P3). Hence the TMH is deleted and *hybO* is immediately followed by a His_6 tag (bold region) and a double stop codon (ttatta) region. The resulting PCR product is ΔTat ΔTM *hybO* His_6 double stop (*hybO*-OP).

Plasmid pQE-80L (Qiagen) is a high copy number plasmid containing a bacteriophage T5 promoter for *E. coli* RNA polymerase, with an embedded *lac* operator. At the N-terminus there is a His_6 tag sequence. From previous studies (Lukey et al., 2010) it was shown that Hyd-2 can be purified using a C-terminal his-tag, and so the N-terminal His_6 sequence was removed from the plasmid: amplification using Primers 3 and 4 (Fig. 4) creates linearized plasmid with regions overlapping with *hybO*-OP. The forward primer (P3, Fig. 4) was of sequence 5′-CCTAGGGCGGCCGCGTCGACC *TGCAGCCAAG CTTAATTAGCTGAGC*-3′, where the underlined region overlaps with the C-terminal region of *hybO*-OP and the italicized region overlaps with pQE-80L. The reverse primer (P4, Fig. 4) was of sequence 5′-ACCTGTTCCTCTGTGACTAGTGCGGCCGCGAGCTC *GAATT CTGTGTGAAATTGTTATCCGCTCACAATTG*-3′, where the underlined region overlaps with the N-terminal region of *hybO*-OP, and the italicized region overlaps with pQE-80L upstream of its N-terminal His_6 tag. Following amplification of *hybO* and linearization of the plasmid, the two fragments were assembled via Gibson Assembly to form the pO^C plasmid.

2.1.1.1 Equipment

- PCR machine
- Standard agarose gel equipment
- Incubator (shaking and static)
- Water bath
- Nanodrop spectrophotometer

2.1.1.2 Reagents and buffers

- Ultrapure water (DNase/RNase-Free, Thermo Fisher Scientific)
- Q5® High-fidelity DNA polymerase kit (New England Biolabs) including GC enhancer is used for all PCR reactions
- Deoxynucleotide (dNTP) solution in an equimolar dATP, dCTP, dGTP, and dTTP mix at 10 m*M* each (New England Biolabs)

- Oligonucleotide primers (100 µ*M* stock, Thermo Fisher Scientific) in ultrapure water at −20°C
- pQE-80L plasmid (Qiagen)
- LB (Miller) broth (Melford)
- Plasmid extraction and PCR purification are achieved using the protocols as described for the Econospin® All-In-One system (Epoch Life Science)
- Gibson Assembly kit (New England Biolabs) used as per manufacturer's instructions
- High efficiency strain NEB®5-alpha competent cell kit including SOC media (New England Biolabs)
- Ampicillin (Cambridge Bioscience Ltd.) 100 mg/mL (1000 ×) stock in 70% ethanol, stored at − 20°C

2.1.1.3 Procedures

- PCR 1—A single colony of FTH013 cells is suspended in 100 µL ultrapure water and 0.5–1 µL is used as DNA template. Primers 1 and 2 (Fig. 4, and above) are used at a final concentration of 10 µ*M* each. The Q5® polymerase protocol is followed according to manufacturer's instructions.
- PCR 2—A single colony of NEB®5-alpha cells transformed with pQE-80L is first grown for 16 h in LB/Ampicillin at 37°C, 200 rpm. Plasmid is extracted using the Econospin system and used as DNA template (0.5–1 µL). Primers 3 and 4 (Fig. 4, and above) are used at a final concentration of 10 µ*M* each. Q5® polymerase protocol is followed according to manufacturer's instructions.
- Samples (10 µL) of PCR 1 and 2 are analyzed for purity and product size by agarose gel electrophoresis and the resulting products of size 967 bp (PCR 1) and 4721 bp (PCR 2) are PCR purified using the Econospin system.
- The Gibson Assembly protocol is followed as per manufacturer's instructions in terms of concentrations of fragments (measured by nanodrop), and the reaction is performed in minimal volume (4–10 µL) at 50°C for 4 h.
- NEB®5-alpha cells are transformed with the Gibson Assembly reaction products using the standard heat-shock method: cells (50 µL) are thawed on ice and DNA added (5 µL maximum). After mixing by swirling with a pipette tip, the cells are incubated on ice for 30 min. Heat-shock, at 42°C for 30 s, is quenched by cooling on ice for 2 min. Recovery is performed

in 1 mL SOC media at 37°C for 1.5–4 h. Single colonies are grown on LB/agar with 100 μg/mL ampicillin at 37°C for 16 h.

- Single colonies are transferred to LB/ampicillin media (5–10 mL, 100 μg/mL) at 37°C, 200 rpm for 16 h.
- Plasmid is extracted using the Econospin system and analyzed for sequence integrity (Source Bioscience, Oxford).

2.1.2 Design of overexpression strain HJ001-hyp

Using *E. coli* strain MC4100 (Casadaban & Cohen, 1979) as the parent strain, strain "HJ001-hyp" was created by deletion of several genes to maximize overproduction of the HybOC catalytic core (Table 1): Δ*hyaB* (Hyd-1 large subunit) with Δ*hycE* (Hyd-3 large subunit) creates a strain known as IC010 (Deplanche et al., 2010), and the additional deletion of *hybO* and *hybA* creates strain IC011. Finally, the deletion of the iron–sulfur cluster regulator system gene (*iscR*, see earlier), known to suppress Hyd-2 expression and FeS cluster biosynthesis (Akhtar & Jones, 2008), creates strain HJ001 (Table 1). Additionally, strain HJ001-hyp includes the *R. eutropha*-derived synthetic operon (optimized for *E. coli* codon preference, Fig. 3) integrated into the chromosome in place of the nonessential *tatD* gene (Fig. 5), i.e., *tatD::hypA1B1F1CDEX* (Kelly et al., 2015; Lamont & Sargent, 2016).

Creation of the *tatD::hypA1B1F1CDEX* allele required several plasmid constructs (Fig. 5). In this study the p-Bluescript II KS^+-based (Stratagene) plasmids "pFAT122" which contains a 551-bp fragment immediately

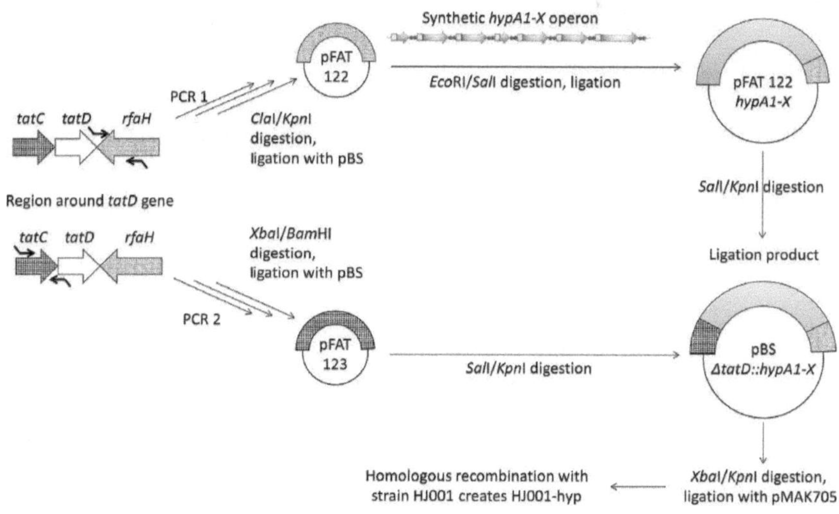

Fig. 5 The design and construction of the overproduction strain HJ001-hyp.

downstream of the *tatD* gene, and "pFAT123" which contains a 557-bp fragment immediately upstream of *tatD* were used as starting constructs (Wexler et al., 2000). The *hypA1-X* operon (Fig. 3) was cloned into pFAT 122. Separately, pFAT 123 and pFAT 122-*hypA1-X* were digested with *Sal*I and *Kpn*I and ligated together. The resulting plasmid has *hypA1-X* flanked by fragments of the downstream *rfaH* and upstream *tatC* genes. Transfer of the *tatD::hypA1-X* allele to plasmid pMAK705 was achieved following *Xba*I/*Kpn*I digestion and ligation.

The pMAK *tatD::hypA1-X* plasmid is used to transform strain HJ001 (Table 1) and the *tatD::hypA1-X* allele is transferred to the chromosome using a well-established technique referred to as the "pMAK protocol" (Hamilton, Aldea, Washburn, Babitzke, & Kushner, 1989): the pMAK705 plasmid has an antibiotic resistance cassette (chloramphenicol) and a temperature-sensitive origin of replication (pSC101) with a permissive temperature of 30°C and nonpermissive temperature at 44°C. Integration of the *hypA1-X* operon in place of *tatD* can be achieved by virtue of the upstream and downstream flanking regions of the *rfaH* and *tatC* genes (Fig. 5). Following transformation of HJ001 with pMAK *tatD::hypA1-X*, homologous recombination between the plasmid and the chromosome results in cointegrates which can be selected for by growth in chloramphenicol at 44°C. Subsequent growth of cointegrates at 30°C frees the plasmid from the chromosome in a second recombination, and is cured from the cells by nonselective growth at 44°C. Single colonies are replica-patched and grown at 30°C in the presence and absence of chloramphenicol. Chloramphenicol sensitive colonies are screened by PCR (flanking the *tatD* gene) and/or DNA sequencing.

2.2 Purification of native Hyd-2 and overproduced Hyd-2: A comparison

2.2.1 Equipment

- Autoclave
- Incubator (static/shaking)
- Schott bottles (5 L, Fisher Scientific)
- Refrigerating centrifuge
- Ultracentrifuge
- Cell disruptor system (Constant Systems)
- Cell homogenizer/tissue grinder with a PTFE pestle and cylindrical mortar (Fisher Scientific)
- ÄKTA purification system with duel wavelength absorbance detector

- His-trap HP Ni-affinity column (GE Healthcare)
- His-trap FF Crude Ni-affinity column (GE Healthcare)
- Centrifugal concentrators, 50,000 molecular weight cutoff (MWCO), polyethersulfone (PES) membrane (Vivaspin)
- PD-10 desalting column (GE Healthcare)
- Standard SDS-PAGE equipment

2.2.2 Reagents and buffers

- LB (Miller) broth (Melford), 25 g/L, autoclaved
- Sodium fumarate (Alfa Aesar): filter sterilized (0.22 μm)
- Glycerol (Fisher Scientific), 50% (v/v, autoclaved)
- IPTG (Cambridge Bioscience Ltd.), filter sterilized (0.22 μm)
- Ampicillin (Cambridge Bioscience Ltd.), filter sterilized (0.22 μm)
- All buffers are stored at 4°C
- Buffer 1: Tris 100 m*M*, EDTA 1 m*M*, NaCl 50 m*M*, DNAse (0.003% (w/v)), and lysozyme (0.012% (w/v)), pH 7.5
- Buffer 2: Tris 100 m*M*, NaCl 350 m*M*, pH 7.5
- Triton X-100 solution: 30% Triton X-100 (v/v) in Buffer 2
- Protease inhibitor cocktail tablets added to cell suspension from strain HJ001-hyp + pO^C: Complete, mini EDTA-free protease inhibitor cocktail, (Roche) 1 tablet per 30 L culture
- Buffer A1: Tris 20 m*M*, NaCl 350 m*M*, imidazole 60 m*M*, Triton X-100 (0.02% (v/v)), dithiothreitol 1 m*M*, pH 7.2
- Buffer A2: Tris 50 m*M*, NaCl 350 m*M*, imidazole 50 m*M*, dithiothreitol 1 m*M*, pH 7.2
- Buffer B1: Tris 20 m*M*, NaCl 350 m*M*, imidazole 750 m*M*, Triton X-100 (0.02% (v/v)), dithiothreitol 1 m*M*, pH 7.2
- Buffer B2: Tris 50 m*M*, NaCl 350 m*M*, imidazole 500 m*M*, dithiothreitol 1 m*M*, pH 7.2
- Buffer C1: Tris 50 m*M*, NaCl 350 m*M*, Triton X-100 0.02% (v/v), 10% glycerol (v/v), dithiothreitol 1 m*M*, pH 7.2
- Buffer C2: Tris 50 m*M*, NaCl 350 m*M*, 10% glycerol (v/v), dithiothreitol 1 m*M*, pH 7.2
- Buffer C3: Tris 20 m*M*, NaCl 150 m*M*, DDM (*n*-Dodecyl β-D-maltoside) 0.02% (w/v), 10% glycerol (v/v), dithiothreitol 1 m*M*, pH 7.2

2.2.3 Procedures

The different procedural steps for production of Hyd-2 from each strain are summarized in Table 2.

Table 2 Protein production and purification strategies for Hyd-2 from strain FTH013 and HJ001-Hyp + pO^C

Step	FTH013 (native Hyd2)	HJ001-hyp + pO^C (Hyd-2-OP)
1 Preculture	A single colony in LB (5 mL), 37°C, 200 rpm, 7 h	A single colony in 5 mL LB/ampicillin (100 μg/mL), 37°C 200 rpm, 7 h
2 Inoculation	Inoculate 5-L bottle of LB media supplemented with glycerol (0.5% (v/v)) and sodium fumarate (0.4% (w/v)). Top up to brim (~6 L total volume) incubate at 37°C, no shaking	Inoculate 5-L bottle of LB media supplemented with glycerol (0.5% (v/v)) and sodium fumarate (0.4% (w/v)) and ampicillin (100 μg/mL). Top up to brim (~6 L total volume) incubate at 37°C, no shaking
3 Induction	n/a	At OD_{600} of 0.2–0.3, induce pO^C expression with IPTG (1 m*M*)
4 Harvest	Harvest cells by centrifugation at 4°C, 3250 rcf for 12 min, 9 h after inoculation	Harvest cells by centrifugation at 4°C, 3250 rcf for 12 min, 9 h after induction
5 Resuspend	Resuspend cell pellet in Buffer 1	Resuspend cell pellet in Buffer 1
6 Cell lysis	Lyse cells at 20 kPSI (×2) and centrifuge (4°C, 2800 rcf for 30 min) to remove unlysed cells	Add protease inhibitor cocktail tablet(s). Lyse cells at 20 kPSI (×2)
7 Pellet membrane	Ultracentrifuge for 60 min	n/a
8 Resuspend membrane	Resuspend membrane in Buffer 2	n/a
9 Homogenize membrane	Mechanically homogenize membrane by hand then increase volume of homogenized membrane with Buffer 2, allow to stir for 1 h	n/a
10 Solubilize membrane	Add Triton X-100 solution to membrane suspension to a final concentration of 1% (v/v). Stir for 40 min at 4°C	n/a

Table 2 Protein production and purification strategies for Hyd-2 from strain FTH013 and HJ001-Hyp + pO^C—cont'd

Step	FTH013 (native Hyd2)	HJ001-hyp + pO^C (Hyd-2-OP)
11 Remove debris	After removing unlysed cells, ultracentrifuge the lysate for 60 min	Ultracentrifuge for 45 min (will pellet unlysed cells and membrane)
12 Ni column	Add imidazole (40 m*M*) and DTT (1 m*M*) to supernatant solution. Load onto His-trap HP column preequilibrated with Buffer A1. Wash with Buffer A1	Add imidazole (50 m*M*) and DTT (1 m*M*) to supernatant solution. Load onto His-trap FF Crude column preequilibrated with Buffer A2. Wash with Buffer A2
13 Elute from Ni column	Elute Hyd-2 over an imidazole gradient from 60 to 750 m*M* over 16–24 column volumes. Collect fractions	Elute Hyd-2 over an imidazole gradient from 50 to 500 m*M* over 16–24 column volumes. Collect fractions
14 SDS-PAGE	Samples of each fraction are separated by denaturing electrophoresis	
15 Concentrate	Pure fractions are pooled and concentrated by centrifugation	
16 PD-10 column	For electrochemical analyses, concentrated protein is buffer-exchanged into Buffer C1	Concentrated protein is buffer-exchanged into Buffer C2 for electrochemical analyses or C3 for structural analysis

All concentrations are final values.

2.2.4 Notes

- Inoculation: The entire 5 mL preculture is used to inoculate each 5-L bottle. To achieve anaerobic conditions bottles are filled to the brim with LB (~6 L) and lids left loose to release gaseous products.
- Cells are resuspended by agitation in Buffer 1 (3.5 mL/L of culture) and stored at −80°C or used immediately. Resuspended/thawed cells should remain on ice/4°C for all further steps.
- Cells are lysed by two passes through the cell disruptor system (Constant Systems) at 20 kPSI, and collected on ice.
- Ultracentrifugation steps: All steps for both strains are carried out at 4°C, 208,000 rcf.
- FTH013 membranes are resuspended in a minimal volume of Buffer 2, transferred to the tissue grinder, and ground gently to create a

homogenous solution. The volume of the homogenized membranes is increased with Buffer 2 by ~3 mL/L of original culture and the suspension is stirred at 4°C for 1 h.

- The volume of the resuspended membrane solution is increased to ~6 mL/L of culture and solubilized using Triton X-100 solution to a final concentration of (1% (v/v)). The solution is stirred for 40 min at 4°C.
- The Ni-affinity columns are first equilibrated with Buffer A1/A2 (Table 2). The absorbance is monitored at 280 nm (aromatic residues) and 420 nm (Fe—S clusters). Unbound proteins are washed from the column with at least 10 column volumes of Buffer A1/A2 or until baseline absorbencies return. Bound protein is eluted from the column by a gradient of imidazole to 100% Buffer B1/B2 (Table 2) over 16–24 column volumes.
- Collected fractions are analysed by SDS-PAGE. Fractions containing the highest ratio of Hyd-2 subunits over contaminating peptides are pooled (Fig. 6A) and concentrated by centrifugation (50,000 MWCO spin concentrator, Vivaspin) at 2800 rcf, 4°C, to a final volume of 0.5–1 mL
- Enzymes for electrochemical analysis are snap-frozen and stored in liquid N_2. Hyd-2-OP for subsequent structural analysis is buffer-exchanged into Buffer C3 without subsequent freezing (see Section 2.3).

2.3 Extension of the Hyd-2 overproduction purification protocol for crystal structure determination

Following Ni-affinity chromatography, Hyd-2-OP is further purified by size-exclusion chromatography (Fig. 6B). This is necessary for crystal growth for structural analyses, achieved using the subsequent Hyd-2-OP at a concentration of 5 mg/mL.

2.3.1 Equipment

- Benchtop centrifuge
- ÄKTA purification system with duel wavelength absorbance detector
- Superdex 200 Increase 10/300 column (GE Healthcare)
- Centrifugal concentrators, 50,000 MWCO with a PES membrane (Vivaspin)
- PD-10 desalting column (GE Healthcare)
- Standard SDS-PAGE equipment
- UV–vis spectrometer for Bradford analysis of protein concentration

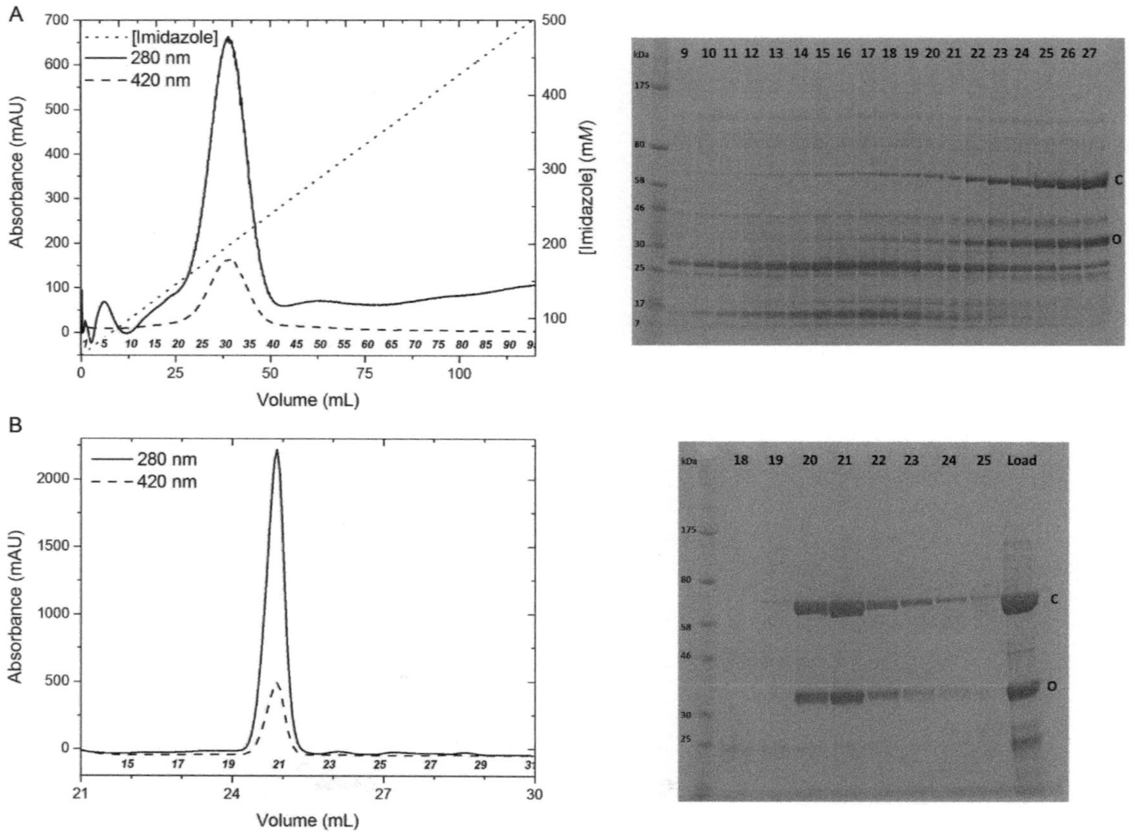

Fig. 6 An example of a typical Hyd-2-OP purification via Ni-affinity chromatography (A) and, for crystallographic analyses via subsequent size-exclusion chromatography (B). The chromatograms indicate the fractions collected in italics. SDS-PAGE gels show samples of each fraction as labeled. HybC is labeled "C" and HybO is labeled "O." In this example, fractions 22–43 were pooled and concentrated after Ni-affinity chromatography, this concentrated sample was then loaded on to the SEC column (Load) of which fractions 20–22 were taken forward to crystallography trials.

2.3.2 Reagents and buffers

- Buffer SEC: Tris 20 m*M*, NaCl 150 m*M*, DDM (*n*-Dodecyl β-D-maltoside, 0.02% (w/v)), 10% glycerol (v/v), dithiothreitol 1 m*M*, pH 7.2
- Bradford reagent (Alfa Aesar)
- Protein standard (bovine serum albumin, 2 mg/mL, Sigma-Aldrich)

2.3.3 Procedures

- Protein is concentrated by centrifugation using a 50,000 MWCO spin concentrator (Vivaspin) at 2800 rcf, 4°C, to a final volume of ~0.5 mL and loaded by injection onto the Superdex column preequilibrated with Buffer SEC.
- Samples of each fraction are analyzed by SDS-PAGE (Fig. 6B) and those containing the purest Hyd-2-OP are pooled.
- Pooled Hyd-2-OP is concentrated using a 50,000 MWCO spin concentrator (Vivaspin) at 2800 rcf, 4°C, to a final concentration of 5 mg/mL, as judged by the method of Bradford (Bradford, 1976), calibrated with protein standard.
- The resulting Hyd-2-OP is used for crystal growth as detailed by Beaton et al., 2018.

3. Biochemical and electrochemical analysis of overproduced Hyd-2 reveals fully formed and active Hyd-2-OP

Rocket immunoelectrophoresis is a nondenaturing technique whereby proteins are loaded into an agarose gel containing an antibody raised specifically against the protein of interest, in this case rabbit anti-Hyd-2 raised against the purified products from strain FTH013. Following electrophoresis, the antibody-precipitated proteins are assessed for H_2 oxidation activity by incubation in benzyl viologen and tetrazolium red in the presence of H_2 (tetrazolium test). Any hydrogenase activity is revealed as a red rocket-shaped "arc." The height of the arc corresponds to the level of protein in the sample.

The amount of Hyd-2-OP produced in HJ001-hyp cells far exceeded that produced in an equivalent native strain (i.e., using strain FTD674 (Beaton et al., 2018) which produces Hyd-2 without the TMH at native expression levels). These results alone suggested that the small subunit HybO had associated with the large subunit HybC partner, since strains without either of these peptides show no hydrogenase activity (Beaton et al.,

2018; Pinske et al., 2011). Solution-based assays for hydrogenase activity were performed for both proton reduction (using methyl viologen) and H_2 oxidation (using benzyl viologen) and the relative rates for Hyd-2 and Hyd-2-OP enzymes were within error, with average rates of ~204 and $8.5\,s^{-1}$, respectively (Beaton et al., 2018).

Protein film electrochemistry techniques rely on the physical adsorption of electroactive proteins to conductive surfaces (Armstrong & Hirst, 2011; Evans & Armstrong, 2014; Evans et al., 2016; Vincent, Parkin, & Armstrong, 2007). Upon application of a voltage, [NiFe]-hydrogenases will either oxidize or reduce the substrate species, depending on the potential applied. The current response is directly proportional to turnover rate. Hydrogen oxidation will be seen as a positive current when a potential above the thermodynamic potential for the $H_2/2H^+$ redox couple is applied. At increasingly high potentials the oxidized inactive resting state known as Ni—B (Ni^{III}-OH^-) will usually form, eventually causing a decrease in current which does not return until a sufficiently negative potential is applied (Armstrong et al., 2016). At increasingly low [H_2], proton reduction is seen as a negative current. The ability to perform proton reduction is dependent on the hydrogenase in question and the conditions of the experiment (Armstrong et al., 2016). Hyd-2 is an example of an "efficient" and reversible electrocatalyst (Armstrong & Hirst, 2011) as it is able to perform H_2 oxidation or proton reduction with minimal overpotential from the thermodynamic potential (Lukey et al., 2010). Proton reduction, however, is subject to product (H_2) inhibition.

The electrochemical measurements of Hyd-2-OP revealed little alteration in the key characteristics of Hyd-2 (Beaton et al., 2018): Hyd-2-OP has no increased onset potential for activity in either direction (Fig. 7). Monitoring the current response to differing concentrations of H_2 allows for $K_M(H_2)$ and $K_i(H_2)$ to be determined. Proton reduction activity increases with decreasing [H_2], i.e., it is product inhibited with a similar $K_i(H_2)$ to Hyd-2 ($240 \pm 11\,\mu M$ compared to $219 \pm 14\,\mu M$ for native Hyd-2). Hydrogen oxidation activity increases with [H_2] and the Michaelis constant is within error for both enzymes ($21 \pm 4\,\mu M$ for native Hyd-2, 23 ± 7 for Hyd-2-OP). The temperature dependence of current across a range of potentials allows for Eyring analysis, and for Hyd-2-OP this yields an activation enthalpy similar to native Hyd-2 (47 ± 2 and 43 ± 3 kJ/mol, respectively).

Taken together, the biochemical and electrochemical analyses of Hyd-2-OP are consistent with the production of fully formed and active enzyme.

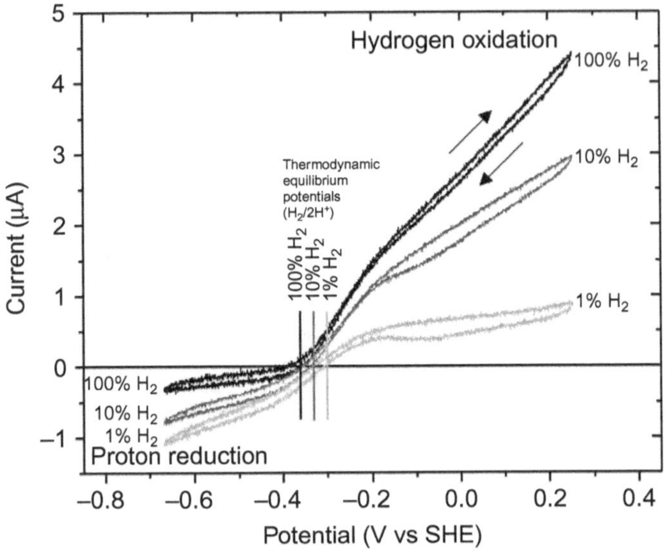

Fig. 7 The electrochemical profile of Hyd-2-OP under differing concentrations of H_2. Conditions: pH 6.0, 30°C, scan rate 5 mV/s, potential scanned from −0.667 to +0.253 V vs the standard hydrogen electrode (SHE), *arrows* indicate the scan direction. The working electrode was rotated at 3000 rpm to avoid mass transport limitation. Gas flow = 1000 standard cubic centimeters per minute, Ar used as carrier gas. The *vertical lines* denote the thermodynamic (equilibrium) potential for each hydrogen concentration in the conditions used, showing clearly that Hyd-2-OP displays efficient reversible behavior.

4. Structural characterization leading to a proposal for the full membrane-bound structure of Hyd-2

The successful overproduction of Hyd-2 allowed for a significant increase in the yield of protein, making crystallography a reality for Hyd-2. Long rod-shaped crystals were produced after 24 h by the sitting drop vapor diffusion technique. Crystals of the as-isolated (aerobic) crystals diffracted to 2.2 Å, whereas dithiothreitol-reduced crystals and H_2-reduced enzyme (exposed to H_2 for 12 h prior to crystallization) diffracted to 1.5 Å. X-ray diffraction conditions and parameters are detailed in Beaton et al. (2018).

As previously suggested by analytical ultracentrifugation (Sawers & Boxer, 1986), Hyd-2-OP is a dimer of heterodimers—i.e., $HybO_2C_2$ (Fig. 8). This is not surprising given the fact that it was isolated from a Δ*hybA*

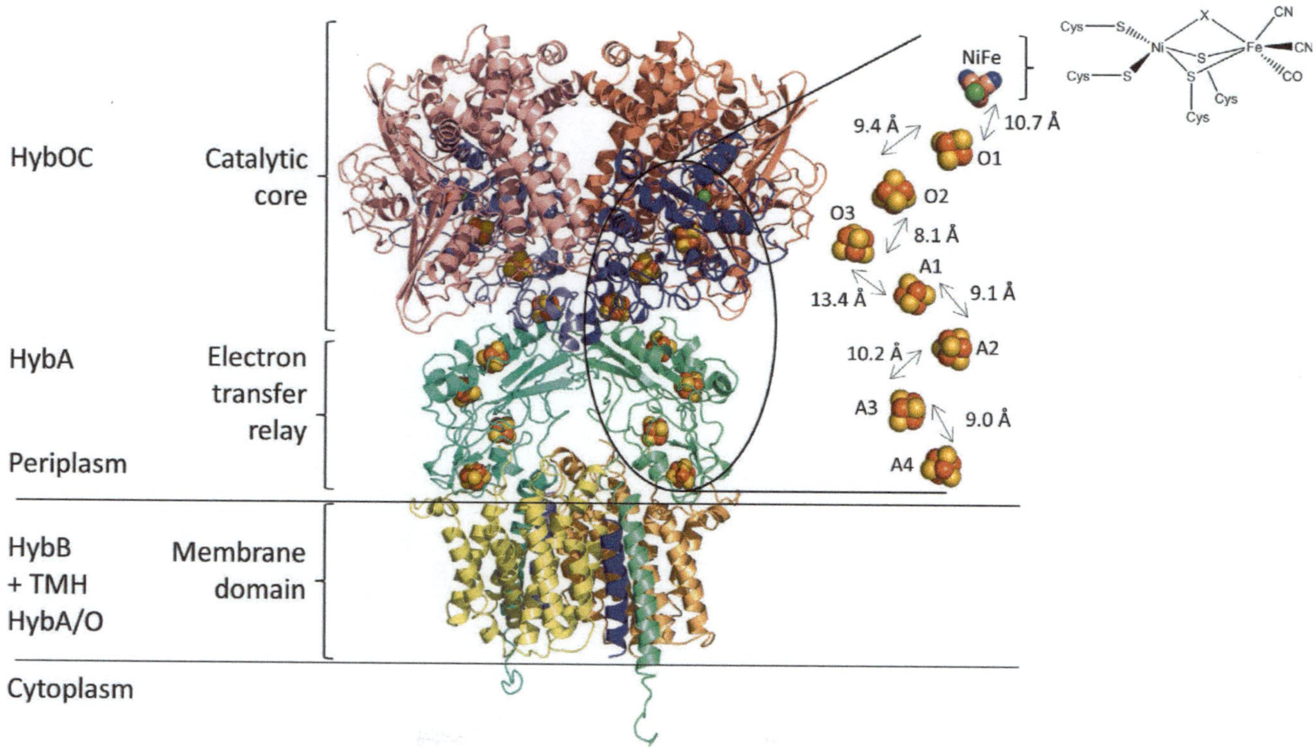

Fig. 8 The proposed structure of the entire Hyd-2 complex based on the crystal structure of the HybOC dimer of heterodimers and the homology models of HybA and HybB. The nearest atom distances between each of the redox centers are shown in Angstroms. Fe—S clusters of HybO are labeled O1–3, and of HybA A1–4. The Chemdraw representation of the NiFe active site includes the four ligating cysteines, two CN^-, and one CO ligand to Fe. Species X, bridging the Ni and Fe atoms is OH^- in the resting inactive oxidized state Ni—B.

strain and from the cell cytoplasm (see Section 2). The distal Fe—S clusters of each HybO unit (O3, Fig. 8) are placed within electron-transferring distance allowing for interheterodimer electron tunneling (Page, Moser, Chen, & Dutton, 1999). Based on the fact that the ferredoxin partner protein is required in whole cells for electron transfer between the quinone pool and the catalytic HybOC core (Pinske et al., 2015), it is likely that the native enzyme in vivo has a $(HybOCA)_2$ arrangement.

HybB is proposed to be a member of the NrfD/DmsC/PsrC family of membrane proteins: NrfD is required for formate-dependent nitrite reduction to ammonia in *E. coli* (Hussain, Grove, Griffiths, Busby, & Cole, 1994; Simon & Kern, 2008); DmsC is involved in dimethyl sulfoxide (DMSO) reduction (Weiner, Shaw, Turner, & Trieber, 1993); and PsrC is involved in polysulfide reduction (Jormakka et al., 2008). This protein family has in common an 8-helix transmembrane segment and a lack of cofactor/heme ligands; instead, menaquinol (MQ) binding pockets have been shown experimentally (Jormakka et al., 2008) localized toward the periplasmic side of the membrane (Rothery, Workun, & Weiner, 2008; Simon & Kern, 2008) near their periplasmic catalytic partners. Only one intact complex from the NrfD/DmsC/PsrC family of proteins has been solved crystallographically—the PsrABC complex (Jormakka et al., 2008), which comprises a dimer of trimers, the majority of the dimer interface being supplied by the membrane integral PsrC protein. PsrC comprises eight transmembrane helices divided into two 4-helix bundles (likely a gene-duplication event), with the MQ binding and oxidation site on the periplasmic side of the membrane. Only one quinone binding pocket is located, in the 4-helix bundle nearest the N-terminus. The distance between the MQ oxidation site and all redox centers is <14 Å, i.e., optimized for fast electron transfer between redox centers (Page et al., 1999). The C-terminal 4-helix bundle is suggested to be involved in proton translocation through a series of conserved residues and water molecules to the MQ pocket in the N-terminal 4-helix bundle.

Homology modeling was performed using the Phyre 2 server (Kelley, Mezulis, Yates, Wass, & Sternberg, 2015) allowing for models of HybA and HybB subunits and their interactions with HybOC to be analyzed. PsrC was used as the template for homology modeling of HybB and gives a structure comprising two 4-helix bundles. To model the HybA ferredoxin subunit of Hyd-2, the FdnH ferredoxin subunit of formate dehydrogenase N from *E. coli* was used (Jormakka, Törnroth, Byrne, & Iwata, 2002). The FdnH subunit is associated with the membrane via a single transmembrane helix and sits between the integral membrane protein FdnI and the catalytic FdnG subunit where formate is oxidized to form CO_2 and protons. FdnH contains four Fe—S clusters which are all positioned at efficient electron transferring distances (Page et al., 1999) from each other and the heme and Fe—S and Mo atom of FdnI and FdnG, respectively.

The homology models of HybA and HybB and the crystal structure of the HybOC dimer of heterodimers allow for a proposed structure of the entire Hyd-2 complex to emerge: the amino acid linker between the HybO subunit and its transmembrane domain is proposed to elevate the catalytic core from the periplasmic side of the membrane and allow the HybA subunit to conduct electrons between the catalytic core and the membrane integral HybB. The proposed model places all the Fe—S clusters within electron tunneling distance of one another (Page et al., 1999) and thus supports the hypothesis that HybA sits in between the catalytic core and HybB. The entire complex of Hyd-2 is thus likely to have a $(HybOCAB)_2$ arrangement.

5. Conclusions

Overproduction of Fe—S-containing proteins is possible in *E. coli* provided the correct machinery is supplied to form and mature protein containing a full complement of Fe—S clusters and other redox centers. Biochemical, structural, and electrochemical characterizations of Hyd-2-OP have provided excellent evidence that this is the case for overproduction of the Hyd-2 catalytic core. The proposed quaternary structure of the entire Hyd-2 complex comprises two copies of each subunit in a physiologically relevant arrangement for efficient electron transfer and conformational proton pumping.

Going forward, it will be important to realize the in vivo structure of Hyd-2 and provide evidence for the conformational changes proposed in the HybB subunit for transmembrane proton pumping.

Acknowledgments

This research was funded by the Biotechnology and Biological Sciences Research Council (BBSRC) Grant BB/N006321/1 awarded to Prof. Fraser A. Armstrong (University of Oxford) and BBSRC EASTBIO Doctoral Training Partnership awards 1510231 and 1280664 awarded to Prof. Frank Sargent (University of Dundee). S.E.B. is grateful to the Dakota Foundation for the award of a Holaday Scholarship.

References

Akhtar, M. K., & Jones, P. R. (2008). Deletion of iscR stimulates recombinant clostridial Fe-Fe hydrogenase activity and H_2-accumulation in Escherichia coli BL21(DE3). *Applied Microbiology and Biotechnology*, *78*(5), 853–862.

Armstrong, F. A., Evans, R. M., Hexter, S. V., Murphy, B. J., Roessler, M. M., & Wulff, P. (2016). Guiding principles of hydrogenase catalysis instigated and clarified by protein film electrochemistry. *Accounts of Chemical Research*, *49*, 884–892.

Armstrong, F. A., & Hirst, J. (2011). Reversibility and efficiency in electrocatalytic energy conversion and lessons from enzymes. *Proceedings of the National Academy of Sciences of the United States of America*, *108*(34), 14049–14054.

Babujee, L., Apodaca, J., Balakrishnan, V., Liss, P., Kiley, P. J., Charkowski, A. O., et al. (2012). Evolution of the metabolic and regulatory networks associated with oxygen availability in two phytopathogenic enterobacteria. *BMC Genomics*, *13*(1), 110–135.

Ballantine, S. P., & Boxer, D. H. (1986). Isolation and characterisation of a soluble active fragment of hydrogenase isoenzyme 2 from the membranes of anaerobically grown Escherichia coli. *European Journal of Biochemistry*, *156*(2), 277–284.

Beaton, S. E., Evans, R. M., Finney, A. J., Lamont, C. M., Armstrong, F. A., Sargent, F., et al. (2018). The structure of hydrogenase-2 from *Escherichia coli*: Implications for H_2-driven proton pumping. *Biochemical Journal*, *475*, 1353–1370.

Böck, A., King, P., Blokesch, M., & Posewitz, M. C. (2006). Maturation of hydrogenases. *Advances in Microbial Physiology*, *51*, 1–71.

Bradford, M. M. (1976). A rapid and sensitive method for the quantitation of microgram quantities of protein utilizing the principle of protein-dye binding. *Analytical Biochemistry*, *72*, 248–254.

Buhrke, T., Bleijlevens, B., Albracht, S. P. J., Friedrich, B., Buhrke, T., Bleijlevens, B., et al. (2001). Involvement of hyp gene products in maturation of the H_2-sensing [NiFe] hydrogenase of *Ralstonia eutropha* involvement of hyp gene products in maturation of the H_2-sensing [NiFe] hydrogenase of *Ralstonia eutropha*. *Journal of Bacteriology*, *183*(24), 7087–7093.

Burgdorf, T., Lenz, O., Buhrke, T., van der Linden, E., Jones, A. K., Albracht, S. P. J., et al. (2005). [NiFe]-hydrogenases of *Ralstonia eutropha* H16: Modular enzymes for oxygen-tolerant biological hydrogen oxidation. *Journal of Molecular Microbiology and Biotechnology*, *10*(2–4), 181–196.

Casadaban, M. J., & Cohen, S. N. (1979). Lactose genes fused to exogenous promoters in one step using a mu-lac bacteriophage: In vivo probe for transcriptional control sequences. *Proceedings of the National Academy of Sciences of the United States of America*, *76*(9), 4530–4533.

Deplanche, K., Caldelari, I., Mikheenko, I. P., Sargent, F., & Macaskie, L. E. (2010). Involvement of hydrogenases in the formation of highly catalytic Pd(0) nanoparticles by bioreduction of Pd(II) using *Escherichia coli* mutant strains. *Microbiology*, *156*(9), 2630–2640.

Dubini, A., Pye, R. L., Jack, R. L., Palmer, T., & Sargent, F. (2002). How bacteria get energy from hydrogen: A genetic analysis of periplasmic hydrogen oxidation in *Escherichia coli*. *International Journal of Hydrogen Energy*, *27*, 1413–1420.

Evans, R. M., & Armstrong, F. A. (2014). Electrochemistry of metalloproteins: Protein film electrochemistry for the study of E. coli [NiFe]-hydrogenase-1. In J. C. Fontecilla-Camps & Y. Nicolet (Eds.), *Metalloproteins—Methods and protocols* (pp. 73–94). New York: Humana Press.

Evans, R. M., Brooke, E. J., Wehlin, S. A. M., Nomerotskaia, E., Sargent, F., Carr, S. B., et al. (2016). Mechanism of hydrogen activation by [NiFe] hydrogenases. *Nature Chemical Biology*, *12*, 46–50.

Ginovska-Pangovska, B., Dutta, A., Reback, M. L., Linehan, J. C., & Shaw, W. J. (2014). Beyond the active site: The impact of the outer coordination sphere on electrocatalysts for hydrogen production and oxidation. *Accounts of Chemical Research*, *47*(8), 2621–2630.

Greening, C., Biswas, A., Carere, C. R., Jackson, C. J., Taylor, M. C., Stott, M. B., et al. (2015). Genomic and metagenomic surveys of hydrogenase distribution indicate H_2 is a widely utilised energy source for microbial growth and survival. *The ISME Journal*, *10*(3), 1–17.

Hambourger, M., Gervaldo, M., Svedruzic, D., King, P. W., Gust, D., Ghirardi, M., et al. (2008). [FeFe]-hydrogenase-catalyzed H2 production in a Photoelectrochemical biofuel cell. *Journal of the American Chemical Society*, *130*(6), 2015–2022.

Hamilton, C. M., Aldea, M., Washburn, B. K., Babitzke, P., & Kushner, S. R. (1989). New method for generating deletions and gene replacements in *Escherichia coli*. *Journal of Bacteriology*, *171*(9), 4617–4622.

Hexter, S. V., Chung, M.-W., Vincent, K. A., & Armstrong, F. A. (2014). Unusual reaction of [NiFe]-hydrogenases with cyanide. *Journal of the American Chemical Society*, *136*(29), 10470–10477.

Hussain, H., Grove, J., Griffiths, L., Busby, S., & Cole, J. (1994). A seven-gene operon essential for formate-dependent nitrite reduction to ammonia by enteric bacteria. *Molecular Microbiology*, *12*(1), 153–163.

Jones, A. K., Sillery, E., Albracht, S. P. J., & Armstrong, F. A. (2002). Direct comparison of the electrocatalytic oxidation of hydrogen by an enzyme and a platinum catalyst. *Chemical Communications*, *8*(2002), 866–867.

Jormakka, M., Törnroth, S., Byrne, B., & Iwata, S. (2002). Molecular basis of proton motive force generation: Structure of formate dehydrogenase. *Science*, *295*(5561), 1863–1868.

Jormakka, M., Yokoyama, K., Yano, T., Tamakoshi, M., Akimoto, S., Shimamura, T., et al. (2008). Molecular mechanism of energy conservation in polysulfide respiration. *Nature Structural & Molecular Biology*, *15*(7), 730–737.

Kelley, L. A., Mezulis, S., Yates, C. M., Wass, M. N., & Sternberg, M. J. E. (2015). The Phyre2 web portal for protein modelling, prediction and analysis. *Nature Protocols*, *10*(6), 845–858.

Kelly, C. L., Pinske, C., Murphy, B. J., Parkin, A., Armstrong, F. A., Palmer, T., et al. (2015). Integration of an [FeFe]-hydrogenase into the anaerobic metabolism of Escherichia coli. *Biotechnology Reports (Amsterdam, Netherlands)*, *8*, 94–104.

Lamont, C. M., & Sargent, F. (2016). Design and characterisation of synthetic operons for biohydrogen technology. *Archives of Microbiology*, *199*, 495–503.

Lubitz, W., Ogata, H., Rüdiger, O., & Reijerse, E. (2014). Hydrogenases. *Chemical Reviews*, *114*(8), 4081–4148.

Lukey, M. J., Parkin, A., Roessler, M. M., Murphy, B. J., Harmer, J., Palmer, T., et al. (2010). How *Escherichia coli* is equipped to oxidize hydrogen under different redox conditions. *Journal of Biological Chemistry*, *285*(6), 3928–3938.

Maier, R. J., Olczak, A., Maier, S., Soni, S., & Gunn, J. (2004). Respiratory hydrogen use by *Salmonella enterica* serovar typhimurium is essential for virulence. *Infection and Immunity*, *72*(11), 6294–6299.

McDowall, J. S., Murphy, B. J., Haumann, M., Palmer, T., Armstrong, F. A., & Sargent, F. (2014). Bacterial formate hydrogenlyase complex. *Proceedings of the National Academy of Sciences of the United States of America*, *111*(38), E3948–E3956.

Olson, J. W., & Maier, R. J. (2002). Molecular hydrogen as an energy source for helicobacter pylori. *Science*, *298*(5599), 1788–1790.

Page, C. C., Moser, C. C., Chen, X., & Dutton, P. L. (1999). Natural engineering principles of electron tunnelling in biological oxidation-reduction. *Nature*, *402*, 47–52.

Pinske, C., Jaroschinsky, M., Linek, S., Kelly, C. L., Sargent, F., & Sawers, R. G. (2015). Physiology and bioenergetics of [NiFe]-hydrogenase 2-catalyzed H_2-consuming and H_2-producing reactions in Escherichia coli. *Journal of Bacteriology, 197*(2), 296–306.

Pinske, C., Krüger, S., Soboh, B., Ihling, C., Kuhns, M., Braussemann, M., et al. (2011). Efficient electron transfer from hydrogen to benzyl viologen by the [NiFe]-hydrogenases of Escherichia coli is dependent on the coexpression of the iron-sulfur cluster-containing small subunit. *Archives of Microbiology, 193*(12), 893–903.

Roessler, M. M. (2011). *EPR investigations of iron-sulfur cluster relays in enzyme.* D. Phil. University of Oxford.

Rothery, R. A., Workun, G. J., & Weiner, J. H. (2008). The prokaryotic complex iron-sulfur molybdoenzyme family. *Biochimica et Biophysica Acta—Biomembranes, 1778*(9), 1897–1929.

Sargent, F. (2016). The model [NiFe]-hydrogenases of Escherichia coli. *Advances in Microbial Physiology, 68*, 433–507.

Sargent, F., Ballantine, S. P., Rugman, P. A., Palmer, T., & Boxer, D. H. (1998). Reassignment of the gene encoding the Escherichia coli hydrogenase 2 small subunit. Identification of a soluble precursor of the small subunit in a hypB mutant. *European Journal of Biochemistry, 255*(3), 746–754.

Sawers, R. G., & Boxer, D. H. (1986). Purification and properties of membrane-bound hydrogenase isoenzyme 1 from anaerobically grown *Escherichia coli* K12. *European Journal of Biochemistry, 156*(2), 265–275.

Simon, J., & Kern, M. (2008). Quinone-reactive proteins devoid of haem b form widespread membrane-bound electron transport modules in bacterial respiration. *Biochemical Society Transactions, 36*(Pt. 5), 1011–1016.

Vignais, P. M., & Billoud, B. (2007). Occurrence, classification, and biological function of hydrogenases: An overview. *Chemical Reviews, 107*(10), 4206–4272.

Vignais, P. M., Billoud, B., & Meyer, J. (2001). Classification and phylogeny of hydrogenases. *FEMS Microbiology Reviews, 25*(4), 455–501.

Vincent, K. A., Parkin, A., & Armstrong, F. A. (2007). Investigating and exploiting the electrocatalytic properties of hydrogenases. *Chemical Reviews, 107*(10), 4366–4413.

Volbeda, A., Darnault, C., Parkin, A., Sargent, F., Armstrong, F. A., & Fontecilla-Camps, J. C. (2013). Crystal structure of the O2-tolerant membrane-bound hydrogenase 1 from Escherichia coli in complex with its cognate cytochrome b. *Structure (London, England: 1993), 21*(1), 184–190.

Weiner, J., Shaw, G., Turner, R., & Trieber, C. (1993). The topology of the anchor subunit of dimethyl sulfoxide reductase of *Escherichia coli. Journal of Biological Chemistry, 268*(5), 3238–3244.

Wexler, M., Sargent, F., Jack, R. L., Stanley, N. R., Bogsch, E. G., Robinson, C., et al. (2000). TatD is a cytoplasmic protein with DNase activity. No requirement for TatD family proteins in sec-independent protein export. *Journal of Biological Chemistry, 275*(22), 16717–16722.

Woolerton, T. W., & Vincent, K. A. (2009). Oxidation of dilute H_2 and H_2/O_2 mixtures by hydrogenases and Pt. *Electrochimica Acta, 54*(22), 5011–5017.

CHAPTER FIVE

O_2-tolerant [NiFe]-hydrogenases of *Ralstonia eutropha* H16: Physiology, molecular biology, purification, and biochemical analysis

Oliver Lenz*, Lars Lauterbach, Stefan Frielingsdorf
Department of Chemistry, Technische Universität Berlin, Berlin, Germany
*Corresponding author: e-mail address: oliver.lenz@tu-berlin.de

Contents

Abstract

Dioxygen-tolerant [NiFe]-hydrogenases are defined by their ability to catalyze the reaction, $H_2 \rightleftharpoons 2H^+ + 2e^-$ even in the presence of O_2. Catalytic and probably also non-catalytic mechanisms protect their active sites from being inactivated by reactive oxygen species, which makes them attractive subjects of investigation from both

Methods in Enzymology, Volume 613
ISSN 0076-6879
https://doi.org/10.1016/bs.mie.2018.10.008

fundamental and applied perspectives. Prominent representatives of the O_2-tolerant [NiFe]-hydrogenases have been isolated from the chemolithoautotrophic model organism *Ralstonia eutropha* H16, which can thrive in a simple mineral medium supplemented with the gases H_2, O_2, and CO_2. In this chapter, we describe methods for cultivation and genetic manipulation of *R. eutropha*, both of which are prerequisites for the reproducible manufacturing of high-quality hydrogenase preparations. The purification procedures for two different O_2-tolerant [NiFe]-hydrogenases from *R. eutropha* are described in detail, as well as the corresponding biochemical procedures used for the determination of the catalytic properties of these sophisticated enzymes.

1. Introduction

In nature, H_2 production as well as H_2 oxidation is catalyzed by metalloenzymes known as hydrogenases, which occur in different structural subtypes and with various physiological functions in their host organisms. Only few hydrogenase exemplars, however, are able to support a chemolithoautotrophic lifestyle based on H_2 as the sole energy source, O_2 as terminal electron acceptor, and CO_2 as the source of carbon. This particular lifestyle has been realized by the so-called Knallgas bacteria and crucially depends on hydrogenases that tolerate the presence of O_2 during H_2 turnover. Probably, the best-characterized member of the Knallgas bacteria is the β-proteobacterium *Ralstonia eutropha* H16 (also called *Cupriavidus necator*). *R. eutropha* contains four O_2-tolerant hydrogenases, all of which carry a catalytic center comprised of nickel and iron atoms, whereby the iron is coordinated by three inorganic ligands, two cyanides, and one carbon monoxide (Lenz, Lauterbach, Frielingsdorf, & Friedrich, 2015).

This chapter describes the engineering, purification, and biochemical characterization of the two energy-generating [NiFe]-hydrogenases from *R. eutropha*, whose catalytic properties are well known and which have already been used for a number of biotechnological applications. The membrane-bound [NiFe]-hydrogenase (MBH) is composed of three subunits (Fig. 1B), a large subunit (HoxG) carrying the catalytic center, a small, electron-transferring subunit (HoxK) with three iron–sulfur clusters, and a membrane-integral cytochrome *b* (HoxZ). The enzyme channels the electrons released upon H_2 oxidation via the quinone pool into the respiratory chain, while the protons are liberated into the periplasmic space, thereby contributing to the generation of a proton motive force. The second H_2-converting enzyme is the NAD^+-reducing [NiFe]-hydrogenase (soluble hydrogenase, SH), which resides in the cytoplasm where it transfers the

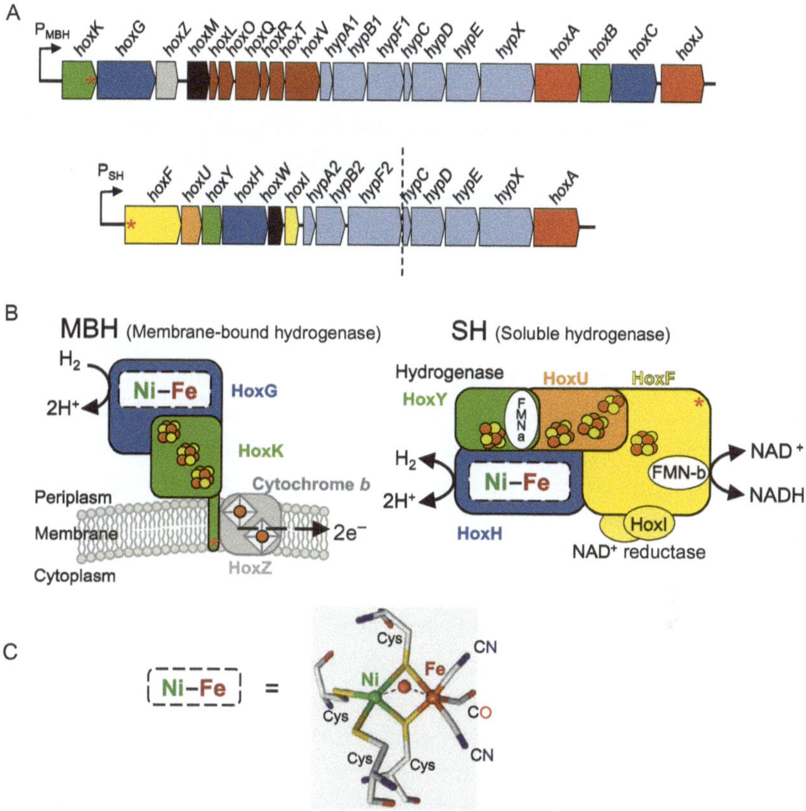

Fig. 1 Cartoon representations of the energy-converting [NiFe]-hydrogenases of *R. eutropha* and the genetic arrangement of the hydrogenase-related genes. (A) Expression of the MBH and SH operons is controlled by the HoxA-dependent MBH and SH promoters (P_{MBH}, P_{SH}). The genes *hoxK, G,* and *Z* encode the MBH subunits while *hoxF, U, Y, H,* and *I* encode the SH constituents (see (B)). Assembly and insertion of the [NiFe] catalytic center (see (C)) is mediated by the products of *hypA1, A2, B1, B2, F1, F2, C, D, E, X,* as well as the hydrogenase-specific protease-encoding genes *hoxM* and *hoxW*. The genes *hoxL, O, Q, R, T,* and *V* encode MBH-specific maturation proteins. Transcriptional control is mediated by the products of HoxA, B, C, and J. HoxB and HoxC form a H_2-sensing regulatory hydrogenase that interacts with the histidine protein kinase HoxJ, which, in turn, forms a two-component regulatory system with the response regulator HoxA. Because of a mutation in *hoxJ* of *R. eutropha* H16, the corresponding HoxJ protein is inactive, leaving HoxA active in hydrogenase gene-specific transcription activation but independent of the presence of H_2. Note that the native SH operon ends with the *hypF2* gene (*dashed line*). For SH overproduction, it has been artificially amended with the *hypC, D, E,* and *X* genes as well as *hoxA* from the MBH operon (see Section 4). *Asterisks* indicate the location of *Strep*-tag II-encoding sequences (in (A)) or peptides (in (B)). (B) Subunit composition and cofactor content of MBH and SH. Iron–sulfur clusters are depicted as spheres and heme groups as diamonds. The structure of the Ni–Fe active site of the MBH is shown in (C). The sphere between Ni and Fe represents the oxygen atom of an OH group that was found to occupy the bridging position in fully oxidized MBH. The Ni–Fe active site of SH is supposed to have a similar structure.

electrons derived from H_2 oxidation directly to NAD^+ to generate reducing power in form of NADH. The heterohexameric SH is composed of a two-subunit hydrogenase module (HoxH and HoxY) and a NAD^+ reductase module, which also comprises two subunits (HoxF and HoxU) in addition to two copies of the nonessential HoxI protein (Fig. 1B).

Knowledge of the genome sequence of *R. eutropha* (Pohlmann et al., 2006), which includes the genetic organization of the hydrogenase-related genes (Fig. 1A) (Schwartz et al., 2003), and the accessibility of *R. eutropha* by standard genetic methods allowed the engineering and purification of recombinant hydrogenase proteins. The purified proteins, in turn, enabled a detailed structural biochemical, electrochemical, and biophysical characterization (Lenz et al., 2015). Purified MBH and SH show maximal H_2 oxidation rates of approximately 240 and 350 s^{-1} and retain more than 50% and 90%, respectively, of the initial activity even in the presence of atmospheric O_2 concentration (Lauterbach & Lenz, 2013; Ludwig, Cracknell, Vincent, Armstrong, & Lenz, 2009), a feature which made them very attractive for biotechnology. In fact, the MBH has been used as the anode catalyst in a biofuel cell that generates electricity from H_2 and O_2 and as a constituent of a photosystem I–hydrogenase hybrid complex that produces H_2 in a light-dependent manner (Krassen et al., 2009; Vincent et al., 2005). The SH represents a proven in vivo and in vitro cofactor recycling catalyst that efficiently regenerates NADH in the course of coupled redox enzyme reactions (Lonsdale et al., 2015; Reeve et al., 2017).

2. Physiology

R. eutropha H16 was isolated from the sludge of a river head in central Germany on the basis of its capacity to grow in a defined mineral medium supplemented with a gas mixture of 60% H_2, 35% O_2, and 5% CO_2 (Wilde, 1962). Besides growing lithoautotrophically, it is also able to use organic compounds as energy and carbon sources. As a typical β-proteobacterium, *R. eutropha* shows a preference for organic acids (succinate, pyruvate, acetate, gluconate) over sugars (Wilde, 1962). Fructose is actually the only sugar that can be metabolized by *R. eutropha* H16 wild type, but genetic engineering has been used to enlarge the spectrum of utilizable sugars to, e.g., mannose and glucose (Sichwart, Hetzler, Bröker, & Steinbüchel, 2011). Hydrogenase synthesis in *R. eutropha* H16 occurs under energy-limiting conditions (Lenz & Friedrich, 1998) and is independent of the presence of H_2 (Friedrich, Friedrich, & Bowien, 1981). Here, we describe the growth conditions resulting in a high cellular hydrogenase content.

2.1 Lithoautotrophic cultivation of *R. eutropha* derivatives

The establishment of gas-based lithoautotrophic growth conditions in a lab requires gas mixing devices that allow the preparation of defined mixtures of H_2, O_2, CO_2, and N_2 (Fig. 2). Cultivation flasks or Petri dishes filled with the basic mineral medium are usually deposited in a thick-walled desiccator (Fig. 2A) which is then evacuated with a vacuum pump and refilled with a safe, noncombustible gas mixture. Cultivations with combustible or explosive H_2/O_2 mixtures must be performed in a certified fermenter (Fig. 2B). Cultures should be routinely incubated at a temperature of 30°C because above 33°C, the hydrogenase gene-specific regulator becomes inactive and hydrogenase synthesis does not occur (Friedrich & Friedrich, 1983).

2.1.1 Equipment

- Erlenmeyer flasks
- Sterile Petri dishes
- Desiccator
- Vacuum pump
- Gas mixing device (with manometer) for H_2, O_2, CO_2, N_2 (Fig. 2A)
- Incubator shaker
- Spectrophotometer

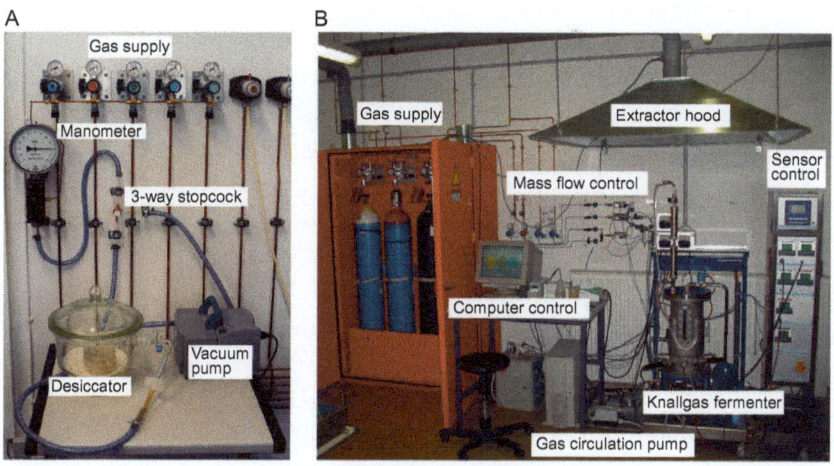

Fig. 2 Devices required for lithoautotrophic cultivation with H_2 and O_2. (A) Gas mixing device for the preparation of defined gas mixtures in desiccators. (B) Fermenter certified for bacterial cell cultivations with explosive H_2–O_2 mixtures. Important parts are labeled.

2.1.2 Buffers and reagents

- 10 × H16 buffer (90 g $Na_2HPO_4 \times 12\ H_2O$, 15 g KH_2PO_4, and 1 L H_2O [adjusted to pH 7.0 with phosphoric acid, if necessary])
- 20% (w/v) NH_4Cl; autoclaved
- 20% (w/v) $MgSO_4 \times 7\ H_2O$; autoclaved
- 1% (w/v) $CaCl_2 \times H_2O$; autoclaved
- 0.5% (w/v) $FeCl_3 \times 6\ H_2O$ (in 0.1 N HCl); sterilized by filtration
- 1 m*M* $NiCl_2$; autoclaved

2.1.3 Procedure

1. For a total volume of 1000 mL AUT medium, mix 100 mL of 10 × H16 buffer with 850 mL H_2O (add 13% (w/v) of Bacto™ agar in case of solid media) and autoclave; subsequently add 10 mL NH_4Cl stock solution and 1 mL each of $MgSO_4$, $CaCl_2$, $FeCl_3$, and $NiCl_2$ stock solution, then fill up to 1000 mL with sterile H_2O. Inoculate with an appropriate amount of preculture.
2. Distribute the AUT medium into baffled Erlenmeyer flasks to 20% of the total volume; in case of solid media pour the agar-containing AUT medium into sterile Petri dishes
3. Place the Erlenmeyer flasks/Petri dishes in a thick-walled desiccator equipped with a perforated ceramic plate and a stopcock, and close it with a lid (Fig. 2A).
4. Remove the air within the desiccator using a vacuum pump and refill with a *safe* gas mixture containing, e.g., 3% H_2, 10% CO_2, 10% O_2, and 77% N_2.
5. Incubate the desiccator at 30°C either in an incubator (solid cultures) or an incubator shaker (flasks must be fixed on the ceramic plate, with, e.g., an adhesive foil) at approximately 120 rpm.
6. Collect the cells when the desired optical density of the liquid culture is reached; check the optical density at 436 nm (OD_{436nm}) with a spectrophotometer.

2.1.4 Notes

1. Precultures should be grown to the late stationary phase in AUT medium supplemented with 0.4% (w/v) fructose (FN medium). Precultures grown at 37°C do not contain any hydrogenase and are therefore well suited for determining growth curves of mutant strains with diminished hydrogenase activities in comparison with the wild-type strain.
2. For the sake of safety, the H_2 concentration (in the presence of O_2) must be kept below 4% (v/v). As H_2 is continuously consumed by the cells,

the gas atmosphere in the desiccator should be frequently exchanged. The gas exchange frequency depends on the culture volume and density relative to the desiccator volume.

3. From the literature, flammable or even explosive H_2/O_2 mixtures have very often been used for lithoautotrophic cultivation of *R. eutropha*. Such cultivations must be done in fermenters that are explicitly approved for such purpose (e.g., as described in Kohlmann et al., 2011) (Fig. 2B).
4. *R. eutropha* H16 is a known producer of polyhydroxyalkanoates, e.g., polyhydroxybutyrate (PHB), which, depending on nutrient availability, can accumulate in high amounts in the cells (Pohlmann et al., 2006). To prevent PHB formation, which actually significantly lowers hydrogenase synthesis, nitrogen and phosphate sources must be supplied in sufficient amounts to liquid cultures. Additional nitrogen supplementation is especially necessary for long-term cultures aiming at high optical densities ($OD_{436nm} > 20$).
5. A strong brownish color of agar plates due to complex formation of Bacto™ agar with phosphate anions in the H16 buffer can be prevented by autoclaving the agar in a suitable amount of water separately from the H16 buffer.

2.2 Heterotrophic cultivation of *R. eutropha* derivatives

In *R. eutropha* H16, the H_2-sensing pathway responsible for H_2-dependent hydrogenase gene expression is disrupted because of a single point mutation in a gene encoding a histidine protein controlling the function of the hydrogenase gene-specific response regulator (Lenz & Friedrich, 1998). Thus, MBH and SH gene expression in strain H16 and its derivatives responds mainly to the global signal of cellular energy depletion. Energy depletion occurs, e.g., in the stationary phase during growth on preferred substrates or upon growth with poor substrates such as glycerol. Within the last decades, heterotrophic growth conditions resulting in a very high cellular hydrogenase content have been continuously optimized. Here, we describe the currently used protocol, i.e., growth under microaerobic conditions on fructose and glycerol (Goris et al., 2011).

2.2.1 Equipment

- Erlenmeyer flasks
- Sterile Petri dishes
- Incubator shaker
- Spectrophotometer

2.2.2 Buffers and reagents

See Section 2.1.2. In addition you will need:

- 40% (w/v) Fructose; sterilized by filtration
- 40% (w/v) Glycerol; autoclaved

2.2.3 Procedure

1. For a total volume of 1000 mL modified fructose–glycerol–ammonium (FGN_{mod}) medium, mix 100 mL of 10 × H16 buffer with 850 mL H_2O (add 13% (w/v) of Bacto™ agar in case of solid media) and autoclave; subsequently add 10 mL NH_4Cl stock solution and 1 mL of each of the $MgSO_4$, $CaCl_2$, $FeCl_3$, and $NiCl_2$ stock solutions. Furthermore, add 1.25 mL fructose stock solution and 10 mL glycerol stock solution. Fill up to 1000 mL with sterile H_2O.
2. Fill baffled Erlenmeyer flasks to 80% of the total volume with FGN medium and inoculate with an appropriate amount of preculture.
3. Incubate the cultures at 30°C in an incubator shaker at approximately 120 rpm.
4. Collect the cells when the desired optical density of the liquid culture is reached; check the optical density at 436 nm with a spectrophotometer. Hydrogenase-producing cultures should have an optical density at 436 nm of 10–12, which is reached after ca. 7–10 days of cultivation (Fig. 3).

Fig. 3 Heterotrophic cultivation of *R. eutropha* derivatives for high-level [NiFe]-hydrogenase synthesis. Shown are FGN_{mod}-based 4-L cultures in (baffled) 5-L Erlenmeyer flasks deposited on adhesive tape, in a temperature-controlled rotary shaker.

2.2.4 Notes

1. Precultures should be grown to the late stationary phase in AUT medium supplemented with 0.4% (w/v) fructose (FN medium). Precultures grown at 37°C do not contain any hydrogenase.
2. The cultivation temperature must not exceed 33°C, otherwise hydrogenase gene expression will cease.
3. Solid media for heterotrophic cultivation should be prepared with FN medium. Pour the agar-containing FN medium into sterile Petri dishes.
4. In case of growing cultures for SH purification (see Section 4.1), supplement the medium (1:1000) with trace element solution SL6 (Atlas, 2010), $ZnCl_2$ solution (1 μM final concentration), and twice the usual amount of $FeCl_3$ (32 μM final concentration). After 4 days, add another 1-mL portion of $NiCl_2$ stock solution per 1000 mL of culture.

3. Molecular biology

DNA can be transferred through the cell wall into the cytoplasm of *R. eutropha* by two different techniques. Plasmid transfer to *R. eutropha* by conjugation was first used in 1981 (Friedrich, Hogrefe, & Schlegel, 1981) and was subsequently simplified by using *E. coli* S17-1 as the donor, in which transfer helper functions are provided from chromosomally integrated transfer genes of the broad-host-range plasmid RP4 (Simon, Priefer, & Pühler, 1983). Much later, the first electrotransformation of *R. eutropha* H16 was reported (Tee et al., 2017). Similarly, the initially applied random mutagenesis methods, which were based on mutagenic agents and transposons (Friedrich et al., 1981; Srivastava, Urban, & Friedrich, 1982), were later substituted by site-directed mutagenesis techniques allowing marker-less modifications of the *R. eutropha* genome (Lenz, Schwartz, Dernedde, Eitinger, & Friedrich, 1994; Xiong et al., 2018).

3.1 Conjugative plasmid transfer to *R. eutropha*

Here, we describe the transfer of mobilizable plasmids from the donor strain *E. coli* S17-1, in which the transfer genes (*tra*) of plasmid RP4 are located on the chromosome, into the recipient *R. eutropha* via the "agar spot mating" technique.

3.1.1 Equipment

- Erlenmeyer flasks
- Sterile glass tubes with screw caps

- Centrifuge
- Sterile Petri dishes
- Incubator shaker

3.1.2 Buffers and reagents

- LB medium containing half the amount of NaCl (low-salt LB: LSLB)
- 1 × H16 buffer, see Section 2.1.2
- FN medium, see Section 2.1.4

3.1.3 Procedure

1. Cultivate the donor and recipient strains in 10 mL liquid LSLB medium with the respective antibiotic overnight on an incubator shaker at 30–37°C.
2. Transfer the culture into sterile glass tubes and close the screw cap. Centrifuge for 10 min at 4000 × *g* and room temperature. Remove the supernatants and resuspend the cell pellets in 5 mL sterile H16 buffer. Centrifuge again and resuspend the cell pellets in 1 mL sterile H16 buffer.
3. Mix each 0.2 mL of the two cell suspensions and transfer the mixture by pipetting onto an antibiotic-free LSLB agar plate (spot mating). Incubate for 6–12 h at 37°C.
4. Remove the cell material from the agar surface with a sterile glass pipette and transfer it to a sterile glass tube containing 1 mL sterile H16 buffer. Resuspend the cells and centrifuge for 10 min at 4000 × *g* and room temperature. Resuspend the cell pellet in 5 mL H16 buffer, centrifuge again, and finally resuspend the cell pellet in 1 mL H16 buffer.
5. Plate 0.1 mL samples of appropriate dilution steps (10^{-2}–10^{-3} in H16 buffer for transfer of broad-host-range vectors; 10^{0}–10^{-1} in H16 buffer for selection after homologous recombination, see Section 3.3) on FN agar plates containing the respective antibiotic(s). After 2–3 days of incubation at 37°C, single colonies of transconjugant cells should become visible. Prepare purity plates of 1–2 single colonies.

3.2 Transformation of *R. eutropha* by electroporation

The restriction-modification system of *R. eutropha* H16 is so far unexplored, which hampers efficient transformation. Here, we describe a relatively simple electrotransformation protocol which is routinely used in our laboratory and results in moderate transformation efficiencies for plasmid DNA.

3.2.1 Equipment

- Electroporator
- Refrigerated centrifuge

3.2.2 Buffers and reagents

- Sterile glycerol solution (10% w/v)
- Electroporation cuvettes, 1–2 mm gap
- LSLB containing 0.4% (w/v) fructose

3.2.3 Procedure

1. Set up a preculture of the desired *R. eutropha* derivative in 10 mL liquid LSLB medium. Cultivate overnight by shaking at 37°C.
2. Inoculate 50 mL LSLB medium with 0.5 mL preculture and incubate on a shaker at 37°C until it reaches an OD_{436nm} between 0.2 and 0.5. Harvest the culture by centrifugation at 4000 × *g* and 4°C in a sterile 2-mL centrifugation tube. All the following steps must be performed on ice or at 4°C.
3. Resuspend the cell pellet in 2 mL ice-cold glycerol solution (10% w/v) and centrifuge in a benchtop centrifuge for 12 min at 4000 × *g*. Repeat this step twice.
4. Resuspend the cell pellet in 1 mL 10% glycerol, prepare 50 μL aliquots in 2-mL reaction tubes, and either shock-freeze them in liquid nitrogen (storage at −80°C) or use them immediately for electroporation (step 5).
5. For electroporation, add 500–2000 ng of purified plasmid to 50 μL of competent *R. eutropha* cells, mix by flicking the reaction tube, and transfer to a chilled electroporation cuvette. Place the cuvette into the holder of the electroporation apparatus and shock at 2500 V, 25 μF, und 200 Ω.
6. Dilute the electroporated cells immediately by adding 1 mL LSLB containing 0.4% fructose, and transfer the suspension to a 1.5-mL reaction tube. Shake the cell suspansion for 1.5 h at 350 rpm and 37°C.
7. Centrifuge the suspension for 1 min at 15,000 rpm and decant the supernatant. Resuspend the cells in the remaining liquid and plate everything on a LSLB agar plate containing the corresponding antibiotics. Incubate the plate for 2 days at 37°C until pale-brown colonies are formed.

3.2.4 Notes

1. The plasmid preparation must be salt-free to avoid a short circuit in the electroporator.
2. Storing at −80°C decreases transformation efficiency of the cells by a factor of 10.

3. For higher transformation efficiency, the slightly more complicated protocol described by Tee et al. (2017) can be used.
4. Transformation efficiency can also be increased significantly through deletion of putative restriction endonuclease genes in the *R. eutropha* genome (Xiong et al., 2018).

3.3 Generation of *R. eutropha* mutants by double homologous recombination

The standard technique for site-directed mutagenesis of the *R. eutropha* genome is to introduce unmarked deletions, insertions, and point mutations by a double homologous recombination procedure (Fig. 4). Based on an initial idea by Gay, Le Coq, Steinmetz, Berkelman, and Kado (1985), conditionally lethal vectors have been constructed and optimized for use with *R. eutropha* (Lenz & Friedrich, 1998; Lenz et al., 1994). The mutagenesis principle is based on polysaccharide synthesis of the levansucrase, SacB, from *Bacillus* species. In gram-positive bacteria, SacB synthesizes high-molecular weight β-(2,6)-linked levan from sucrose in the extracellular space. In gram-negative bacteria, SacB executes its function in the periplasm, finally leading to cell disruption. Here, we describe the allele exchange procedure that is carried out with the conditionally lethal, *sacB*-based suicide plasmids developed in our group.

3.3.1 Equipment

- Incubator shaker
- Refrigerated centrifuge

3.3.2 Buffers and reagents

- Sterile H16 buffer
- Liquid LB and Nutrient broth (NB) media
- LSLB agar plates and FN agar plates with and without antibiotics
- LSLB containing 15% (w/v) sucrose

3.3.3 Procedure

1. Clone the DNA fragment of interest of the *R. eutropha* genome into a suitable vector and introduce the desired mutation by standard molecular biological techniques. Choose the size of the fragment so as to leave on both sides of the mutation approximately 500 bp of unaltered *R. eutropha* DNA (necessary for efficient recombination).

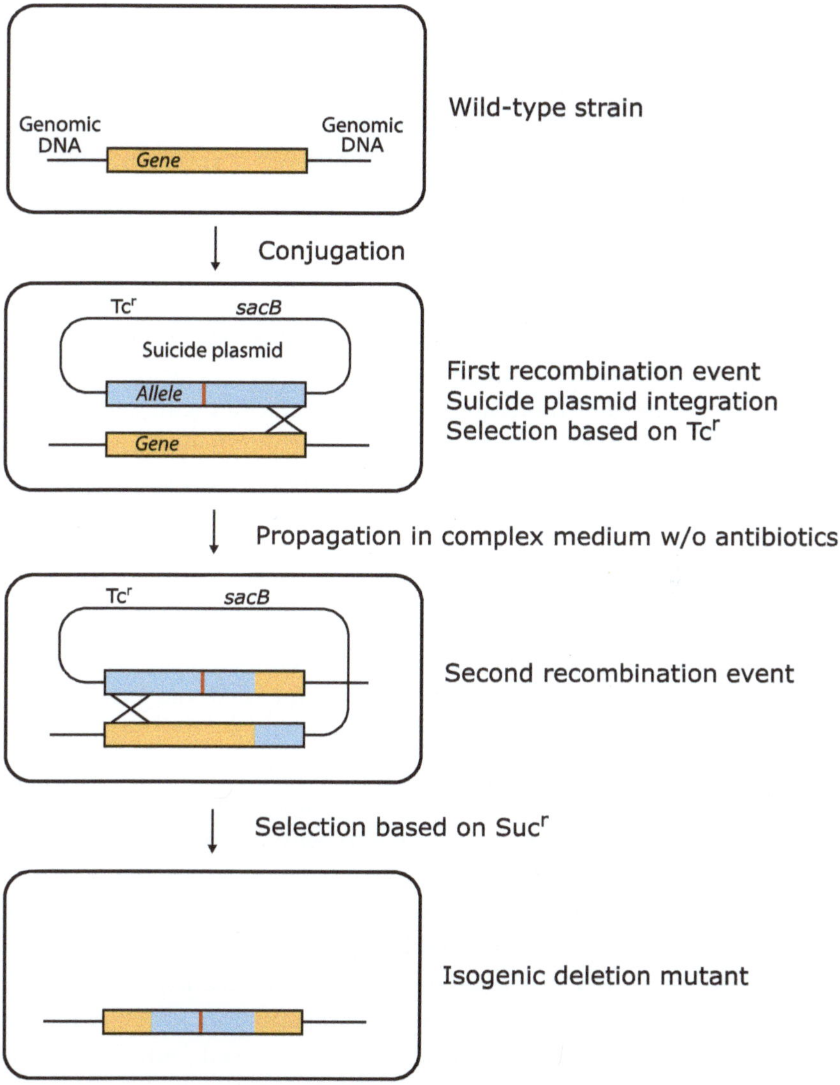

Fig. 4 Allelic exchange strategy based on double homologous recombination. A genomic wild-type gene (*yellow*) is replaced by a plasmid-encoded allele (*blue*) carrying either a base exchange, a deletion, or an insertion (indicated by the *red line*). Selection after the first recombination is based on antibiotic resistance (Tc^r, tetracycline resistance). Selection after the second recombination event is based on sucrose resistance (Suc^r). In the presence of an intact *sacB* gene, sucrose is polymerized to levan which clogs the periplasmic space leading to cell death. Only those cells that have lost the *sacB* gene survive.

2. Transfer the mutagenized DNA fragment to either of the mobilizable suicide plasmids pLO2 (confers kanamycin resistance, Lenz et al., 1994) or pLO3 (confers tetracycline resistance, Lenz & Friedrich, 1998) and transform the resulting plasmids into *E. coli* S17-1.
3. Transfer the recombinant pLO2/3 derivatives by conjugation from *E. coli* S17-1 to the *R. eutropha* strain to be mutagenized (see Section 3.1.1). The pLOX plasmids cannot replicate in *R. eutropha* and must therefore integrate into the genome by homologous recombination. Cells in which the plasmid has been successfully integrated can be selected on FN agar plates containing the appropriate antibiotic(s). Prepare purity plates of 2–4 single colonies.
4. Inoculate 10 mL LSLB medium without antibiotics with 1–4 pure colonies obtained after single homologous recombination and incubate overnight while shaking at 30°C. This step allows the cells to undergo a second homologous recombination event, resulting in either the wild type or the mutagenized DNA sequence with a statistical probability of 50:50.
5. Harvest the cells by centrifugation (10 min at 4000 × *g*, room temperature), remove the supernatant, and resuspend the cells in 5 mL H16 buffer. Centrifuge again and finally resuspended the cell pellet in 1 mL H16 buffer.
6. Plate 0.1 mL of appropriately diluted cell suspension (10^{-2}–10^{-4}) on LSLB agar plates containing 15% (w/v) sucrose and incubate 36–48 h at 30°C. Only cells that have lost the *sacB* gene are able to survive during growth in the presence of sucrose.
7. Transfer cell material of sucrose-resistant single colonies to both antibiotic-free FN agar plates and FN agar plates containing the antibiotic which was used for the first selection round. Incubate the plates for 2–3 days at 30°C. Cells that grow solely on antibiotic-free FN medium should be examined for the presence of the desired mutation by, e.g., colony PCR.

3.3.4 Notes

1. The design of the DNA fragment used for recombination crucially determines the success rate. The DNA fragments neighboring the mutation site should ideally have a similar length. In case of large deletions or insertions (more than 1 kbp in length), the neighboring DNA regions should be extended to 1 kbp or larger.

2. As the first recombination yields clones in which the vector has been integrated in different ways (depending on the extent of the exchanged DNA regions), it might be beneficial to inoculate the LSLB medium for the second recombination event (Section 3.3.3, step 4) with multiple single colonies. This might improve the success rate.
3. The statistical success rate for insertion of the desired mutation after the second recombination event is 50%, but occasionally can be less than 1%, meaning that sometimes more than 100 colonies need to be screened.

3.4 Plasmids replicating in *R. eutropha*

Besides the indigenous and self-transmissible megaplasmid pHG1, which has a size of 452,156 bp and harbors all hydrogenase-related genes (Schwartz et al., 2003), broad-host-range plasmids from different incompatibility groups were shown to replicate in *R. eutropha*. This includes plasmids derived from RP4 (IncP), RSF1010 (IncQ), pSa (IncW), and pBBR1 (incompatibility group to be defined). Without applying antibiotic pressure, however, the plasmids have limited stability over long fermentation periods, with RP4 derivatives showing the highest stability (Gruber, Hagen, Schwab, & Koefinger, 2014). By transplanting the RP4 partition functions (*par*) to, e.g., a RSF1010 derivative, long-term plasmid stability could be improved considerably. The resulting RSF1010-par_{RP4} plasmid was equipped with a range of constitutive (Gruber et al., 2014) and, later, with inducible promoters allowing for IPTG-, cumate- (Gruber et al., 2016), and rhamnose- (Sydow et al., 2017) dependent gene expression at high levels. Another strategy to guarantee accurate plasmid segregation without addition of antibiotics is the use of metabolism-based addiction systems. Systems based on KDPG aldolase, pyrroline-5-carboxylate reductase, and the hydrogenase gene-specific response regulator HoxA have been successfully applied to overproduce (isotopically labeled) cyanophycin and polyhydroxyalkanoate. In these cases, autoinducible *R. eutropha*-derived promoters were employed for gene expression (Budde, Riedel, Willis, Rha, & Sinskey, 2011; Lütte et al., 2012; Voss & Steinbüchel, 2006).

4. Hydrogenase purification

SH and MBH from *R. eutropha* H16 had already been purified in the 1970s (Schink & Schlegel, 1979; Schneider & Schlegel, 1976), but the elaborate procedures consisted of multiple chromatographic steps and were very time-consuming. The genetic amenability of *R. eutropha* H16 has made it feasible to attach peptide tags to hydrogenase structural genes, which, in turn, enabled fast, easy, and mild purification of the *Ralstonia* [NiFe]-hydrogenases by affinity chromatography.

Synthesis of catalytically active [NiFe]-hydrogenase at high levels requires multigene expression. Consequently, operons that comprise all the genes necessary for functional MBH and SH synthesis were constructed and subsequently cloned onto the RP4 derivative, pEDY309 (Kleihues, Lenz, Bernhard, Buhrke, & Friedrich, 2000). The resulting MBH operon on plasmid pGE636 comprised the genes *hoxKGZMLOQRTVhypA1B1F1CDEXhoxA BCJ* whereby the 3′ end of the *hoxK* gene was equipped with a *Strep*-tag II-encoding sequence (Fig. 1A) (Schmidt & Skerra, 2007; Schubert, Lenz, Krause, Volkmer, & Friedrich, 2007). The artificial SH operon on plasmid pGE771 consisted of the genes *hoxFUHYWIhypA2B2F2CDEXhoxA* (Fig. 1A). Here, the *Strep*-tag II-encoding sequence was attached to the 5′ end of *hoxF* (Lauterbach & Lenz, 2013). The respective plasmids were transferred via conjugation to an *R. eutropha* derivative that is devoid of its indigenous megaplasmid pHG1. The resulting transconjugants synthesize either *Strep*-tagged MBH or *Strep*-tagged SH which can be easily purified via *Strep*-Tactin affinity chromatography (Schmidt & Skerra, 2007) as described later.

4.1 Purification of the soluble NAD^+-reducing [NiFe]-hydrogenase (SH)

4.1.1 Equipment

- French press
- Ultracentrifuge
- *Strep*-Tactin column (homemade, filled with *Strep*-Tactin matrix)
- Centrifugal filter units for buffer exchange and concentrating

4.1.2 Buffers and reagents

- Buffer SH-A: 50 m*M* K_iPO_4 buffer, pH 7.0, containing 5% glycerol and 5 m*M* NAD^+ and a protease inhibitor cocktail (EDTA-free protease inhibitor, Roche)
- Buffer SH-B: 50 m*M* K_iPO_4 buffer, pH 7.0, containing 5% glycerol and 5 m*M* NAD^+
- Buffer SH-C: 50 m*M* K_iPO_4 buffer, pH 7.0, containing 5% glycerol

- Buffer SH-D: 50 m*M* K_iPO_4 buffer, pH 7.1, containing 5% glycerol, 5 m*M* desthiobiotin
- Regeneration buffer: 0.5 *M* NaOH

4.1.3 Procedure

1. Cultivate the *R. eutropha* transconjugant harboring plasmid pGE771 in FGN_{mod} medium to an OD_{436nm} of 10–12 as described in Section 2.2. Collect the cells by centrifugation for 20 min at 11,500 × *g* and 4°C. Remove the supernatant and freeze the cell pellet in liquid nitrogen. Store the frozen cells at −80°C.
2. Flush the frozen cells for 10 min with Ar gas. Remove dioxygen from the resuspension buffer SH-A by bubbling for at least 10 min with N_2 or Ar.
3. Resuspend cells in two volumes (vol/wt) of resuspension buffer SH-A, and flush the suspension for 2 min with Ar.
4. Disrupt the cells by two passages at 124.1 MPa through a chilled French press cell, and transfer the suspension into an Ar-flushed ultracentrifugation tube.
5. Flush the tube again with Ar, close the lid, and centrifuge for 45 min at 100,000 × *g*.
6. Equilibrate the *Strep*-Tactin column (1 mL bed volume per 5 mL soluble extract) with buffer SH-B.
7. Pour the soluble extract obtained after ultracentrifugation immediately onto the column and let it run completely through the resin (collect the flow-through when appropriate).
8. Wash the column immediately with 3 column bed volumes (CV) of buffer SH-B.
9. Wash the column with 7 CV of buffer SH-C in order to remove NAD^+.
10. Elute the *Strep*-tagged SH with 4 CV of buffer SH-D.
11. If required, concentrate the SH solution with a centrifugal filter unit (e.g., Amicon Ultra-15, 100 kDa exclusion size, Merck).
12. Regenerate the *Strep*-Tactin column matrix with 5 CV of regeneration buffer and immediately wash the column with ≥5 CV buffer SH-C.

4.1.4 Notes

1. Cell extract preparation and all purification steps should be performed on ice or at 4°C.

2. The procedure to be described results in the isolation of a homodimer of heterohexameric SH (HoxHYFUI$_2$). Higher pH and higher ionic strength results in dissociation of the HoxI subunits. By using 50 m*M* Tris–HCl, pH 8.0, 150 m*M* KCl, 5% glycerol as the buffer basis pH 8, the heterotetrameric HoxHYFU complex can be enriched.
3. If the SH is purified with Tris–HCl buffer at high pH, the buffer should finally be exchanged with Buffer SH-C in which the enzyme shows higher long-term stability.
4. The presence of NAD^+ in buffers SH-A and SH-B supports full oxidation of the SH cofactors and therefore prevents formation of reactive oxygen species.
5. Argon is denser than air and is therefore preferred over N_2 for establishing (semi)anaerobic conditions.
6. Check the pH of buffer SH-D after adding desthiobiotin and readjust, if necessary.
7. SH activity is Na^+-sensitive. Avoid the presence of sodium ions during SH purification.
8. Soluble extracts of *R. eutropha* H16 derivatives contain at least two proteins that are not related to hydrogenase and bind specifically to the *Strep*-Tactin matrix (Jones, Lenz, Strack, Buhrke, & Friedrich, 2004). The amount of these contaminants in the final eluate depends on the production rate of the protein of interest.

4.2 Purification of membrane-bound [NiFe]-hydrogenase

4.2.1 Equipment

- French press
- Centrifuge (refrigerated)
- Ultracentrifuge
- Potter-Elvehjem or Dounce homogenizer
- *Strep*-Tactin column (homemade, filled with *Strep*-Tactin matrix)
- Centrifugal filter units for buffer exchange and concentrating

4.2.2 Buffers and reagents

- Buffer MBH-A: 50 m*M* K_iPO_4; pH 7.1
- Buffer MBH-B: 50 m*M* K_iPO_4, 150 m*M* NaCl; pH 7.1
- Buffer MBH-C: 50 m*M* K_iPO_4, 150 m*M* NaCl, 10% (v/v) glycerol, 3 m*M* desthiobiotin; pH 5.5
- Buffer MBH-D: 50 m*M* K_iPO_4, 150 m*M* NaCl, 10% (v/v) glycerol; pH 5.5

- Regeneration buffer: 0.5 *M* NaOH
- Ferricyanide stock solution: 500 m*M* $K_3[Fe(CN)_6]$
- Digitonin stock solution: 5% (w/v) Digitonin (water soluble, Serva)
- Protease inhibitor tablets: Roche complete EDTA-free or similar
- DNase I (Roche)

4.2.3 Procedure for heterodimeric MBH

1. Cultivate the *R. eutropha* transconjugant harboring plasmid pGE636 in FGN_{mod} medium to an OD_{436nm} of 10–12 as described in Section 2.2. Harvest the cells by centrifugation for 10 min at 11,500 × *g*. Optional: Freeze cells in liquid nitrogen for long-term storage at −80°C.
2. Resuspend cells in 3.5 mL buffer MBH-A per gram (wet weight) of cells. Add protease inhibitor to the final working concentration recommended by the manufacturer. Add a small amount (tip of a spatula) of DNase I. Stir until there are no lumps left.
3. Disrupt the cells by two passages through a prechilled French pressure cell at 124.1 MPa.
4. Centrifuge the cell lysate for 20 min at 4000 × *g* and 4°C. Carefully decant the supernatant into a prechilled beaker and discard the pellet.
5. Add prechilled ferricyanide stock solution to a final concentration of 50 m*M* ferricyanide and mix briefly.
6. Centrifuge the suspension for 45 min at 100,000 × *g* and 4°C. Carefully discard the supernatant.
7. Using a spatula, transfer the brown pellet layers to a prechilled beaker and add 15 mL buffer MBH-B per g of membrane and 1/3 of the recommended final concentration of protease inhibitor.
8. Resuspend the membrane using a Potter-Elvehjem/Dounce homogenizer to generate a homogeneous lump-free suspension.
9. Centrifuge the suspension for 35 min at 100,000 × *g* and 4°C. Discard the supernatant and transfer the brown pellet layers to a prechilled vessel. At this step, the membrane can be either flash frozen in liquid N_2 for long-term storage at −80°C or kept overnight on ice for subsequent MBH purification.
10. Per 1 g of membrane add 5 mL buffer MBH-B containing protease inhibitor at twice the recommended concentration. Resuspend the membrane using a homogenizer until there are no lumps left. Add 5 mL buffer MBH-B including 4% (v/v) Triton X-114 per 1 g of membrane and gently stir for 2 h at 4°C.

11. Centrifuge at 100,000 × *g* for 25 min at 4°C. Carefully decant the solubilized membrane extract into a prechilled vessel and discard the pellet.
12. Equilibrate a *Strep*-Tactin-column with 3.5 CV of buffer MBH-B.
13. Pour the solubilized membrane extract onto the column matrix.
14. Wash the column matrix with 10 CV of buffer MBH-B.
15. Elute MBH with 3 CV of buffer MBH-C and collect the eluate.
16. Regenerate the *Strep*-Tactin column matrix with 5 CV of regeneration buffer and immediately wash the column with ≥5 CV buffer MBH-B.
17. Use a centrifugal filter unit with a molecular weight cutoff of 50 kDa to concentrate the eluted protein to 10 g/L, dilute 10-fold with buffer MBH-D and concentrate again to 10 g/L.
18. In order to remove protein precipitate that may appear during concentration, transfer the concentrate to a prechilled tube, centrifuge for 5 min at maximum speed in a benchtop centrifuge, and transfer the supernatant to another prechilled tube. Analyze a fraction of the sample for purity by SDS-PAGE (Fig. 5). The prepared MBH solution can be flash frozen in liquid N_2 and stored at −80°C or in liquid N_2.

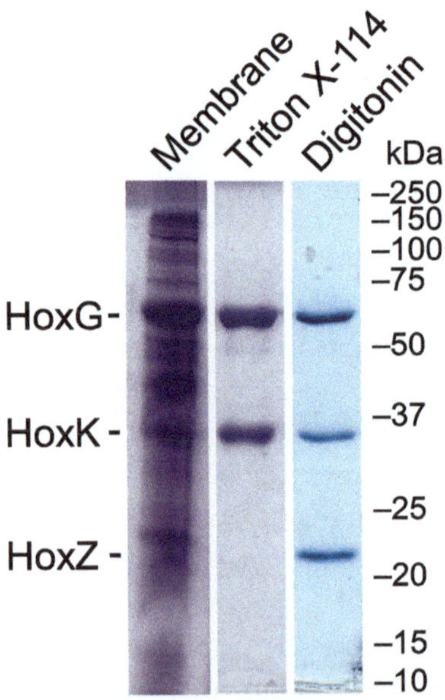

Fig. 5 SDS-PAGE gel showing the band pattern of purified heterodimeric and heterotrimeric MBH. Heterodimeric and heterotrimeric MBHs were isolated using the detergents Triton X-114 and Digitonin, respectively. 10 μg protein of extract and 2 μg of purified protein were separated on 12.5% acrylamide SDS-PAGE. Molecular masses are given in kDa, and the positions of the MBH subunits are indicated.

4.2.4 Procedure for heterotrimeric MBH

1. Follow the instructions in Section 4.2.3 and apply the following changes:
2. Instead of using Triton X-114, solubilize the membrane with digitonin (1.5% (w/v) final concentration).
3. In order to keep the membrane-integral part of the heterotrimeric MBH complex in solution, digitonin needs to be added to all MBH buffers with a final concentration of 0.01% (w/v).
4. Concentrate the eluate by using a centrifugal filtration unit with a molecular weight cutoff of 100 kDa to avoid concentration of the detergent.
5. Apply the concentrated eluate to size-exclusion chromatography using a Superdex 200 column in order to remove, by size exclusion, heterodimeric HoxGK and single HoxZ proteins that are always present to some degree due to complex disintegration (Frielingsdorf, Schubert, Pohlmann, Lenz, & Friedrich, 2011). Analyze a fraction of the sample for purity via SDS-PAGE (Fig. 5).

4.2.5 Notes

1. Ferricyanide addition is critical during membrane preparation. This oxidant removes remaining electrons from the quinone pool and redox-active membrane proteins, thereby preventing oxidative damage of metal centers during purification under oxic conditions.
2. PHB (Pohlmann et al., 2006) appears after ultracentrifugation as white material at the bottom of the membrane pellet and does not need to be included in further steps.
3. In case the membrane fraction will be directly subjected to ferricyanide-sensitive spectroscopic analyses, wash the membrane at least two more times.
4. Always perform membrane resuspension in the absence of detergent, to avoid foam formation.
5. Avoid membrane lumps during solubilization, since the detergent only has access to the lump surface.
6. About 25 mL solubilized membrane extract can be applied to 1 mL *Strep*-Tactin high-capacity matrix.
7. The pH of K_iPO_4 buffer changes upon dilution and NaCl addition. Check (and adjust) the final buffer pH, because the *Strep*-tag II binding efficiency severely drops below pH 7.0.
8. Washing step: when using homemade gravity-flow columns, let the buffer drain completely before adding more buffer, in order to secure that impurities are washed away and not just diluted in the headspace of the column.

9. Ni-NTA columns can be used in combination with His-tagged MBH following the same protocol(s). Affinity purification is then performed in the presence of 20 m*M* imidazole (in buffers MBH-A and -B) and elution is carried out with 250 m*M* imidazole (in buffer MBH-C). Affinity chromatography with Ni-NTA, however, yields His-tagged MBH that is less pure and has significantly lower catalytic activity compared to *Strep*-tagged MBH.
10. Freezing leads to disintegration of the heterotrimeric MBH complex, *so do not freeze.*

5. Biochemical analyses

5.1 H_2-dependent reduction of NAD^+

The quantitative activity determination of SH is based on the H_2-dependent reduction of NAD^+. The formation of the reduced form, NADH, is recorded spectrophotometrically.

5.1.1 Equipment

- UV–vis spectrophotometer, temperature controlled
- Water bath with temperature control
- Cannula (length: 120 mm, diameter 0.8 mm) for gas introduction
- Cannula (length: 25 mm, diameter: 0.6 mm) for gas release
- Cuvette (3 mL volume) closable with a rubber septum (Fig. 6A)
- Hamilton syringe for solution/enzyme addition through the cuvette septum
- Gas washing bottle with screw-capped sampling opening (Fig. 6B)
- Gas regulators with needle valves for fine tuning the gas flow.

5.1.2 Buffers and reagents

- Activity buffer (AB): 50 m*M* Tris–HCl, pH 8.0
- Stock solutions: 50 m*M* NAD^+; 200 μ*M* flavin mononucleotide (FMN); 50 m*M* dithiothreitol (DTT) (or Tris(2-carboxyethyl)phosphine, TCEP)
- Purified SH protein, SH-containing soluble extract or whole cells ($OD_{436nm} = 30$)
- H_2, N_2, and O_2 (grade 5.0)

5.1.3 Procedure

1. Bubble approximately 100 mL AB in a 500-mL gas washing bottle (Fig. 6B) with H_2 gas for at least 30 min at 30°C.

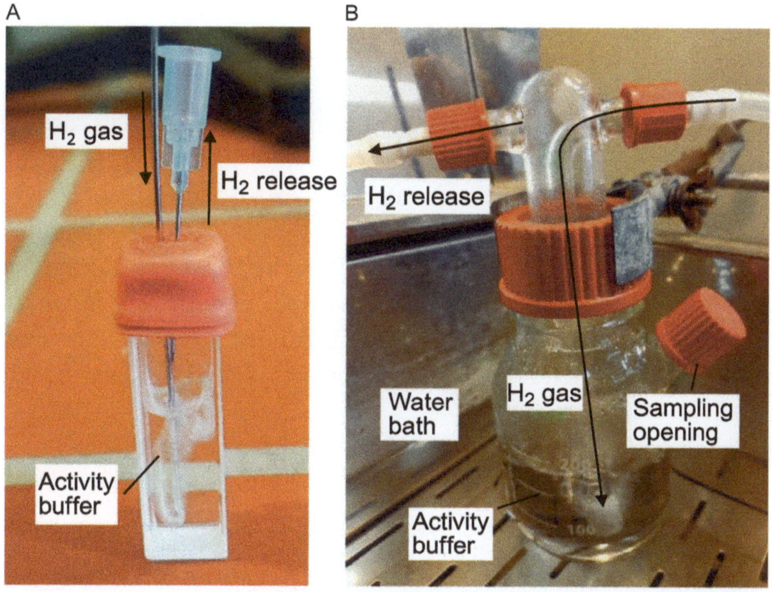

Fig. 6 Setup for hydrogenase activity measurement. (A) Cuvette for photometrical activity measurement. (B) Preparation of H_2-saturated activity buffer in a special gas washing bottle with screw-capped side connections.

2. Pipette 2 mL H_2-saturated activity buffer into the 3-mL cuvette. Add 10, 40, and 40 μL of the FMN, DTT/TCEP, and NAD^+ stock solutions, respectively, and close the cuvette with a rubber septum. Bubble with H_2 for an additional 2 min using a second cannula as exhaust tube (Fig. 6A).
3. Start the reaction by adding 2–50 μL of enzyme solution through the rubber septum using a Hamilton syringe, and place the cuvette immediately in the cuvette holder of a UV–vis spectrophotometer thermostatted at 30°C.
4. Record SH-catalyzed H_2-driven NAD^+ reduction by detecting the increase in NADH concentration at a wavelength of 365 nm.
5. Quantify the enzyme activity by relating the calculated slope of the activity curve to the cuvette/sample dimensions, the extinction coefficient ($\varepsilon_{NADH(365nm)} = 3.3\ mM^{-1}\ cm^{-1}$), and the protein concentration. One unit of specific enzyme activity corresponds to 1 μmol NADH produced per min and 1 mg of protein.

5.1.4 Notes

1. As the pH is temperature dependent, titrate the activity buffer at 30°C. Avoid sodium ions in the assay, because they inhibit SH activity.

2. The absorbance maximum of NADH lies at 340 nm. The absorbance increase, however, remains linear for longer time periods when monitored at 365 nm.
3. SH activity can also be monitored in cell suspensions. In this case the cell suspension needs to be first concentrated to an OD_{436nm} of 30. A volume of 30 μL of the concentrated cell suspension (corresponding to 3.16 mg/mL) is then added to the assay mixture described in Section 5.1.3, step 2, with the exception that NAD^+ is omitted. After bubbling with H_2, 10 μL of 0.5% (wt/vol) cetyltrimethylammonium bromide (CTAB) is added to permeabilize the cell wall. The reaction is initiated 60 s later by the addition of NAD^+ (Friedrich, Heine, Finck, & Friedrich, 1981).
4. For evaluating O_2 tolerance of SH derivatives, various mixtures of activity buffers saturated with either of O_2, N_2, and H_2 are pipetted into the cuvette, and the headspace is filled with the appropriate gas mixtures. NADH is O_2-stable, so that SH activity can be monitored in the presence of O_2.
5. FMN and DTT are added to the activity assay in order to maintain full activity of purified SH protein. They have no beneficial effect to SH activity in soluble extracts or cell suspensions. DTT can be exchanged with TCEP, which is more stable in the presence of O_2 and does not form aggregates with Ni^{2+} ions.

5.2 H_2-dependent reduction of redox dyes

In many cases the natural electron acceptor cannot be used or is unknown, hence, artificial redox dyes need to be used to assay the catalytic activity of hydrogenases or any kind of redox enzyme. Unfortunately, it cannot be predicted which redox dye works well with a given enzyme. Therefore, a certain variety of redox dyes needs to be tested for their compatibility. The redox dyes ideally cover a wide range of redox midpoint potentials.

5.2.1 Equipment

- See Section 5.1.1

5.2.2 Buffers and reagents

- Activity buffer: 50 m*M* K_iPO_4, pH 5.5 ($AB_{5.5}$) and pH 7.0 (AB_7)
- 50 m*M* Methylene blue

5.2.3 Procedure

1. Activity measurements using redox dyes are essentially carried out as described in Section 5.1.
2. For methylene blue reduction mediated by purified MBH, fill the cuvette with 1.9 mL H_2-saturated $AB_{5.5}$ (use AB_7 if membrane extract is assayed), add methylene blue to a final concentration of 200 μ*M*, close the cuvette with a rubber septum, and bubble the solution for 2 min with H_2 gas.
3. Add 1.5 μg of purified MBH using a Hamilton syringe and follow the reaction in a spectrophotometer at 570 nm (Supplementary Movie 1 in the online version at https://doi.org/10.1016/bs.mie.2018.10.008; with kind permission of biotechnologie.tv, full content is accessible in episode 93 via YouTube channel biotechgermany, https://youtu.be/rjrZ8YcOy8Q).
4. Quantify the enzyme activity by relating the calculated slope of the activity curve to the cuvette/sample dimensions, the extinction coefficient ($\varepsilon = 13.1\ \mathrm{m}M^{-1}\ \mathrm{cm}^{-1}$), and the protein concentration. One unit of specific enzyme activity corresponds to 1 μmol H_2 oxidized (corresponding to 1 μmol methylene blue reduced) per min per 1 mg of protein.

5.2.4 Notes

1. Most [NiFe]-hydrogenases as well as MBH derivatives with low activity require reductive activation before entering the catalytic H_2/H^+ cycle. The corresponding lag phase in activity tests can be considerably shortened by adding dithionite at 80 μ*M* final concentration to the activity assay.
2. For uncharacterized hydrogenases find a redox dye that shows good reactivity with your enzyme (see, e.g., Podzuweit, Schneider, & Knüttel, 1987; Schäfer, Friedrich, & Lenz, 2013).
3. Hydrogenase activity crucially depends on pH, and thus the pH optimum should be determined. Wide range buffer systems should be used to avoid effects of single buffer components on catalytic activity (Carmody, 1961; McIlvaine, 1921).
4. Enzyme activities in the hydrogenase field are usually given in U/mg where one U equals the conversion of 1 μmol H_2 per minute. When using redox dyes one has to consider that upon reduction these dyes can accept one electron (e.g., methyl viologen), two electrons (e.g., methylene blue), or even four electrons (e.g., nitroblue tetrazolium chloride) per molecule, i.e., the number of electrons accepted by the

redox dye has to be taken into account when redox dye reduction rates are related to H_2 oxidation rates. Furthermore, we would encourage researchers to give also the turnover frequency or rather the catalytic constant k_{cat} in addition to the specific activity, since k_{cat} is independent of the enzyme's molecular mass and thus allows for easier comparison of enzymes.

5.3 H_2 evolution measurements with modified Clark electrode

A modified Clark electrode allows for continuous recording of reaction kinetics for both H_2 production and consumption. The Clark electrode consists of a working platinum anode and a counter Ag/AgCl cathode, which are immersed in a 3 *M* KCl/10 m*M* HCl electrolyte solution at a potential of +0.6 V. The working electrode is covered with a gas-permeable Teflon membrane. The resulting electrolysis current is linearly dependent on the H_2 concentration in the reaction mixture.

5.3.1 Equipment

- Inversely polarized Clark-type electrode (Sweet, Houchins, Rosen, & Arp, 1980), temperature controlled (e.g., DW2/2 electrode chamber; Hansatech, UK)
- Cigarette rolling paper (ultra-thin)
- Teflon membrane (12.5 μm)

5.3.2 Buffers and reagents

- 500 m*M* Methyl viologen
- 1 *M* Dithionite
- 50 m*M* Tris–HCl, pH 8.0 (SH)
- 50 m*M* K_iPO_4, pH 5.5 (MBH)

5.3.3 Procedure

1. Calibrate the modified Clark electrode by determining the current, first with 1.3 mL N_2-saturated buffer (0% H_2) and then with 1.2 mL N_2-saturated buffer supplemented with 100 μL of H_2-saturated buffer (7.7% H_2).
2. For activity measurements, fill the chamber with 1.3 mL N_2-saturated buffer and add methyl viologen and dithionite to final concentrations of 5 and 7.7 m*M*, respectively.
3. Add 5–10 μg SH or 1.5 μg MBH to the chamber and determine the slope of the linear current increase. One unit of hydrogenase activity is defined as the amount of enzyme which catalyzes the conversion of 1 μmol of H_2 per min.

5.3.4 Notes

1. Prepare the dithionite stock solution in N_2-saturated buffer, because it is O_2 sensitive.
2. In step 2, transfer the N_2-saturated buffer gently to the chamber in order to prevent O_2 contamination. Alternatively, bubble the reaction mixture for 5 min with N_2 prior to enzyme addition.
3. Cover the modified Clark electrode with new cellulose (cigarette rolling paper) and Teflon membranes (at the top) 1 day before the measurement, in order to allow sufficient equilibration time.
4. Avoid touching the Teflon membrane.
5. Physiological electron donors, such as NADH (50 m*M* final concentration) for SH, can be used instead of the DT/MV couple.

5.4 H_2 evolution measurements by gas chromatography

5.4.1 Equipment

- Gas chromatograph (e.g., Shimadzu GC-2014AT) equipped for the detection of small gases with Hayesep N columns and a 13 × molecular sieve column with Argon as the gas carrier according to specification A0608005
- Septum-sealed vials (5–10 mL)
- Gastight Hamilton syringes

5.4.2 Buffers and reagents

- 500 m*M* Methyl viologen
- 1 M Dithionite
- 50 m*M* Tris–HCl, pH 8.0 (SH)
- 50 m*M* K_iPO_4, pH 5.5 (MBH)

5.4.3 Procedure

1. Calibrate the GC with different volumes of pure H_2 gas
2. Perform the H_2-production assay similar as described in Sections 5.2 and 5.3 by applying the following changes: Fill the vial with activity buffer to 10% of the total volume (at maximum) and close the vial with a rubber septum.
3. Bubble the assay mixture with N_2 for additional 5 min in order to remove residual O_2. Puncture the rubber septum with a cannula, which serves as a gas exit tube (see Section 5.1.3).
4. Start the reaction by adding enzyme. Perform the reaction while shaking, in order to ensure adequate mass transfer between liquid and gas phase.
5. Remove 100 μL samples from the gas phase with a gas-tight Hamilton syringe and apply them to gas chromatography.

5.5 Catalytic hydrogen–deuterium exchange measurements

The initial H_2 splitting step of the catalytic cycle can be followed by measuring the catalytic H/D exchange reaction, which is independent of long-range electron transfer kinetics (Cournac, Guedeney, Peltier, & Vignais, 2004; Hartmann et al., 2018). In the absence of an electron acceptor, only the first step of the catalytic cycle is performed, namely, the splitting of H_2 into $H^+ + H^-$. If the reaction takes place in D_2O, the liberated H^+ exchanges with D^+ and finally recombines with H^- to form HD:

$$H_2 + D_2O \rightleftharpoons HD + HDO$$

Analogously, the H^+ of an HD molecule can be exchanged for D^+ to yield D_2:

$$HD + D_2O \rightleftharpoons D_2 + HDO$$

Both HD and D_2 formation can be assayed mass spectrometrically by following m/z of 2, 3, and 4 referring to H_2, HD, and D_2, respectively.

5.5.1 *Equipment*

- Mass spectrometer equipped for analysis of small gases (e.g., OmniStar™ GSD 301, Pfeiffer Vakuum, Germany)
- Reaction chamber (e.g., modified DW2/2 electrode chamber, Hansatech Instruments, UK). We have equipped a DW2/2 chamber with a home-made baseplate that is connected via two Ultra-Torr fittings (Swagelok, Germany) to the mass spectrometer vacuum line (Fig. 7). The baseplate is sealed toward the reaction chamber by two custom product O-rings, a membrane support disc (Frit, SS, 2 μm, semi-prep, ERC GmbH, Germany), and a Teflon membrane (0.0125 mm, Hansatech, UK).
- Magnetic stirrer

5.5.2 *Buffers and reagents*

- 50 m*M* K_iPO_4 in D_2O, pD 5.5 (D_2O, 99.8% atom D, Carl-Roth, Germany)
- Prepare buffer in D_2O by setting up the buffer with water followed by lyophilization or dehydration and subsequent rehydration in D_2O.
- H_2 gas, grade 5.0
- D_2 gas, grade 2.7

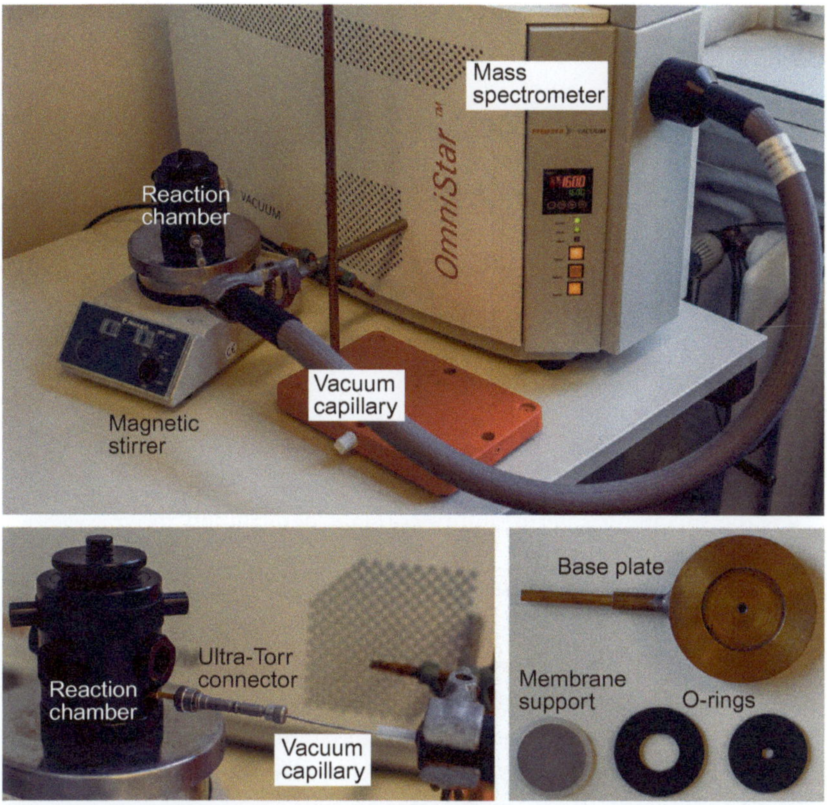

Fig. 7 Setup for catalytic H/D exchange measurements. A modified Hansatech DW2/2 cell is connected via a homemade baseplate and two Swagelok Ultra-Torr fittings to the heated vacuum line of a Pfeiffer Vacuum mass spectrometer. The baseplate and membrane support are sealed using custom product O-rings.

5.5.3 Procedure

1. Set up mass spectrometer and general settings as described previously (Frielingsdorf et al., 2014).
2. Use a small magnetic bar for continuous mixing of the solution.
3. Connect the reaction chamber to the vacuum line and wait for stable baseline.
4. Calibration: introduce 2 mL of buffer to the reaction chamber and bubble with H_2 gas until saturation is reached. Use the raw values obtained with 0% H_2 and 100% H_2 for system calibration. Stop bubbling the reaction chamber and record the change in signal with time. Fit the data to $c(t) = c_0 \cdot e^{-kt}$ and determine k which is a specific constant for H_2 permeability across the Teflon membrane area. Proceed as described earlier

using D_2. Since HD gas is not commercially available, calculate the mean value of the *k* constants determined for H_2 and D_2 to define the HD-specific *k* constant.

5. Measurement: introduce 2 mL of buffer to the reaction chamber, bubble the buffer with H_2 gas until saturation. Stop bubbling and close the reaction chamber. Take care to dip the lid slightly into the liquid phase to prevent gas exchange with the atmosphere of a potential headspace. Inject the hydrogenase sample using a Hamilton syringe and follow the formation of HD and D_2. Convert the raw data into gas concentrations. Correct the gas concentrations for the gas consumption by the mass spectrometer using equation

$$c(t)_{\text{corrected}} = c(t)_{\text{observed}} + \int k \cdot c(t)\, dt.$$

k is the specific gas permeability constant determined above.

5.5.4 Notes

1. The magnetic bar as well as the syringe used for bubbling the solution can disrupt the Teflon membrane leading to liquid entering the mass spectrometer. For mass spectrometer protection a cooling trap or a heated capillary should be integrated into the system.
2. Thoroughly clean all parts of the reaction chamber after each measurement to prevent carryover of active enzyme.

6. Summary and conclusion

Heterologous production of [NiFe]-hydrogenases is still challenging, because their maturation relies on the dedicated function of at least seven specific auxiliary proteins. Biosynthesis of O_2-tolerant [NiFe]-hydrogenases requires even more maturases, which are responsible for metal center protection under oxic conditions (Fritsch, Lenz, & Friedrich, 2013). Thus, the latter enzymes are usually isolated from their native hosts. The aerobic H_2 oxidizer *Ralstonia eutropha* probably harbors the most prominent O_2-tolerant [NiFe]-hydrogenases, and is nonpathogenic, genetically amenable, and easily cultivatable. These advantageous characteristics allow the convenient isolation of O_2-tolerant [NiFe]-hydrogenases and derivatives thereof, which can be easily obtained by genetic engineering.

The cultivation conditions, cell extract preparations, and purification procedures we have described have been designed to obtain high amounts of pure and homogenous preparations of MBH and SH from *R. eutropha*. Crucial for homogeneity is the initial preparation of the cell extracts which includes the treatment of the proteins with oxidants (ferricyanide, NAD^+) in a quasi-cellular context (Goris et al., 2011; Lauterbach & Lenz, 2013). In living cells, O_2 tolerance of [NiFe]-hydrogenases requires the redox-communication of the enzymes with their natural electron acceptors/donors, i.e., the oxidized/reduced cytochrome *b* in case of MBH and NAD^+/NADH in the case of SH (Horch et al., 2010; Radu, Frielingsdorf, Evans, Lenz, & Jeuken, 2014). This communication is interrupted during purification. Thus, the addition of suitable oxidants ensures complete oxidation of the hydrogenases prior to oxic purification, preventing the enzymes from reacting with O_2 and generating reactive oxygen species that would damage the metal centers.

Apart from the biochemical methods for hydrogenase characterization described in this chapter, a range of further analytical techniques have been used to investigate, in particular, the various metal centers. Their use and the corresponding data analysis rely on sufficient amounts of homogeneous enzyme where the metal centers reside in defined redox states. These methods include—among others—Fourier-transform infrared spectroscopy (Horch et al., 2015), Raman spectroscopy (Siebert et al., 2013), electron paramagnetic resonance spectroscopy (Löwenstein, Lauterbach, Teutloff, Lenz, & Bittl, 2015), protein film electrochemistry (Cracknell, Wait, Lenz, Friedrich, & Armstrong, 2009), as well as X-ray crystallography (Frielingsdorf et al., 2014; Fritsch et al., 2011), all of which have contributed in complementary ways to obtaining a deep understanding of the catalytic processes occurring in O_2-tolerant [NiFe]-hydrogenases.

References

Atlas, R. M. (2010). *Handbook of microbiological media* (4th ed.). . Hoboken: CRC Press. Retrieved from *http://lib.myilibrary.com/detail.asp?ID=250412*.

Budde, C. F., Riedel, S. L., Willis, L. B., Rha, C., & Sinskey, A. J. (2011). Production of poly(3-hydroxybutyrate-co-3-hydroxyhexanoate) from plant oil by engineered *Ralstonia eutropha* strains. *Applied and Environmental Microbiology*, 77(9), 2847–2854. https://doi.org/10.1128/AEM.02429-10.

Carmody, W. R. (1961). An easily prepared wide range buffer series. *Journal of Chemical Education*, *38*(11), 559. https://doi.org/10.1021/ed038p559.

Cournac, L., Guedeney, G., Peltier, G., & Vignais, P. M. (2004). Sustained photoevolution of molecular hydrogen in a mutant of *Synechocystis* sp. strain PCC 6803 deficient in the type I NADPH-dehydrogenase complex. *Journal of Bacteriology*, *186*(6), 1737–1746. https://doi.org/10.1128/JB.186.6.1737-1746.2003.

Cracknell, J. A., Wait, A. F., Lenz, O., Friedrich, B., & Armstrong, F. A. (2009). A kinetic and thermodynamic understanding of O_2 tolerance in NiFe-hydrogenases. *Proceedings of*

the National Academy of Sciences of the United States of America, 106(49), 20681–20686. https://doi.org/10.1073/pnas.0905959106.
Friedrich, C. G., & Friedrich, B. (1983). Regulation of hydrogenase formation is temperature sensitive and plasmid coded in *Alcaligenes eutrophus*. *Journal of Bacteriology, 153*(1), 176–181.
Friedrich, C. G., Friedrich, B., & Bowien, B. (1981). Formation of enzymes of autotrophic metabolism during heterotrophic growth of *Alcaligenes eutrophus*. *Microbiology, 122*(1), 69–78. https://doi.org/10.1099/00221287-122-1-69.
Friedrich, B., Heine, E., Finck, A., & Friedrich, C. G. (1981). Nickel requirement for active hydrogenase formation in *Alcaligenes eutrophus*. *Journal of Bacteriology, 145*(3), 1144–1149.
Friedrich, B., Hogrefe, C., & Schlegel, H. G. (1981). Naturally occurring genetic transfer of hydrogen-oxidizing ability between strains of *Alcaligenes eutrophus*. *Journal of Bacteriology, 147*(1), 198–205.
Frielingsdorf, S., Fritsch, J., Schmidt, A., Hammer, M., Löwenstein, J., Siebert, E., et al. (2014). Reversible [4Fe3S] cluster morphing in an O_2-tolerant [NiFe] hydrogenase. *Nature Chemical Biology, 10*(5), 378–385. https://doi.org/10.1038/NCHEMBIO.1500.
Frielingsdorf, S., Schubert, T., Pohlmann, A., Lenz, O., & Friedrich, B. (2011). A trimeric supercomplex of the oxygen-tolerant membrane-bound [NiFe]-hydrogenase from *Ralstonia eutropha* H16. *Biochemistry, 50*(50), 10836–10843. https://doi.org/10.1021/bi201594m.
Fritsch, J., Lenz, O., & Friedrich, B. (2013). Structure, function and biosynthesis of O_2-tolerant hydrogenases. *Nature Reviews Microbiology, 11*(2), 106–114. https://doi.org/10.1038/nrmicro2940.
Fritsch, J., Scheerer, P., Frielingsdorf, S., Kroschinsky, S., Friedrich, B., Lenz, O., et al. (2011). The crystal structure of an oxygen-tolerant hydrogenase uncovers a novel iron-sulphur centre. *Nature, 479*(7372), 249–252. https://doi.org/10.1038/nature10505.
Gay, P., Le Coq, D., Steinmetz, M., Berkelman, T., & Kado, C. I. (1985). Positive selection procedure for entrapment of insertion sequence elements in gram-negative bacteria. *Journal of Bacteriology, 164*(2), 918–921.
Goris, T., Wait, A. F., Saggu, M., Fritsch, J., Heidary, N., Stein, M., et al. (2011). A unique iron-sulfur cluster is crucial for oxygen tolerance of a NiFe-hydrogenase. *Nature Chemical Biology*, 7(5), 310–318. https://doi.org/10.1038/nchembio.555.
Gruber, S., Hagen, J., Schwab, H., & Koefinger, P. (2014). Versatile and stable vectors for efficient gene expression in *Ralstonia eutropha* H16. *Journal of Biotechnology, 186*, 74–82. https://doi.org/10.1016/j.jbiotec.2014.06.030.
Gruber, S., Schwendenwein, D., Magomedova, Z., Thaler, E., Hagen, J., Schwab, H., et al. (2016). Design of inducible expression vectors for improved protein production in *Ralstonia eutropha* H16 derived host strains. *Journal of Biotechnology, 235*, 92–99. https://doi.org/10.1016/j.jbiotec.2016.04.026.
Hartmann, S., Frielingsdorf, S., Ciaccafava, A., Lorent, C., Fritsch, J., Siebert, E., et al. (2018). O_2-tolerant H_2 activation by an isolated large subunit of a [NiFe] hydrogenase. *Biochemistry, 57*, 5339–5349. https://doi.org/10.1021/acs.biochem.8b00760.
Horch, M., Lauterbach, L., Mroginski, M. A., Hildebrandt, P., Lenz, O., & Zebger, I. (2015). Reversible active site sulfoxygenation can explain the oxygen tolerance of a NAD^+-reducing NiFe hydrogenase and its unusual infrared spectroscopic properties. *Journal of the American Chemical Society, 137*(7), 2555–2564. https://doi.org/10.1021/ja511154y.
Horch, M., Lauterbach, L., Saggu, M., Hildebrandt, P., Lendzian, F., Bittl, R., et al. (2010). Probing the active site of an O_2-tolerant NAD^+-reducing NiFe-hydrogenase from *Ralstonia eutropha* H16 by in situ EPR and FTIR spectroscopy. *Angewandte Chemie (International Ed. in English), 49*(43), 8026–8029. https://doi.org/10.1002/anie.201002197.

Jones, A. K., Lenz, O., Strack, A., Buhrke, T., & Friedrich, B. (2004). NiFe hydrogenase active site biosynthesis: Identification of Hyp protein complexes in *Ralstonia eutropha*. *Biochemistry*, *43*(42), 13467–13477. https://doi.org/10.1021/bi048837k.

Kleihues, L., Lenz, O., Bernhard, M., Buhrke, T., & Friedrich, B. (2000). The H_2 sensor of *Ralstonia eutropha* is a member of the subclass of regulatory [NiFe] hydrogenases. *Journal of Bacteriology*, *182*(10), 2716–2724. https://doi.org/10.1128/JB.182.10.2716-2724.2000.

Kohlmann, Y., Pohlmann, A., Otto, A., Becher, D., Cramm, R., Lütte, S., et al. (2011). Analyses of soluble and membrane proteomes of *Ralstonia eutropha* H16 reveal major changes in the protein complement in adaptation to lithoautotrophy. *Journal of Proteome Research*, *10*(6), 2767–2776. https://doi.org/10.1021/pr101289v.

Krassen, H., Schwarze, A., Friedrich, B., Ataka, K., Lenz, O., & Heberle, J. (2009). Photosynthetic hydrogen production by a hybrid complex of photosystem I and NiFe-hydrogenase. *ACS Nano*, *3*(12), 4055–4061. https://doi.org/10.1021/nn900748j.

Lauterbach, L., & Lenz, O. (2013). Catalytic production of hydrogen peroxide and water by oxygen-tolerant NiFe-hydrogenase during H_2 cycling in the presence of O_2. *Journal of the American Chemical Society*, *135*(47), 17897–17905. https://doi.org/10.1021/ja408420d.

Lenz, O., & Friedrich, B. (1998). A novel multicomponent regulatory system mediates H_2 sensing in *Alcaligenes eutrophus*. *Proceedings of the National Academy of Sciences of the United States of America*, *95*(21), 12474–12479. https://doi.org/10.1073/pnas.95.21.12474.

Lenz, O., Lauterbach, L., Frielingsdorf, S., & Friedrich, B. (2015). Oxygen-tolerant hydrogenases and their biotechnological potential. In M. Rögner (Ed.), *Biohydrogen* (pp. 61–88). Berlin, Boston: Walter de Gruyter GmbH & Co. KG.

Lenz, O., Schwartz, E., Dernedde, J., Eitinger, M., & Friedrich, B. (1994). The *Alcaligenes eutrophus* H16 *hoxX* gene participates in hydrogenase regulation. *Journal of Bacteriology*, *176*(14), 4385–4393.

Lonsdale, T. H., Lauterbach, L., Honda Malca, S., Nestl, B. M., Hauer, B., & Lenz, O. (2015). H_2-driven biotransformation of n-octane to 1-octanol by a recombinant *Pseudomonas putida* strain co-synthesizing an O_2-tolerant hydrogenase and a P450 monooxygenase. *Chemical Communications (Cambridge, England)*, *51*(90), 16173–16175. https://doi.org/10.1039/c5cc06078h.

Löwenstein, J., Lauterbach, L., Teutloff, C., Lenz, O., & Bittl, R. (2015). Active site of the NAD^+-reducing hydrogenase from *Ralstonia eutropha* studied by EPR spectroscopy. *The Journal of Physical Chemistry. B*, *119*(43), 13834–13841. https://doi.org/10.1021/acs.jpcb.5b04144.

Ludwig, M., Cracknell, J. A., Vincent, K. A., Armstrong, F. A., & Lenz, O. (2009). Oxygen-tolerant H_2 oxidation by membrane-bound NiFe hydrogenases of *Ralstonia* species. Coping with low level H_2 in air. *Journal of Biological Chemistry*, *284*(1), 465–477. https://doi.org/10.1074/jbc.M803676200.

Lütte, S., Pohlmann, A., Zaychikov, E., Schwartz, E., Becher, J. R., Heumann, H., et al. (2012). Autotrophic production of stable-isotope-labeled arginine in *Ralstonia eutropha* strain H16. *Applied and Environmental Microbiology*, *78*(22), 7884–7890. https://doi.org/10.1128/AEM.01972-12.

McIlvaine, T. C. (1921). A buffer solution for colorimetric comparison. *Journal of Biological Chemistry*, *49*(1), 183–186.

Podzuweit, H. G., Schneider, K., & Knüttel, H. (1987). Comparison of the membrane-bound hydrogenases from *Alcaligenes eutrophus* H16 and *Alcaligenes eutrophus* type strain. *Biochimica et Biophysica Acta—Biomembranes*, *905*(2), 435–446. https://doi.org/10.1016/0005-2736(87)90473-1.

Pohlmann, A., Fricke, W. F., Reinecke, F., Kusian, B., Liesegang, H., Cramm, R., et al. (2006). Genome sequence of the bioplastic-producing "Knallgas" bacterium *Ralstonia eutropha* H16. *Nature Biotechnology*, *24*(10), 1257–1262. https://doi.org/10.1038/nbt1244.

Radu, V., Frielingsdorf, S., Evans, S. D., Lenz, O., & Jeuken, L. J. C. (2014). Enhanced oxygen-tolerance of the full heterotrimeric membrane-bound NiFe-hydrogenase of *Ralstonia eutropha*. *Journal of the American Chemical Society*, *136*(24), 8512–8515. https://doi.org/10.1021/ja503138p.

Reeve, H. A., Ash, P. A., Park, H., Huang, A., Posidias, M., Tomlinson, C., et al. (2017). Enzymes as modular catalysts for redox half-reactions in H_2-powered chemical synthesis: From biology to technology. *The Biochemical Journal*, *474*(2), 215–230. https://doi.org/10.1042/BCJ20160513.

Schäfer, C., Friedrich, B., & Lenz, O. (2013). Novel, oxygen-insensitive group 5 [NiFe]-hydrogenase in *Ralstonia eutropha*. *Applied and Environmental Microbiology*, *79*(17), 5137–5145. https://doi.org/10.1128/AEM.01576-13.

Schink, B., & Schlegel, H. G. (1979). The membrane-bound hydrogenase of *Alcaligenes eutrophus*. I. Solubilization, purification, and biochemical properties. *Biochimica et Biophysica Acta (BBA)—Enzymology*, *567*(2), 315–324. https://doi.org/10.1016/0005-2744(79)90117-7.

Schmidt, T. G. M., & Skerra, A. (2007). The *Strep*-tag system for one-step purification and high-affinity detection or capturing of proteins. *Nature Protocols*, *2*(6), 1528–1535. https://doi.org/10.1038/nprot.2007.209.

Schneider, K., & Schlegel, H. G. (1976). Purification and properties of soluble hydrogenase from *Alcaligenes eutrophus* H16. *Biochimica et Biophysica Acta (BBA)—Enzymology*, *452*(1), 66–80. https://doi.org/10.1016/0005-2744(76)90058-9.

Schubert, T., Lenz, O., Krause, E., Volkmer, R., & Friedrich, B. (2007). Chaperones specific for the membrane-bound NiFe-hydrogenase interact with the Tat signal peptide of the small subunit precursor in *Ralstonia eutropha* H16. *Molecular Microbiology*, *66*(2), 453–467. https://doi.org/10.1111/j.1365-2958.2007.05933.x.

Schwartz, E., Henne, A., Cramm, R., Eitinger, T., Friedrich, B., & Gottschalk, G. (2003). Complete nucleotide sequence of pHG1: A *Ralstonia eutropha* H16 megaplasmid encoding key enzymes of H_2-based lithoautotrophy and anaerobiosis. *Journal of Molecular Biology*, *332*(2), 369–383. https://doi.org/10.1016/S0022-2836(03)00894-5.

Sichwart, S., Hetzler, S., Bröker, D., & Steinbüchel, A. (2011). Extension of the substrate utilization range of *Ralstonia eutropha* strain H16 by metabolic engineering to include mannose and glucose. *Applied and Environmental Microbiology*, 77(4), 1325–1334. https://doi.org/10.1128/AEM.01977-10.

Siebert, E., Horch, M., Rippers, Y., Fritsch, J., Frielingsdorf, S., Lenz, O., et al. (2013). Resonance Raman spectroscopy as a tool to monitor the active site of hydrogenases. *Angewandte Chemie (International Ed in English)*, *52*(19), 5162–5165. https://doi.org/10.1002/anie.201209732.

Simon, R., Priefer, U., & Pühler, A. (1983). A broad host range mobilization system for in vivo genetic engineering: Transposon mutagenesis in Gram negative bacteria. *Nature Biotechnology*, *1*(9), 784. https://doi.org/10.1038/nbt1183-784.

Srivastava, S., Urban, M., & Friedrich, B. (1982). Mutagenesis of *Alcaligenes eutrophus* by insertion of the drug-resistance transposon *Tn*5. *Archives of Microbiology*, *131*(3), 203–207.

Sweet, W. J., Houchins, J. P., Rosen, P. R., & Arp, D. J. (1980). Polarographic measurement of H_2 in aqueous solutions. *Analytical Biochemistry*, *107*(2), 337–340. https://doi.org/10.1016/0003-2697(80)90393-0.

Sydow, A., Pannek, A., Krieg, T., Huth, I., Guillouet, S. E., & Holtmann, D. (2017). Expanding the genetic tool box for *Cupriavidus necator* by a stabilized L-rhamnose inducible plasmid system. *Journal of Biotechnology*, *263*, 1–10. https://doi.org/10.1016/j.jbiotec.2017.10.002.

Tee, K. L., Grinham, J., Othusitse, A. M., González-Villanueva, M., Johnson, A. O., & Wong, T. S. (2017). An efficient transformation method for the bioplastic-producing

"Knallgas" bacterium *Ralstonia eutropha* H16. *Biotechnology Journal*, *12*(11) https://doi.org/10.1002/biot.201700081.

Vincent, K. A., Cracknell, J. A., Lenz, O., Zebger, I., Friedrich, B., & Armstrong, F. A. (2005). Electrocatalytic hydrogen oxidation by an enzyme at high carbon monoxide or oxygen levels. *Proceedings of the National Academy of Sciences of the United States of America*, *102*(47), 16951–16954. https://doi.org/10.1073/pnas.0504499102.

Voss, I., & Steinbüchel, A. (2006). Application of a KDPG-aldolase gene-dependent addiction system for enhanced production of cyanophycin in *Ralstonia eutropha* strain H16. *Metabolic Engineering*, *8*(1), 66–78. https://doi.org/10.1016/j.ymben.2005.09.003.

Wilde, E. (1962). Untersuchungen über Wachstum und Speicherstoffsynthese von *Hydrogenomonas*. *Archives of Microbiology*, *43*(2), 109–137. https://doi.org/10.1007/BF00406429.

Xiong, B., Li, Z., Liu, L., Zhao, D., Zhang, X., & Bi, C. (2018). Genome editing of *Ralstonia eutropha* using an electroporation-based CRISPR-Cas9 technique. *Biotechnology for Biofuels*, *11*, 172. https://doi.org/10.1186/s13068-018-1170-4.

CHAPTER SIX

Cytoplasmic and membrane-bound hydrogenases from *Pyrococcus furiosus*

Chang-Hao Wu, Dominik K. Haja, Michael W.W. Adams*
Department of Biochemistry and Molecular Biology, University of Georgia, Athens, GA, United States
*Corresponding author: e-mail address: adamsm@uga.edu

Contents

Abstract

Hydrogenases catalyze the simplest of chemical reactions, the reversible interconversion of protons, electrons, and hydrogen gas. These enzymes have potential to be utilized for several biotechnological applications, such as in vitro hydrogen production from renewable materials and in enzyme-based fuel cells for electricity generation. Based on the metal content of their catalytic sites, hydrogenases are classified as either [NiFe], [FeFe], or mononuclear Fe enzymes, and [NiFe] hydrogenases are further categorized into five groups based on the sequences of the catalytic subunits. This chapter describes recombinant engineering strategies, purification procedures, and catalytic properties of two distinct types of [NiFe] hydrogenase from *Pyrococcus furiosus*, a

Methods in Enzymology, Volume 613
ISSN 0076-6879
https://doi.org/10.1016/bs.mie.2018.10.009

microorganism with an optimal growth temperature of 100°C. These enzymes are termed soluble hydrogenase I (SHI, group 3) and membrane-bound hydrogenase (MBH, group 4). The two hydrogenases were affinity-tagged to facilitate their purification and the purified enzymes have been used for biochemical, mechanistic, and structural analyses. The results have provided us with new insights into how catalysis by SHI is achieved, which could also lead to the development of catalysts for economic hydrogen production, and knowledge of how MBH couples hydrogen gas production to conservation of energy in the form of an ion gradient. The methods described in this chapter provide the basis for these studies.

1. Introduction

Hydrogenases catalyze the simplest of reactions, the reversible interconversion of hydrogen gas, protons, and electrons: $2H^+ + 2e^- \leftrightarrow H_2$. They have several biotechnological applications involving both H_2 production and H_2 oxidation. For instance, electrons from H_2 oxidation can be used for the reduction of the nicotinamide cofactors, thereby providing a clean energy source for NAD(P)H-dependent enzymes in chemical synthesis (Reeve et al., 2017). In addition, H_2 oxidation by hydrogenase can be utilized in enzyme-based fuel cells for electricity generation where hydrogen gas is the energy source (Chenevier et al., 2013). Hydrogen gas is a renewable and efficient energy carrier with $3\times$ the energy content per unit mass compared to fossil fuels (Veziroglu & Barbir, 1992). Hydrogenases provide an opportunity for environmentally friendly and renewable methods for H_2 production, in contrast to the current method for the mass production of hydrogen gas, which still primarily relies on fossil fuels (Lee, Vermaas, & Rittmann, 2010). Indeed, a cell-free synthetic pathway containing more than 10 enzymes, including a H_2-evolving hydrogenase, provides a promising, alternative, and renewable method for converting renewable sugars to hydrogen gas (Kim et al., 2017; Zhang, Evans, Mielenz, Hopkins, & Adams, 2007; Zhang, Sun, & Ma, 2017). This in vitro cell-free system is particularly attractive as it can achieve the maximum yield of 12 H_2 per C_6 sugar, $3\times$ the maximum yield that can be obtained by in vivo cellular fermentations (≤ 4 H_2 per C_6 sugar). In addition, fundamental studies of hydrogenases provide an opportunity for the design and development of artificial catalysts that are more stable than native enzymes and based on iron and/or nickel catalytic sites. These would replace the current catalysts that rely on expensive metals such as platinum to catalyze H_2 production (Chenevier et al., 2013; Mondal & Dey, 2017).

Based on the metal contained in the catalytic sites, hydrogenases can be classified into three types: [NiFe] hydrogenases, [FeFe] hydrogenases, and mononuclear Fe hydrogenases (Lubitz, Ogata, Rudiger, & Reijerse, 2014; Sondergaard, Pedersen, & Greening, 2016; Vignais & Billoud, 2007). The [NiFe] hydrogenases are widespread in bacteria and archaea and have been extensively studied (Lubitz et al., 2014). They can be further categorized into five groups (designated 1–5) based on the sequences of their catalytic subunits and these groups generally represent five different physiological functions (Sondergaard et al., 2016; Vignais, Billoud, & Meyer, 2001; Vignais & Colbeau, 2004). In this chapter we focus on one group 3 and one group 4 hydrogenase. The group 3 enzymes are bidirectional and cofactor coupled and the focus here is on a group 3b hydrogenase that reversibly oxidizes H_2 and reduces $NADP^+$. The group 4 enzymes are respiratory H_2-evolving hydrogenases that are membrane-bound and couple proton reduction to ion pumping. Both of these enzymes have been purified from the microorganism *Pyrococcus furiosus*.

P. furiosus grows optimally at 100°C (Fiala & Stetter, 1986) and contains three [NiFe] hydrogenases. Two are cytoplasmic (SHI and SHII, both group 3b) and one is membrane-bound (MBH, group 4). *P. furiosus* has an unusual Embden–Meyerhof glycolytic pathway (Kengen et al., 1994) in which the reductant reduces the iron–sulfur redox protein ferredoxin (Fd) rather than NAD^+ (Mukund & Adams, 1995). MBH oxidizes the reduced Fd generated from glycolysis and reduces protons to H_2, and also generates a sodium ion gradient that is further used for ATP generation by ATP synthase (Pisa, Huber, Thomm, & Muller, 2007). It is thought that the H_2 produced by MBH is oxidized by SHI and SHII to produce NAD(P)H, although the exact physiological roles of these enzymes are in doubt due to the lack of a growth phenotype in mutational studies (Lipscomb et al., 2011). All three hydrogenases from *P. furiosus* have been purified and characterized. SHI and SHII were obtained from native biomass while SHI was also genetically modified so that it could be purified from *P. furiosus* by affinity chromatography (Bryant & Adams, 1989; Chandrayan et al., 2012; Chandrayan, Wu, McTernan, & Adams, 2015; Ma, Weiss, & Adams, 2000). MBH was also engineered for affinity-based purification (McTernan et al., 2015, 2014). In this chapter, we describe the engineering strategies that led to the overexpression of SHI and MBH in *P. furiosus* as well as facilitating their purification. The catalytic properties of these enzymes are also discussed.

2. Expression strategy

The original purification of SHI from native *P. furiosus* biomass involved multistep column chromatography, and the yield (2.5 mg/100 g of cells, wet weight) was too low for detailed characterization or applied studies. Therefore, the four-gene operon (PF0891-PF0894) encoding heterotetrameric (βγδα) SHI was overexpressed using a highly active constitutive promoter (P_{slp}, for the gene encoding the S-layer protein) in *P. furiosus* and a 9x-histidine tag was placed at the N-terminus of the β subunit (PF0891), resulting in the His-SHI strain as shown in Fig. 1 (Chandrayan et al., 2015).

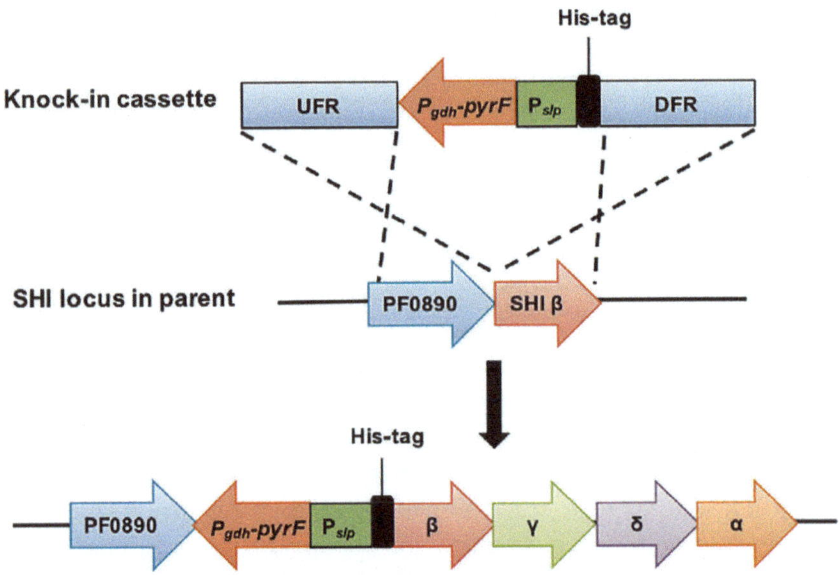

Fig. 1 Scheme for the construction of the affinity-tagged SHI strain. The knock-in cassette contains the upstream flanking region (UFR), homologous to PF0890, and downstream flanking region (DFR), homologous to PF0891. The selectable marker *pyrF* is driven by the glutamate dehydrogenase promoter (P_{gdh}), and the constitutive promoter of S-layer protein gene (P_{slp}) was used to overexpress SHI operon. A 9x-histidine tag was also placed at the N-terminus of SHIβ (PF0891). The arrangement of the SHI locus after homologous recombination in the genetically traceable strain COM1 is also shown. *Modified from Chandrayan, S. K., Wu, C. H., McTernan, P. M., & Adams, M. W. (2015). High yield purification of a tagged cytoplasmic [NiFe]-hydrogenase and a catalytically-active nickel-free intermediate form.* Protein Expression and Purification, 107, *90–94.*

The knock-in cassette contained the homologous recombination upstream flanking region (UFR) and downstream flanking region (DFR), the selection marker *pyrF* (PF1114) driven by the promoter for the glutamate dehydrogenase gene (PF1602), the promoter of the S-layer protein gene (PF1399), and the histidine tag. This cassette was assembled using overlapping PCR (Bryksin & Matsumura, 2010) and was placed upstream of the first gene in the SHI operon (PF0891) in the genetically tractable *P. furiosus* strain COM1 (Lipscomb et al., 2011).

A similar strategy was used to engineer an affinity purification tag on to MBH, as shown in Fig. 2. However, MBH is encoded by an operon of 14 genes (*mbhA-N*) where the first 9 encode proteins with multiple helices that are predicted to be membrane-bound. Hence, instead of placing the tag upstream of the first gene in the *mbh* operon, the knock-in cassette was placed upstream of *mbhJ*, the first gene predicted not to be integrally membrane bound. The location of the affinity tag was shown to be important for purifying MBH. Specifically, when the affinity tag was on the N-terminus of MbhJ and MBH was solubilized from the membranes and subjected to affinity chromatography, 27% of the MBH activity (H_2 evolution from an artificial electron donor, see below) was recovered. This compares with only 2% when the affinity tag was at the C-terminus of MbhN, which is encoded by the last gene in the operon (McTernan et al., 2014). However, the insertion of the affinity tag onto MbhJ, together with a new promoter (P_{slp}), splits the MBH operon such that two transcripts are produced, one (*mbhA-I*) driven by the native promoter (P_{mbh}) and one (*mbhJ-N*) driven by P_{slp} (McTernan et al., 2015). Transcriptional analysis showed that the expression level of the transcript driven by P_{slp} was twice that of the transcript driven P_{mbh}. Assuming both transcripts are translated to protein with similar efficiency, twice as much MbhJ-N would be produced compared to MbhA-I. Hence, it was anticipated that two forms of MBH should be purified from the recombinant *P. furosus* strain. As shown in Fig. 2, an intact 14-subunit MbhA-N holoenzyme (solubilized or S-MBH) is obtained from the solubilized membrane fraction, while the excess MbhJ-N should yield a 5-subunit subcomplex (cytoplasmic or C-MBH) from the cytoplasm fraction. Importantly, C-MBH contains the catalytic subunit (MbhL) and the subunits (MbhJ and MbhN) containing the three iron–sulfur clusters that transfer electrons to the catalytic site (see Fig. 2); thus C-MBH should be catalytic active in in vitro assays. The purification procedures for each of these two forms of MBH are described below.

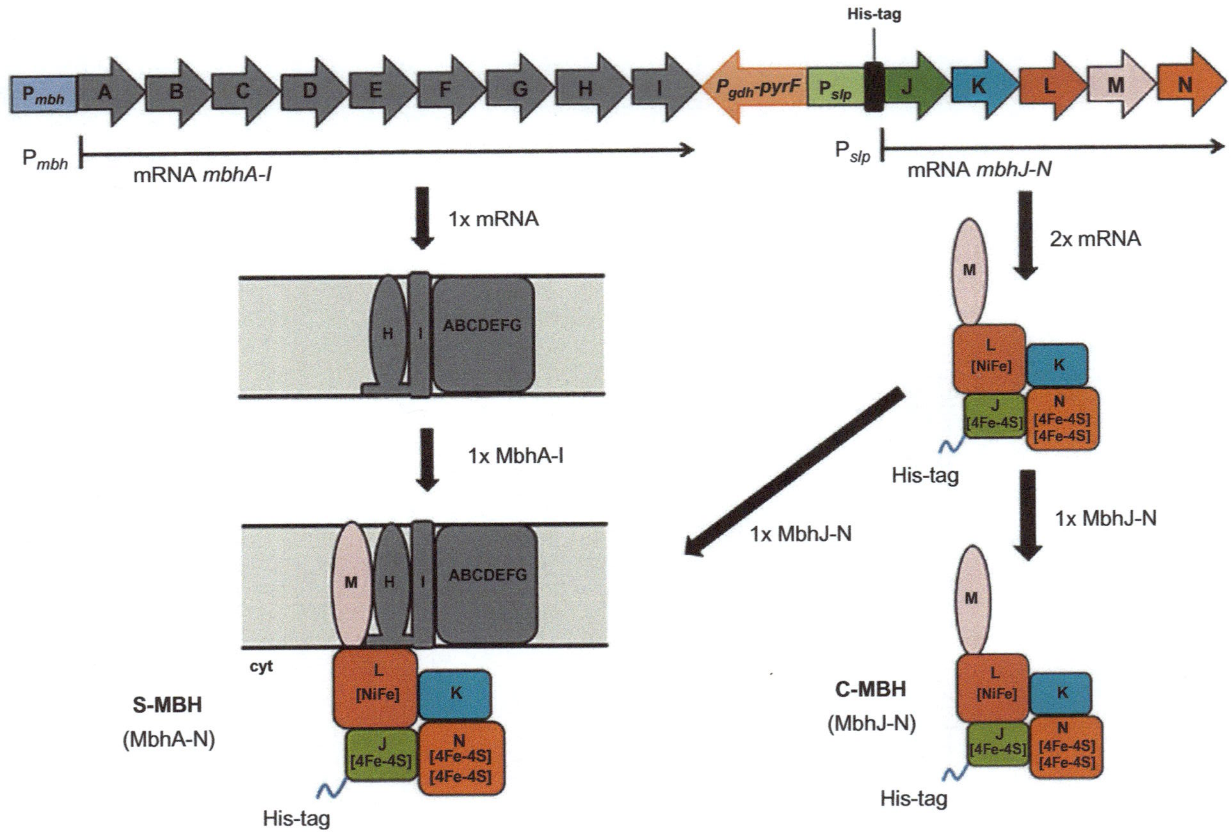

Fig. 2 The operon-splitting strategy for affinity-tagging MBH. The expression level of the second transcript driven by the constitutive promoter P_{slp} (*mbhJ-N*) is twice that of the first transcript driven by MBH native promoter P_{mbh} (*mbhA-N*). Two forms of MBH, S-MBH and C-MBH, could be purified from this strains. *Modified from McTernan, P. M., Chandrayan, S. K., Wu, C. H., Vaccaro, B. J., Lancaster, W. A., & Adams, M. W. (2015). Engineering the respiratory membrane-bound hydrogenase of the hyperthermophilic archaeon* Pyrococcus furiosus *and characterization of the catalytically active cytoplasmic subcomplex.* Protein Engineering, Design & Selection: PEDS, 28, *1–8.*

3. Large-scale growth of recombinant *P. furiosus* strains

The *P. furiosus* strains containing affinity-tagged MBH (His-MBH) and SHI (His-SHI) strains (Chandrayan et al., 2015; McTernan et al., 2014) were grown at 90°C in a 20-L fermenter (DCI-Biolafitte) using the medium containing (per liter) 1 × base salts, 1 × trace minerals ($NiCl_2{\cdot}6H_2O$ was adjusted to a final concentration of 1 μ*M*), 10 μ*M* sodium tungstate, 5 g yeast extract, 5 g casein hydrolysate, 5 g maltose, 0.5 g cysteine hydrochloride, 0.5 g sodium sulfide, and 1 m*M* potassium phosphate buffer, the pH being adjusted to 6.8. The 5 × base salt stock is composed of (per liter) 140 g NaCl, 17.5 g $MgSO_4{\cdot}7H_2O$, 13.5 g $MgCl_2{\cdot}6H_2O$, 1.65 g KCl, 1.25 g NH_4Cl, and 0.7 g $CaCl_2{\cdot}2H_2O$. The 1000 × trace mineral stock contains (per liter) 0.1% (v/v) concentrated HCl, 0.5 g Na_4EDTA, 2 g $FeCl_3$, 0.05 g H_3BO_3, 0.05 g $ZnCl_2$, 0.03 g $CuCl_2{\cdot}2H_2O$, 0.05 g $MnCl_2{\cdot}4H_2O$, 0.05 g $(NH_4)_2MoO_4$, 0.05 g $AlK(SO_4){\cdot}2H_2O$, 0.05 g $CoCl_2{\cdot}6H_2O$, and 0.05 g $NiCl_2{\cdot}6H_2O$. The strains were grown with agitation at 150 rpm and constant flushing of 20% (v/v) CO_2 and 80% N_2 for ~14 h, while the pH was maintained at 6.8 by the addition of 10% (w/v) sodium bicarbonate. Growth was monitored by OD_{600} and cells were harvested at late exponential phase by continuous centrifugation at 31,000 rpm (Sharples). The harvested cells were flash-frozen with liquid nitrogen and stored at −80°C.

4. Purification of SHI

4.1 Equipment

- Coy Lab anaerobic chamber
- Beckman Coulter Optima L-90K Ultracentrifuge
- Bio-Rad NGC Chromatography Systems
- GE Healthcare 5-mL His-Trap Crude FF column
- GE Healthcare 5-mL Q HP column
- Bio-Rad Mini-PROTEAN Tetra Cell

4.2 Buffers and reagents

- Lysis buffer: 25 m*M* sodium phosphate, pH 7.5, containing 1 m*M* dithiothreitol (DTT) and 50 μg/mL DNase I
- Buffer A: 25 m*M* sodium phosphate, pH 7.5, containing 300 m*M* NaCl, 10 m*M* imidazole, and 1 m*M* DTT

- Buffer B: 25 mM sodium phosphate, pH 7.5, containing 300 mM NaCl, 500 mM imidazole, and 1 mM DTT
- Buffer C: 50 mM Tris–HCl pH 7.5 containing 1 mM sodium dithionite (DT)
- Buffer D: 50 mM Tris–HCl pH 7.5 containing 1 M NaCl and 1 mM DT

4.3 Procedure

The purification procedure is carried out anaerobically. All solutions are thoroughly degassed and subsequently maintained under a positive pressure of Ar, and all liquid transfers are made via needles and syringes. The frozen cells are lysed osmotically with lysis buffer (5 mL buffer per gram wet weight cells) and stirred for 1 h at room temperature in an anaerobic chamber. For His-SHI purification (Chandrayan et al., 2015), the cell lysate is centrifuged at 100,000 × g for 1 h and the supernatant is loaded directly onto a 5-mL His-Trap Crude FF column using the Bio-Rad NGC Chromatography System. The column is equilibrated with Buffer A before applying the sample containing 2% Buffer B. The column is washed with 10-column volumes of Buffer A, and the bound protein is eluted with a 20-column volume gradient from 0% to 100% Buffer B. The SHI is eluted when ~30% of the gradient has been applied. The active fractions are pooled and loaded onto a 5-mL Q HP column equilibrated with Buffer C, while diluting the sample 10-fold with Buffer C. The column is washed with 10-column volumes of Buffer C, and the bound enzyme is eluted with a 20-column volume gradient from 0% to 100% Buffer D. The intact heterotetrameric SHI elutes when ~30% of the gradient has been applied. The purified protein is analyzed by SDS-PAGE using a Bio-Rad Mini-PROTEAN TGX Stain-Free Gels, 4%–15% gradient.

5. Purification of C-MBH and S-MBH

5.1 Equipment

- Coy Lab anaerobic chamber
- Beckman Coulter Optima L-90K Ultracentrifuge
- Bio-Rad NGC Chromatography Systems
- GE Healthcare 5 mL His-Trap Crude FF column
- GE Healthcare 5 mL Q HP column
- GE Healthcare 1 mL His-Trap HP column
- GE Healthcare Superose 6 10/300 GL column
- Invitrogen Novex Mini-Cell

5.2 Buffers and reagents

- Lysis buffer: 25 m*M* sodium phosphate, pH 7.5, containing 1 m*M* DTT and 50 μg/mL DNase I
- Buffer A: 25 m*M* sodium phosphate, pH 7.5, containing 300 m*M* NaCl and 1 m*M* DTT
- Buffer B: 25 m*M* sodium phosphate, pH 7.5, containing 300 m*M* NaCl, 500 m*M* imidazole and 1 m*M* DTT
- Buffer C: 50 m*M* Tris–HCl pH 7.5 containing 1 m*M* DT
- Buffer D: 50 m*M* Tris–HCl pH 7.5 containing 1 *M* NaCl and 1 m*M* DT
- Buffer E: 25 m*M* sodium phosphate pH 7.5, containing 300 m*M* NaCl, 1 m*M* DTT, and 0.02% (v/v) Triton X-100
- Buffer F: 25 m*M* sodium phosphate pH 7.5, containing 300 m*M* NaCl, 500 m*M* imidazole, 1 m*M* DTT, and 0.02% (v/v) Triton X-100
- Buffer G: 50 m*M* Tris–HCl, pH 8.2, containing 300 m*M* NaCl, 2 m*M* DT, and 0.02% (v/v) Triton X-100
- Wash buffer: 50 m*M* EPPS, pH 8.0, containing 5 m*M* $MgCl_2$, 50 m*M* NaCl, 10% (v/v) glycerol, 2 m*M* DTT, and 0.1 m*M* PMSF
- Resuspension buffer: 50 m*M* Tris–HCl, pH 8.0, containing 5 m*M* $MgCl_2$, 50 m*M* NaCl, 5% (v/v) glycerol, 1 m*M* DTT, and 0.1 m*M* PMSF

5.3 Procedure

Both purification procedures are carried out anaerobically. Frozen cells are lysed osmotically with lysis buffer (5 mL buffer per gram wet weight cells) and stirred for 1 h at room temperature in an anaerobic chamber. The cell lysate is centrifuged at 1,000,000 × g for 1 h, and the supernatant is collected in order to purify the cytoplasmic subcomplex (MbhJ-N, C-MBH) (McTernan et al., 2015), while the membrane-containing pellet is saved to solubilize and purify the intact holoenzyme form (MbhA-N, S-MBH) (McTernan et al., 2014). The cytoplasmic extract is loaded directly onto a 5-mL His-Trap Crude FF column preequilibrated with Buffer A. The column is then washed with 10-column volumes of Buffer A, and the bound enzyme elutes with a 20-column volume gradient from 0% to 100% Buffer B. C-MBH is eluted when ~15% of the gradient has been applied. The active fractions are pooled and loaded onto a 5-mL Q HP column (GE Healthcare), equilibrated with Buffer C, while diluting the sample 10-fold with Buffer C. The column is washed with 10-column volumes of Buffer C and the bound enzyme is eluted with a 20-column volume gradient from 0% to 100% Buffer D. C-MBH elutes as ~20% of the gradient is applied and the active fractions are analyzed by SDS-PAGE using a precast Invitrogen NuPAGE 4%–12% Bis-Tris Gel.

The membrane-containing pellet, along with the milky layer containing the S-layer protein, is resuspended in wash buffer. The mixture is centrifuged at 100,000 × *g* for 1 h, and the supernatant is discarded after centrifugation. This washing step is carried out 3 ×. The washed membranes are then resuspended using a Pyrex tissue grinder with resuspension buffer, and 10% (v/v) Triton X-100 is used to solubilize the membranes with gentle stirring at 4°C for 16 h. The S-layer protein is retained at each step, which improves the overall yield of purified S-MBH after purification. The solubilized sample is centrifuged at 100,000 × *g* for 1 h and the supernatant is collected and applied to a 5-mL His-Trap Crude FF column, preequilibrated with Buffer E, while diluting it 10-fold with Buffer E. The column is washed with 10-column volumes of Buffer E and the bound protein is eluted with a 20-column volume gradient from 0% to 100% Buffer F. S-MBH elutes as ~25% of the gradient is applied. The active fractions are concentrated by reloading them on to a 1-mL His-Trap HP column, while diluting them fivefold with Buffer E. A 30-column volume gradient from 0% to 100% Buffer F is applied, and S-MBH elutes when ~30% of the gradient has been applied. The S-MBH sample is further purified using a Superose 6 10/300 GL column equilibrated with Buffer G. For the sample prepared for cryo-EM studies, *n*-dodecyl β-D-maltoside (DDM) is used to solubilize S-MBH and for the purification because Triton X-100 is not compatible with the methods used for structure determination. DDM, 3% (w/v), is used for the solubilization and 0.03% (w/v) DDM is included in all buffers during the purification (Yu et al., 2018). The purified enzyme is analyzed by SDS-PAGE using a precast Invitrogen NuPAGE 4%–12% Bis-Tris Gel.

6. Assays

6.1 Equipment

- Agilent Technologies 6850 Network GC system
- Agilent Technologies Cary 100 UV–vis spectrophotometer with a Peltier-based temperature controller

6.2 Buffers and reagents

- 100 m*M* *N*-(2-hydroxyethyl)-piperazine-*N*′-3-propanesulfonic acid (EPPS), pH 8.4
- 50 m*M* Tris–HCl pH 8.0 containing 300 m*M* NaCl and 2 m*M* DT

- 1 *M* sodium pyruvate
- 20 m*M* coenzyme A
- 40 m*M* thiamine pyrophosphate (TPP)
- 200 m*M* $MaCl_2$
- 100 m*M* methyl viologen (MV)
- 100 m*M* DT
- 200 m*M* NADPH

6.3 Procedure

6.3.1 Hydrogen evolution activity

Hydrogenase activity can be conveniently measured by H_2 production using MV reduced by sodium DT as the electron donor. In this assay, the 2 mL mixture in an 8-mL anaerobically sealed vial containing 100 m*M* EPPS, pH 8.4, 1 m*M* MV and 10 m*M* DT is preincubated by shaking at 150 rpm at 80°C for 1 min. The enzyme is added to the assay vials and the mixture is incubated for 6 min. The hydrogen gas produced in the headspace is measured using an Agilent Technologies 6850 Network GC. One unit of hydrogenase activity is defined as 1 μmol H_2 produced per minute. The physiological substrate, NADPH, can also be used to assay SHI in an NADPH-dependent H_2 production activity, where 1 m*M* NADPH is used in place of DT-reduced MV as the electron donor. For the assay of MBH with *P. furiosus* ferredoxin (Fd) as the electron donor (Schut, Menon, & Adams, 2001), reduced Fd is continuously generated by the reaction of *P. furiosus* pyruvate ferredoxin oxidoreductase (POR) (Kim, Brereton, Verhagen, & Adams, 2001). In this assay, the 2 mL mixture contains 100 m*M* EPPS, pH 8.4, 10 m*M* sodium pyruvate, 0.2 m*M* coenzyme A, 0.4 m*M* TPP, 2 m*M* $MgCl_2$, 2 m*M* DTT, 30 μg/mL POR, and 100 μg/mL Fd. After preincubation at 80°C for 1 min, the reaction is initiated by addition of the enzyme, and the H_2 that is produced is measured by GC.

6.3.2 Hydrogen oxidation activity

MV is also used as an electron acceptor in the H_2 oxidation assay, but this reaction is followed by measuring the reduction of MV spectrophotometrically. The headspace of the anaerobically stoppered cuvette containing 2 mL 100 m*M* EPPS, pH 8.4, is replaced with 100% H_2 gas, then after preincubation at 80°C, 1 m*M* MV is added to the cuvette and the reaction is initiated with the addition of enzyme. The reduction of MV is measured by the increase in the absorbance at 600 nm using a molar coefficient of 8.25 $mM^{-1}\ cm^{-1}$. One unit of H_2 oxidation activity is defined as 1 μmol H_2 oxidized or 2 μmol MV reduced per minute.

6.3.3 Oxygen-stability and thermostability

The stability of the different hydrogenases is measured using the MV-linked H_2-evolution assay. Oxygen sensitivity is determined by exposing the enzymes (100 μg/mL in a buffer consisting of 50 m*M* Tris, pH 8.0, 300 m*M* NaCl and 2 m*M* DT) to air with shaking at 30 rpm, while thermostability is examined by incubating anaerobically sealed vessels at 90°C. Samples are taken every 2 h to measure the residual H_2-evolution activity.

7. Catalytic properties of SHI and MBH

SHI catalyzes the reversible reaction of H_2 and $NADP^+$ in vitro, and the activities are summarized in Table 1. SHI has fivefold higher H_2 oxidation (940 U/mg) activity than H_2 evolution activity (190 U/mg) and has a high affinity for both hydrogen (K_m 20 μ*M*) and $NADP^+$ (K_m 37 μ*M*). The predicted physiological function of SHI is therefore to use H_2 as

Table 1 Properties of SHI and MBH (S-MBH and C-MBH)

Property	SHI	S-MBH	C-MBH
MV-linked H_2 evolution activity (U/mg) at 80°C	190	19	0.3
NADPH-linked H_2 evolution activity (U/mg) at 80°C	1.5	—	—
H_2 oxidation using MV (U/mg) at 80°C	940	0.8	0.02
POR-Ferredoxin-linked H_2 evolution activity (U/mg) at 80°C	—	0.14	0.004
Half-life ($t_{1/2}$) at 90°C under argon (h)	14	1	8
Half-life ($t_{1/2}$) at 25°C under air (h)	21	15	13
Optimal temperature measured by MV-linked H_2 evolution (°C)[a]	95	95	60

[a]Measured from 40°C to 95°C.

Numbers were adapted from van Haaster, D. J., Silva, P. J., Hagedoorn, P. L., Jongejan, J. A., & Hagen, W. R. (2008). Reinvestigation of the steady-state kinetics and physiological function of the soluble NiFe-hydrogenase I of *Pyrococcus furiosus*. *Journal of Bacteriology, 190*, 1584–1587. Chandrayan, S. K., McTernan, P. M., Hopkins, R. C., Sun, J., Jenney, F. E., Jr., & Adams, M. W. (2012). Engineering hyperthermophilic archaeon *Pyrococcus furiosus* to overproduce its cytoplasmic [NiFe]-hydrogenase. *Journal of Biological Chemistry, 287*, 3257–3264. McTernan, P. M., Chandrayan, S. K., Wu, C. H., Vaccaro, B. J., Lancaster, W. A., Yang, Q., et al. (2014). Intact functional fourteen-subunit respiratory membrane-bound [NiFe]-hydrogenase complex of the hyperthermophilic archaeon *Pyrococcus furiosus*. *Journal of Biological Chemistry, 289*, 19364–19372. McTernan, P. M., Chandrayan, S. K., Wu, C. H., Vaccaro, B. J., Lancaster, W. A., & Adams, M. W. (2015). Engineering the respiratory membrane-bound hydrogenase of the hyperthermophilic archaeon *Pyrococcus furiosus* and characterization of the catalytically active cytoplasmic subcomplex. *Protein Engineering, Design & Selection: PEDS, 28*, 1–8.

the electron donor to generate NADPH (van Haaster, Silva, Hagedoorn, Jongejan, & Hagen, 2008). However, no growth phenotype was observed for the SHI deletion mutant, so the true function of SHI in vivo is still unknown (Lipscomb et al., 2011; Schut, Nixon, Lipscomb, Scott, & Adams, 2012). The half-lives of SHI at 90°C (anaerobically) and under air (at 23°C), as measured by the loss of H_2 evolution activity at 80°C, are 14 and 21 h, respectively (Table 1). The optimal temperature for H_2 evolution activity was $\geq$95°C when assayed in the range from 40°C to 95°C (Bryant & Adams, 1989). Based on an electrochemical study, it was reported that SHI, a group 3 hydrogenase, has a unique (although as yet uncharacterized) mechanism of oxygen resilience that is distinct from that of the so-called oxygen-tolerant group 1 hydrogenases. These enzymes have unique iron–sulfur clusters, crucial for their oxygen tolerance, that are not present in SHI (Kwan et al., 2015).

Two forms of MBH have been purified from the operon-split His-MBH strain and the properties of these two forms are summarized in Table 1 (McTernan et al., 2015, 2014). The S-MBH holoenzyme has a 40-fold higher activity than the C-MBH subcomplex in both H_2 evolution and H_2 oxidation assays, using either the artificial electron donor MV or the physiological electron donor Fd. It was expected that the subcomplex would be less thermostable than the intact holoenzyme but this was not the case. C-MBH and S-MBH had half-lives at 90°C of 8 and 1 h, respectively. The instability of S-MBH could be due to the presence of Triton X-100, which was used to solubilize membranes and was present throughout the S-MBH purification. Triton X-100 has a cloud point around 70°C, above which micelles would form a detergent-rich aggregate that could destabilize the enzyme.

8. Summary and conclusions

Herein, the methods of engineering and purifying SHI and MBH from *P. furiosus* have been described and the properties of the enzymes have been briefly discussed. The availability of these two classes of [NiFe] hydrogenase has provided an opportunity for their detailed characterization and, in the case of SHI, its application. For example, SHI has been used recently to study its catalytic mechanism by use of nanosecond transient infrared and visible absorbance spectroscopy, resulting in the identification of new intermediates in the catalytic cycle for NiFe-hydrogenases (Greene, Vansuch, Wu, Adams, & Dyer, 2016; Greene, Wu, McTernan, Adams, & Dyer, 2015;

Greene, Wu, Vansuch, Adams, & Dyer, 2016). In addition, the structure of MBH was very recently determined by cryo-EM, and this represented the first structure of a group 4 hydrogenase (Yu et al., 2018). The cryo-EM structure not only provides insights into the catalytic reaction but also into the evolutionary relationship of MBH with Complex I of the aerobic respiratory chain. The close relationship between group 4 hydrogenases and Complex I has been known for decades (Albracht, 1993; Friedrich & Weiss, 1997; Schut et al., 2016). The structure of S-MBH, purified using the methods described herein, provides atomic level evidence for the first time of the close relationship between these two respiratory systems. In addition to these fundamental studies, SHI has been used in several applications for in vitro hydrogen and NADPH production (Wu, McTernan, Walter, & Adams, 2015). Overall, the ability to purify SHI and MBH from *P. furiosus* is not only beneficial for artificial catalyst development but also for direct applications using these enzymes and eventually in the development of economically viable industrial hydrogen production based on renewable feedstocks (Mondal & Dey, 2017).

Acknowledgments

This work was supported by a Grant (DE-FG05-95ER20175 to M.W.W.A.) from the Division of Chemical Sciences, Geosciences and Biosciences, Office of Basic Energy Sciences of the Department of Energy.

References

Albracht, S. P. (1993). Intimate relationships of the large and the small subunits of all nickel hydrogenases with two nuclear-encoded subunits of mitochondrial NADH: Ubiquinone oxidoreductase. *Biochimica et Biophysica Acta, 1144*, 221–224.

Bryant, F. O., & Adams, M. W. (1989). Characterization of hydrogenase from the hyperthermophilic archaebacterium, *Pyrococcus furiosus*. *Journal of Biological Chemistry, 264*, 5070–5079.

Bryksin, A. V., & Matsumura, I. (2010). Overlap extension PCR cloning: A simple and reliable way to create recombinant plasmids. *BioTechniques, 48*, 463–465.

Chandrayan, S. K., McTernan, P. M., Hopkins, R. C., Sun, J., Jenney, F. E., Jr., & Adams, M. W. (2012). Engineering hyperthermophilic archaeon *Pyrococcus furiosus* to overproduce its cytoplasmic [NiFe]-hydrogenase. *Journal of Biological Chemistry, 287*, 3257–3264.

Chandrayan, S. K., Wu, C. H., McTernan, P. M., & Adams, M. W. (2015). High yield purification of a tagged cytoplasmic [NiFe]-hydrogenase and a catalytically-active nickel-free intermediate form. *Protein Expression and Purification, 107*, 90–94.

Chenevier, P., Mugherli, L., Darbe, S., Darchy, L., DiManno, S., Tran, P. D., et al. (2013). Hydrogenase enzymes: Application in biofuel cells and inspiration for the design of noble-metal free catalysts for H-2 oxidation. *Comptes Rendus Chimie, 16*, 491–505.

Fiala, G., & Stetter, K. O. (1986). *Pyrococcus furiosus* sp. nov represents a novel genus of marine heterotrophic archaebacteria growing optimally at 100 degrees C. *Archives of Microbiology*, *145*, 56–61.

Friedrich, T., & Weiss, H. (1997). Modular evolution of the respiratory NADH:ubiquinone oxidoreductase and the origin of its modules. *Journal of Theoretical Biology*, *187*, 529–540.

Greene, B. L., Vansuch, G. E., Wu, C. H., Adams, M. W., & Dyer, R. B. (2016). Glutamate gated proton-coupled electron transfer activity of a [NiFe]-hydrogenase. *Journal of the American Chemical Society*, *138*, 13013–13021.

Greene, B. L., Wu, C. H., McTernan, P. M., Adams, M. W., & Dyer, R. B. (2015). Proton-coupled electron transfer dynamics in the catalytic mechanism of a [NiFe]-hydrogenase. *Journal of the American Chemical Society*, *137*, 4558–4566.

Greene, B. L., Wu, C. H., Vansuch, G. E., Adams, M. W., & Dyer, R. B. (2016). Proton inventory and dynamics in the Nia-S to Nia-C transition of a [NiFe] hydrogenase. *Biochemistry*, *55*, 1813–1825.

Kengen, S. W., de Bok, F. A., van Loo, N. D., Dijkema, C., Stams, A. J., & de Vos, W. M. (1994). Evidence for the operation of a novel Embden-Meyerhof pathway that involves ADP-dependent kinases during sugar fermentation by *Pyrococcus furiosus*. *Journal of Biological Chemistry*, *269*, 17537–17541.

Kim, C., Brereton, P. S., Verhagen, M. F., & Adams, M. W. (2001). Ferredoxin from *Pyrococcus furiosus*. *Methods in Enzymology*, *334*, 30–40.

Kim, J. E., Kim, E. J., Chen, H., Wu, C. H., Adams, M. W. W., & Zhang, Y. H. P. (2017). Advanced water splitting for green hydrogen gas production through complete oxidation of starch by in vitro metabolic engineering. *Metabolic Engineering*, *44*, 246–252.

Kwan, P., McIntosh, C. L., Jennings, D. P., Hopkins, R. C., Chandrayan, S. K., Wu, C. H., et al. (2015). The [NiFe]-hydrogenase of *Pyrococcus furiosus* exhibits a new type of oxygen tolerance. *Journal of the American Chemical Society*, *137*, 13556–13565.

Lee, H. S., Vermaas, W. F., & Rittmann, B. E. (2010). Biological hydrogen production: Prospects and challenges. *Trends in Biotechnology*, *28*, 262–271.

Lipscomb, G. L., Stirrett, K., Schut, G. J., Yang, F., Jenney, F. E., Jr., Scott, R. A., et al. (2011). Natural competence in the hyperthermophilic archaeon *Pyrococcus furiosus* facilitates genetic manipulation: Construction of markerless deletions of genes encoding the two cytoplasmic hydrogenases. *Applied and Environmental Microbiology*, *77*, 2232–2238.

Lubitz, W., Ogata, H., Rudiger, O., & Reijerse, E. (2014). Hydrogenases. *Chemical Reviews*, *114*, 4081–4148.

Ma, K., Weiss, R., & Adams, M. W. (2000). Characterization of hydrogenase II from the hyperthermophilic archaeon *Pyrococcus furiosus* and assessment of its role in sulfur reduction. *Journal of Bacteriology*, *182*, 1864–1871.

McTernan, P. M., Chandrayan, S. K., Wu, C. H., Vaccaro, B. J., Lancaster, W. A., & Adams, M. W. (2015). Engineering the respiratory membrane-bound hydrogenase of the hyperthermophilic archaeon *Pyrococcus furiosus* and characterization of the catalytically active cytoplasmic subcomplex. *Protein Engineering, Design & Selection*, *28*, 1–8.

McTernan, P. M., Chandrayan, S. K., Wu, C. H., Vaccaro, B. J., Lancaster, W. A., Yang, Q., et al. (2014). Intact functional fourteen-subunit respiratory membrane-bound [NiFe]-hydrogenase complex of the hyperthermophilic archaeon *Pyrococcus furiosus*. *Journal of Biological Chemistry*, *289*, 19364–19372.

Mondal, B., & Dey, A. (2017). Development of air-stable hydrogen evolution catalysts. *Chemical Communications*, *53*, 7707–7715.

Mukund, S., & Adams, M. W. (1995). Glyceraldehyde-3-phosphate ferredoxin oxidoreductase, a novel tungsten-containing enzyme with a potential glycolytic role in the hyperthermophilic archaeon *Pyrococcus furiosus*. *Journal of Biological Chemistry*, *270*, 8389–8392.

Pisa, K. Y., Huber, H., Thomm, M., & Muller, V. (2007). A sodium ion-dependent A1AO ATP synthase from the hyperthermophilic archaeon *Pyrococcus furiosus*. *The FEBS Journal*, *274*, 3928–3938.

Reeve, H. A., Ash, P. A., Park, H., Huang, A. L., Posidias, M., Tomlinson, C., et al. (2017). Enzymes as modular catalysts for redox half-reactions in H-2-powered chemical synthesis: From biology to technology. *The Biochemical Journal*, *474*, 215–230.

Schut, G. J., Menon, A. L., & Adams, M. W. (2001). 2-keto acid oxidoreductases from *Pyrococcus furiosus* and *Thermococcus litoralis*. *Methods in Enzymology*, *331*, 144–158.

Schut, G. J., Nixon, W. J., Lipscomb, G. L., Scott, R. A., & Adams, M. W. (2012). Mutational analyses of the enzymes involved in the metabolism of hydrogen by the hyperthermophilic archaeon *Pyrococcus furiosus*. *Frontiers in Microbiology*, *3*, 163.

Schut, G. J., Zadvornyy, O., Wu, C. H., Peters, J. W., Boyd, E. S., & Adams, M. W. (2016). The role of geochemistry and energetics in the evolution of modern respiratory complexes from a proton-reducing ancestor. *Biochimica et Biophysica Acta*, *1857*, 958–970.

Sondergaard, D., Pedersen, C. N., & Greening, C. (2016). HydDB: A web tool for hydrogenase classification and analysis. *Scientific Reports*, *6*, 34212.

van Haaster, D. J., Silva, P. J., Hagedoorn, P. L., Jongejan, J. A., & Hagen, W. R. (2008). Reinvestigation of the steady-state kinetics and physiological function of the soluble NiFe-hydrogenase I of *Pyrococcus furiosus*. *Journal of Bacteriology*, *190*, 1584–1587.

Veziroglu, T. N., & Barbir, F. (1992). Hydrogen—The wonder fuel. *International Journal of Hydrogen Energy*, *17*, 391–404.

Vignais, P. M., & Billoud, B. (2007). Occurrence, classification, and biological function of hydrogenases: An overview. *Chemical Reviews*, *107*, 4206–4272.

Vignais, P. M., Billoud, B., & Meyer, J. (2001). Classification and phylogeny of hydrogenases. *FEMS Microbiology Reviews*, *25*, 455–501.

Vignais, P. M., & Colbeau, A. (2004). Molecular biology of microbial hydrogenases. *Current Issues in Molecular Biology*, *6*, 159–188.

Wu, C. H., McTernan, P. M., Walter, M. E., & Adams, M. W. (2015). Production and application of a soluble hydrogenase from *Pyrococcus furiosus*. *Archaea*, *2015*, 912582.

Yu, H., Wu, C. H., Schut, G. J., Haja, D. K., Zhao, G., Peters, J. W., et al. (2018). Structure of an ancient respiratory system. *Cell*, *173*, 1636–1649.

Zhang, Y. H., Evans, B. R., Mielenz, J. R., Hopkins, R. C., & Adams, M. W. (2007). High-yield hydrogen production from starch and water by a synthetic enzymatic pathway. *PLoS One*, *2*, e456.

Zhang, Y. P., Sun, J., & Ma, Y. (2017). Biomanufacturing: History and perspective. *Journal of Industrial Microbiology & Biotechnology*, *44*, 773–784.

CHAPTER SEVEN

Characterization of the [NiFeSe] hydrogenase from *Desulfovibrio vulgaris* Hildenborough

Sónia Zacarias[a], Marisela Vélez[b], Marcos Pita[b], Antonio L. De Lacey[b,*], Pedro M. Matias[a,c,*], Inês A.C. Pereira[a,*]

[a]Instituto de Tecnologia Química e Biológica António Xavier, Universidade Nova de Lisboa, Oeiras, Portugal
[b]Instituto de Catálisis y Petroleoquímica, CSIC, Madrid, Spain
[c]iBET, Instituto de Biologia Experimental e Tecnológica, Oeiras, Portugal
*Corresponding authors: e-mail address: alopez@icp.csic.es; matias@itqb.unl.pt; ipereira@itqb.unl.pt

Contents

Methods in Enzymology, Volume 613
ISSN 0076-6879
https://doi.org/10.1016/bs.mie.2018.10.003

Abstract

The [NiFeSe] hydrogenases are a subgroup of the well-characterized family of [NiFe] hydrogenases, in which a selenocysteine is a ligand to the nickel atom in the binuclear NiFe active site instead of cysteine. These enzymes display very interesting catalytic properties for biological hydrogen production and bioelectrochemical applications: high H_2 production activity, bias for H_2 evolution, low H_2 inhibition, and some degree of O_2 tolerance.

Here we describe the methodologies employed to study the [NiFeSe] hydrogenase isolated from the sulfate-reducing bacteria *D. vulgaris* Hildenborough and the creation of a homologous expression system for production of variant forms of the enzyme.

1. Introduction

[NiFe] hydrogenases are the most abundant and well-characterized group of hydrogenases. Their active site contains a NiFe binuclear center bound to the protein chain by four cysteine residues, two of which bridge the two metals while the other two are bound terminally to the Ni. The [NiFeSe] hydrogenases are a subclass of the [NiFe] hydrogenases in which a selenocysteine (Sec) replaces cysteine as one of the nickel ligands. Sec is codified by the same codon as "stop" (UGA), and a dedicated cellular machinery is needed to recode this codon as "selenocysteine" instead of "stop" and to insert Sec correctly into the protein: this means that the use of Sec instead of Cys is obviously very costly to the cell (Papp, Lu, Holmgren, & Khanna, 2007). The evolutionary advantage of using Sec over Cys is associated not only with a higher catalytic efficiency but most importantly with the ability of Sec to react with O_2 in a readily reversible manner and thus resist oxidative inactivation (Maroney & Hondal, 2018; Reich & Hondal, 2016). This concept was recently confirmed also for the *D. vulgaris* Hildenborough [NiFeSe] hydrogenase, through production of a variant where the Sec residue was replaced by a Cys. This variant exhibited reduced catalytic activity and a higher sensitivity to O_2 inactivation, confirming the role of selenium in protecting against oxidative damage (Marques et al., 2017). The Sec/Cys hydrogenase variant also displayed a much reduced Ni incorporation, suggesting the direct involvement of Sec in the maturation of the active site.

[NiFeSe] hydrogenases present interesting catalytic properties in comparison with other hydrogenases: (i) high catalytic activities and a bias for H_2 evolution (Baltazar et al., 2011); (ii) lower inhibition by H_2

(Parkin, Goldet, Cavazza, Fontecilla-Camps, & Armstrong, 2008); (iii) some degree of tolerance to O_2, as they do not form inactive Ni(III) states (Teixeira et al., 1987), and can be rapidly reactivated under reducing conditions (Parkin et al., 2008). These properties have been exploited in biocatalytic applications of [NiFeSe] hydrogenases for photo- and electrochemical H_2 production (Caputo et al., 2014; Caputo, Wang, Beranek, & Reisner, 2015; Dong et al., 2018; Lee, Park, Fontecilla-Camps, & Reisner, 2016; Mersch et al., 2015; Reisner, Fontecilla-camps, & Armstrong, 2009; Reisner, Powell, Cavazza, Fontecilla-Camps, & Armstrong, 2009; Rüdiger et al., 2010; Sakai, Mersch, & Reisner, 2013; Tapia et al., 2016) and also for electrochemical ATP synthesis (Gutiérrez-Sanz et al., 2016).

To date [NiFeSe] hydrogenases have been reported only in sulfate-reducing bacteria and methanogenic archaea (Baltazar et al., 2011). From these groups only a few [NiFeSe] hydrogenases have been isolated, namely, those from the sulfate-reducing bacteria *Desulfovibrio salexigens* (Teixeira et al., 1986), *Dm. baculatum* (Teixeira et al., 1987), *D. vulgaris* Hildenborough (Valente et al., 2005), and *D. vulgaris* Miyazaki (Nonaka, Nguyen, Yoon, & Ogo, 2013), and from the methanogens *Methanococcus vannielii* (Yamazaki, 1982) and *Methanococcus voltae* (Muth, Morschel, & Klein, 1987).

In this chapter, we describe the methods employed to express, purify, and characterize forms of the [NiFeSe] hydrogenase from *D. vulgaris* Hildenborough, which is an anaerobic sulfate-reducing organism that can use H_2 as energy source or produce H_2 fermentatively or during syntrophic growth. The *D. vulgaris* Hildenborough genome encodes seven hydrogenases of the [FeFe] and [NiFe] families (Pereira et al., 2011). Four of these are periplasmic: the [FeFe] hydrogenase (Hyd), two [NiFe] hydrogenases (HynAB-1 and HynAB-2), and a [NiFeSe] hydrogenase (HysAB), whereas three lie within or face cytoplasm, namely, the two membrane-associated energy-conserving hydrogenases, Ech and Coo, and a soluble [FeFe] hydrogenase (Pereira et al., 2011).

The *D. vulgaris* Hildenborough [NiFeSe] hydrogenase (HysAB) is a membrane-associated periplasmic-facing protein, consisting of two subunits, a large subunit (HysA) with molecular mass 63 kDa and a small subunit (HysB) with molecular mass 35 kDa (Valente et al., 2005). The membrane association is achieved through a lipidic group linked to Cys4 at the N-terminus of the large subunit. HysA was the first example reported of a bacterial lipoprotein where the signal peptide is limited to a conserved four-residue lipobox (Fig. 1), since the other regions present in typical

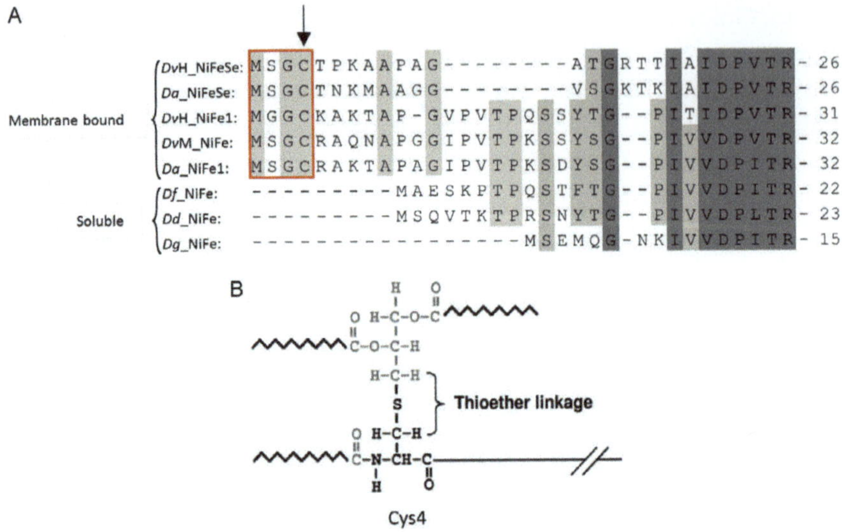

Fig. 1 (A) Sequence alignment of the large subunit N-terminus of several *Desulfovibrionaceae* hydrogenases. The conserved four-residue lipobox is highlighted by the *red box*, and the cysteine, where the lipid group will bind, by an *arrow*. Membrane-associated hydrogenases: *D. vulgaris* Hildenborough [NiFeSe] (*DvH*_NiFeSe); *D. alaskensis* [NiFeSe] (*Da*_NiFeSe); *D. vulgaris* Hildenborough [NiFe]1 (*DvH*_NiFe1); *D. vulgaris* Miyazaki [NiFe] (*DvM*_NiFe); *D. alaskensis* [NiFe]1 (*Da*_NiFe1). Soluble hydrogenases: *D. fructosovorans* [NiFe] (*Df*_NiFe); *D. desulfuricans* ATTC 27774 [NiFe] (*Db*_NiFe); *D. gigas* [NiFe] (*Dg*_NiFe). (B) Representation of the *N*-acyl-*S*-diacylglyceryl group bound to Cys4.

lipoprotein signal peptides, involved in recognition and transport by the general secretory pathway (Sec pathway), are absent (Valente et al., 2007). The reason behind this difference is that HysA is translocated by the Tat pathway in complex with HysB (Rodrigue et al., 1999; Valente et al., 2007), which contains a standard Tat signal peptide. Thus, HysA was the first example of a bacterial lipoprotein translocated through the Tat and not the Sec pathway (Valente et al., 2007).

The main physiological role of HysAB is in H_2 oxidation, with the periplasmic type-I cytochrome c_3 (TpIc_3) as the physiological electron acceptor (Pereira, Romão, Xavier, Le Gall, & Teixeira, 1998; Valente et al., 2005). In solution assays this enzyme is biased toward H_2 production, presenting very high activities: 4787 U mg^{-1} (7008 s^{-1}) and 6908 U mg^{-1} (10,132 s^{-1}) (Table 1), in the presence of detergent or phospholipids, respectively.

Table 1 Catalytic activities for *D. vulgaris* Hildenborough [NiFeSe] hydrogenase

	[NiFeSe]$_m$ hydrogenase	**[NiFeSe]$_s$ hydrogenase**	**[NiFeSe]$_r$ hydrogenase**
H_2 evolution	4787 U mg^{-1} (7008 s^{-1})[a] 6908 U mg^{-1} (10,132 s^{-1})[b]	860 U mg^{-1} (1261 s^{-1})	5640 U mg^{-1} (8272 s^{-1})
H_2 uptake	900 U mg^{-1} (1320 s^{-1})	N.D.	3037 U mg^{-1} (4454 s^{-1})
H_2 photoproduction	N.D.	N.D.	672 U mg^{-1} (986 s^{-1})
Isotopic D_2–H^+ exchange activity			
HD + H_2 evolution	140 U mg^{-1} (205 s^{-1})	N.D.	1300 U mg^{-1} (2000 s^{-1})

H_2 evolution and uptake activities for [NiFeSe]$_m$ hydrogenase in the presence of detergent[a] and phospholipids[b] (Marques, Coelho, Pereira, & Matias, 2013; Valente et al., 2005). H_2 evolution activity for [NiFeSe]$_s$ hydrogenase (Valente et al., 2005). H_2 evolution activity for [NiFeSe]$_r$ hydrogenase (Marques et al., 2017) and H_2-photoproduction activity (Tapia et al., 2016). Isotopic exchange activities for [NiFeSe]$_m$ hydrogenase in the presence of detergent (Gutiérrez-Sanz et al., 2013) and [NiFeSe]$_r$ hydrogenase (Marques et al., 2017). N.D., not determined.

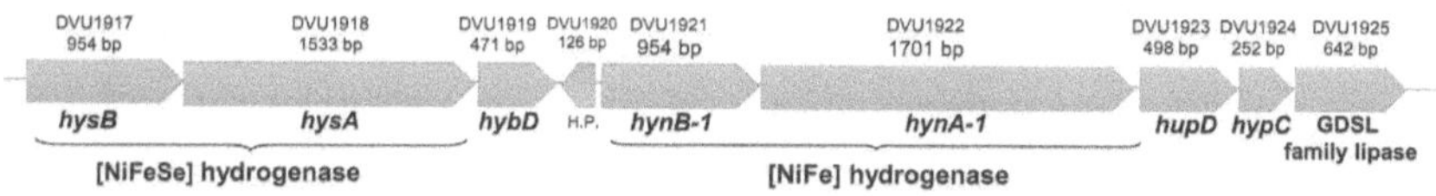

Fig. 2 The *D. vulgaris* Hildenborough gene locus containing the [NiFeSe] hydrogenase genes is approximately 8 kb long. It comprises the genes for the small and large subunits of [NiFeSe] hydrogenase (*hysB* and *hysA*) and [NiFe]1 hydrogenase (*hynB* and *hynA*), a set of maturation genes (*hybD, hupD, hypC*), a hypothetical protein (H.P.), and a gene coding for an GPLS family lipase.

Besides the membrane-associated form of the [NiFeSe] hydrogenase (herein named [NiFeSe]$_m$ hydrogenase), a soluble form ([NiFeSe]$_s$ hydrogenase) also exists in the cell, which lacks the first 11 residues of the large subunit and the hydrophobic group. This soluble form may be produced by cleavage of the lipidic group in [NiFeSe]$_m$ hydrogenase by a lipase encoded in the same locus (Fig. 2) as the [NiFeSe] hydrogenase genes (Valente et al., 2007), or by autoproteolytic degradation, since its presence increases in aged preparations of the [NiFeSe]$_m$ hydrogenase (Valente et al., 2005). The reported catalytic activities for native [NiFeSe]$_s$ hydrogenase (860 U mg^{-1} for H_2 production) (Marques et al., 2013; Valente et al., 2005) were always much lower than those of [NiFeSe]$_m$ hydrogenase.

Recently, a recombinant version of the soluble [NiFeSe] hydrogenase ([NiFeSe]$_r$ hydrogenase) was created (Marques et al., 2017). It contains a strep-tag in the N-terminus of the large subunit to facilitate purification and is homologously expressed in *D. vulgaris* Hildenborough, given the known specificity of the maturation proteins and selenocysteine insertion machinery. The [NiFeSe]$_r$ hydrogenase displays a high H_2-production activity of $5640 \pm 260\,U\,mg^{-1}$ ($8272 \pm 381\,s^{-1}$) (Table 1) which is comparable with the activities reported for the native membrane form (Valente et al., 2005). This result suggests that the low activities obtained for the native soluble form of the enzyme were mainly caused by the long purification protocol, rather than by the loss of the lipidic N-terminal tail of enzyme.

The structure of *D. vulgaris* Hildenborough [NiFeSe] hydrogenase was solved in 2010 (Fig. 3A) (Marques, Coelho, De Lacey, Pereira, & Matias, 2010) and, as observed for the *D. baculatum* enzyme (Garcin et al., 1999), revealed an overall arrangement very similar to the standard [NiFe] hydrogenases, with major differences restricted to the active site and the presence of a medial [4Fe4S] cluster in the small subunit, rather than a [3Fe4S] cluster. The structure of the *D. vulgaris* Hildenborough [NiFeSe] hydrogenase was the first structure reported for an oxidized hydrogenase of the [NiFe] family that did not contain an oxide species bridging the metals in the active site (Marques et al., 2010). However, evidence of O_2 damage was present in the terminal Cys75 ligand of Ni, which was oxidized to a sulfenate/sulfinate. Also, the proximal FeS cluster was partially oxidized to $Fe_4S_3O_2$ (Fig. 3B), but this oxidative modification was reversed upon reduction (Marques et al., 2010, 2013); it has also been observed in the structure of [NiFe] hydrogenase of *D. desulfuricans* ATCC (Matias et al., 2001). The active site of the oxidized *D. vulgaris* Hildenborough [NiFeSe] hydrogenase is found in three different conformations (Fig. 3C). In two of them the Se atom is bound to an exogenous sulfur atom (conformers I and II), while the third has a direct Ni–Se bond (conformer III) and corresponds to the conformation of the reduced enzyme. The extra sulfur atom in conformers I and II induces a Sec conformation that shields the NiFe site from contact with oxygen, preventing inactivation associated with bridging oxide ligands (Marques et al., 2013, 2017).

Recently, it was proposed that the sulfur atom at the active site results from the sulfide produced from the bacterial metabolism, which may be involved in repair of oxidative damage at the active site (Maroney & Hondal, 2018). This is supported by the fact that in the reduced form of the enzyme the sulfide group can occupy a cavity in the enzyme (Garcin et al., 1999; Marques et al., 2017).

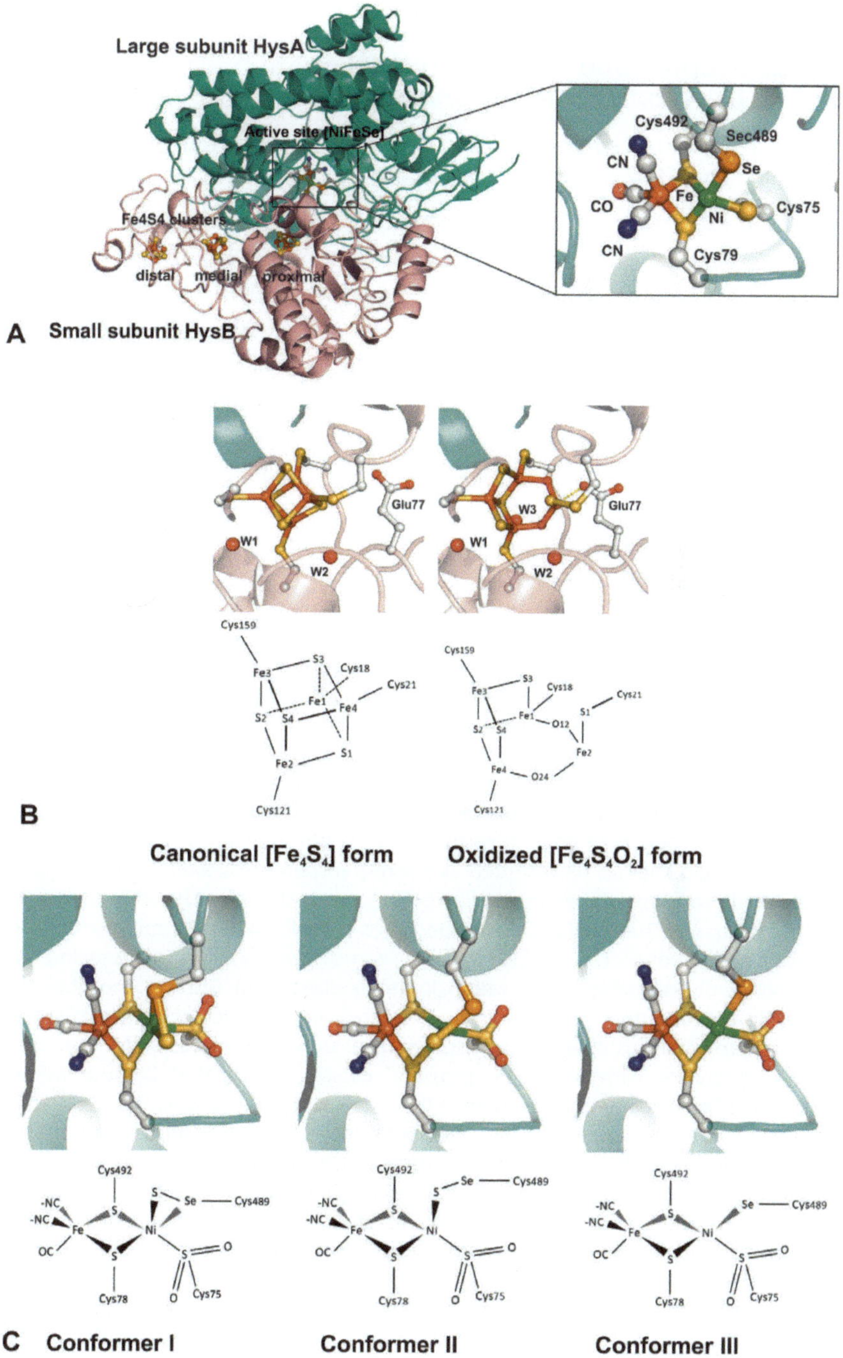

Fig. 3 See legend on next page.

2. Preparation of the native [NiFeSe] hydrogenase from *D. vulgaris* Hildenborough

2.1 Growth of *D. vulgaris* Hildenborough for expression of the [NiFeSe] hydrogenase

The strain used is *D. vulgaris* Hildenborough (DSM 644). In this organism the expression of the periplasmic hydrogenases is regulated by the growth substrate, H_2 concentration, and by Ni and Se levels (Caffrey et al., 2007; Valente, Almeida, Pacheco, Saraiva, & Pereira, 2006). Although a higher protein yield can be obtained from cells grown with H_2, the *D. vulgaris* Hildenborough is routinely purified from cells grown in standard lactate–sulfate conditions. To induce the expression of the [NiFeSe] hydrogenase and downregulate the expression of the Hyd and HynAB-1/2 enzymes, selenium must be present in the medium. In this case the [NiFeSe] hydrogenase becomes the dominant hydrogenase, irrespective of the electron donor used (Valente et al., 2006).

Fig. 3 (A) The 3D structure of the *D. vulgaris* Hildenborough [NiFeSe] hydrogenase (taken from PDB 5JSK). *Left*: overall view. The two protein chains are displayed as cartoons; the [NiFeSe] active site, its coordinating ligands, and the three [4Fe4S] clusters are shown in ball-and-stick representation. *Right:* Close-up view of the active site. The [NiFe] binuclear center and the side chains of the coordinating protein ligands are shown in ball-and-stick representation. Atom colors are *brown* for iron, *green* for nickel, *gold* for sulfur, *red* for oxygen, *light gray* for carbon, *blue* for nitrogen, and *orange* for selenium. The HysA protein chain is represented as a semitransparent cartoon. (B) The canonical and the oxidized forms of the proximal [4Fe4S] cluster in the structure of *D. vulgaris* [NiFeSe] hydrogenase. The proximal cluster becomes partially oxidized under aerobic crystallization conditions. This oxidized form is not observed after reducing with H_2/dithionite or anaerobic crystallization. In the oxidized form, Glu77 approaches atom Fe2 within coordinating bond distance. The three water molecules represented surround the proximal cluster and are conserved in all known structures of [NiFe] and [NiFeSe] hydrogenases with a [4Fe4S] proximal cluster coordinated by four cysteine residues. (C) The three Sec conformers in the active site of *D. vulgaris* [NiFeSe] hydrogenase. *Conformer I* (taken from PDB 3ZE6) contains an exogenous sulfur atom (presumably as HS^-) bound to both the Ni and Se atoms, and is predominant in crystals obtained aerobically; *Conformer II* (taken from PDB 3ZE8) also contains an exogenous sulfur atom (presumably as HS^-) bound to both the Ni and Se atoms, and is predominant in crystals reoxidized under air after reduction with H_2/dithionite or anaerobic crystallization; *Conformer III* (taken from PDB 3ZE7) represents the active form of the enzyme and is predominant after crystal reduction with H_2/dithionite or anaerobic crystallization. The three crystal structures were obtained from an aerobically purified enzyme sample, hence the oxidative damage at Cys75.

The medium used is the modified Postgate C medium (Postgate, 1984), with the following composition: 40 m*M* sodium lactate, 18 m*M* Na_2SO_4, 18.5 m*M* NH_4Cl, 1 g/L yeast extract, 3.7 m*M* KH_2PO_4, 1 m*M* sodium citrate tribasic dihydrate, 0.9 m*M* sodium thioglycolate, 0.6 m*M* ascorbic acid, 0.4 m*M* $CaCl_2{\cdot}2H_2O$, 0.2 m*M* $MgSO_4{\cdot}7H_2O$, 26 μ*M* $FeSO_4{\cdot}7H_2O$, 1 μ*M* $NiCl_2{\cdot}6H_2O$, 1 μ*M* $NaSeO_3{\cdot}5H_2O$, and 0.001% (w/v) resazurin (giving a pink color when oxygen is present). The pH is adjusted to 7.2 ± 0.1 with HCl. Cells are grown at 37°C, always in anaerobic conditions, achieved by degassing the medium solution and replacing the atmosphere with N_2.

To grow *D. vulgaris* Hildenborough, 1 mL of a cell suspension freezer stock (stored in vials at −80°C in 10% (v/v) glycerol) is added to 10 mL medium and grown overnight. A 10% (v/v) inoculum is used for all growth transfers. Depending on the amount of protein required, the growth volume can vary from 30 to 300 L. The 30 L growth volume is performed using three 10 L size Schott flasks with a small headspace filled with N_2. The 300 L growth is performed in a bioreactor with N_2 in the headspace.

The initial downstream processes are usually performed under air, with no major impact on the [NiFeSe] hydrogenase activity, given the presence of reducing sulfide in the crude cell extract. Cells are transferred to centrifuge bottles inside the fume hood to avoid inhalation of hydrogen sulfide and collected by centrifuging them at 11,000 *g*, 4°C. Cells can be stored at −80°C for several years without affecting the enzyme activity.

2.2 Purification of the membrane form of the [NiFeSe] hydrogenase

Cells are suspended in 20 m*M* Tris–HCl pH 7.6 with DNase and disrupted by three passages through a French press at 6.9 MPa. The resulting extract is centrifuged at 12,000 *g* for 10 min at 4°C to remove cell debris. The supernatant is then centrifuged at 180,000 *g* for 90 min at 4°C to collect the cell membranes, which can be processed immediately or stored at −80°C. Membranes are first washed and homogenized by suspending them in 20 m*M* Tris–HCl pH 7.6 with 1 m*M* EDTA and DNase followed by ultracentrifugation at 180,000 *g* for 90 min at 4°C. The washed membranes are solubilized in 20 m*M* Tris–HCl at pH 7.6 with 2% (w/v) Zwittergent SB3–12 (SB3–12) in the presence of protease inhibitors. Solubilization is carried out overnight in an ice bath with gentle stirring, with the headspace of the glass vial filled with N_2 to avoid prolonged contact with O_2. The solubilized proteins are separated by ultracentrifugation at 180,000 *g* for 90 min at 4°C, and the pellet can be used for a second solubilization using the same conditions. The extract with solubilized membrane proteins is loaded on a Q-Sepharose HP column (XK 26/10—GE Healthcare) equilibrated with buffer A (20 m*M* Tris–HCl at pH 7.6 plus 0.2% (w/v) SB3–12). The flow rate is 4 mL/min and a stepwise gradient of NaCl concentration is performed (0–350 m*M*, with 50 m*M* steps). Fractions eluted at different NaCl concentrations are concentrated in an Amicon stirred ultrafiltration cell with a 30 kDa cutoff membrane, where the ionic strength is also adjusted, if necessary. The total protein concentration is determined by the Bradford assay and an activity-stained native gel is used to identify those fractions having hydrogenase activity. The $[NiFeSe]_m$ hydrogenase elutes at around 300 m*M* NaCl.

The membrane-bound Ech and Coo hydrogenases of *D. vulgaris* Hildenborough are not detected during purification of the membrane extract: this can be due to their intrinsic instability, a characteristic of the family of multisubunit membrane [NiFe] hydrogenases related to Complex I (Hedderich, 2004), or to a low expression level (Valente et al., 2005).

A single step of ionic exchange chromatography is not sufficient to obtain pure protein, so the fraction of interest from the first chromatographic step is again loaded in a second Q-Sepharose HP column (XK 16/10—GE Healthcare) and the procedures described above are repeated. For the final polishing step, the fraction having activity expected for $[NiFeSe]_m$ hydrogenase is loaded onto a Q-resource column equilibrated with 20 m*M* Tris–HCl at pH 7.6 and 0.1% (w/v) *n*-dodecyl β-D-maltoside (DDM) or 0.2% SB3–12, and a stepwise NaCl gradient is performed (0–350 m*M*, with 50 m*M* steps). This column yields pure $[NiFeSe]_m$ hydrogenase as judged by the SDS-PAGE and activity-stained native gels. The fraction with pure enzyme is concentrated and the ionic strength is adjusted. It is stored in 20 m*M* Tris–HCl buffer pH 7.6 with 0.1% (w/v) DDM or 0.2% SB3–12.

2.3 Purification of the soluble form of the [NiFeSe] hydrogenase

The $[NiFeSe]_s$ hydrogenase can be purified either from the soluble extract obtained by washing the crude membranes (Valente et al., 2005) or from the solubilized membranes (Marques, Coelho, Pereira, & Matias, 2009). The purification of the soluble form follows the steps described above for the membrane form, with the exception that no detergent is used after the first column.

Another way to obtain the $[NiFeSe]_s$ hydrogenase is by incubating the membrane form with a commercial lipase (Valente et al., 2007). This procedure circumvents the low and nonreproducible yields obtained when the $[NiFeSe]_s$ hydrogenase is isolated directly from the membranes. The procedure consists of mixing 50 m*M* $[NiFeSe]_m$ hydrogenase with 5 mg of lipase, in a 20 m*M* Tris–HCl pH 8.0 buffer, at 37°C for 12 h: yields of almost 99% soluble $[NiFeSe]_s$ hydrogenase are obtained (Marques et al., 2013). It should be noted that the lipase acts here as a protease, similar to that which happens in vivo, cleaving off the first 11 residues from the large subunit (Valente et al., 2007).

3. Production of the recombinant *D. vulgaris* Hildenborough [NiFeSe] hydrogenase

3.1 Creation of the deletion mutant for the [NiFeSe] hydrogenase

The growth medium used when genetically manipulating *D. vulgaris* is the MOYLS4-rich medium (Zane, Yen, & Wall, 2010) with the following composition: 60 m*M* sodium lactate, 30 m*M* Na_2SO_4, 8 m*M* $MgCl_2$, 20 m*M* NH_4Cl, 0.6 m*M* $CaCl_2$, 2 m*M* phosphate (K_2HPO_4/NaH_2PO_4), 60 μ*M* $FeCl_2$, 120 μ*M* EDTA, 30 m*M* Tris (pH 7.4), 1 g/L yeast extract, 1 mL/L Thauers vitamin (Brandis & Thauer, 1981), and 6 mL/L trace elements solution with the pH adjusted to 7.2, plus 0.001% (w/v) resazurin (as oxygen indicator). The trace elements solution contains the following concentrations of salts: 2.5 m*M* $MnCl_2{\cdot}4H_2O$, 1.26 m*M* $CoCl_2{\cdot}6H_2O$, 1.47 m*M* $ZnCl_2$, 210 μ*M* $Na_2MoO_4{\cdot}2H_2O$, 320 μ*M* H_3BO_3, 380 μ*M* $NiSO_4{\cdot}2H_2O$, 11.7 μ*M* $CuCl_2{\cdot}2H_2O$, 35 μ*M* $Na_2SeO_3{\cdot}5H_2O$, and 24 μ*M* $Na_2WO_4{\cdot}2H_2O$.

A mutant strain for *hys*A and *hys*B genes (coding for the large and small subunits of [NiFeSe] hydrogenase, respectively) was produced by double homologous recombination in *D. vulgaris* Hildenborough, based on the marker exchange approach as previously described (Keller, Wall, & Chhabra, 2011). For the *hys*AB deletion, the plasmid pMOIP01 was produced by sequence ligation-independent cloning (SLIC) (Li & Elledge, 2007), using the following PCR amplified segments: 1055 bp upstream of *hys*AB (hysAB Up Fw—GCCTTTTGCTGGCCTTTTGCTCACAT GGACAAGGATGAGCCCGTTGTGAA and hysAB Up Rev—AAGACTGTAGCCGTACCTCGAATCTATGC AGTGCCAGCCAA

TAGAGTGAA), 885 bp downstream of hysAB (hysAB Dwn Fw—AATCCGCTCACTAAGTTCATAGACCGGACGCCCATGAT GTTAGGGTTCCAA and *hys*AB Dwn Rev—CGAGGCATTTC TGTCCTGGCTGGGCGTACGCATTACGCACGTATCAT); the kanamycin resistance gene, taken from the pSC27 plasmid (Keller, Bender, & Wall, 2009) (Kan Fw—TAGATTCGAGGTACGGCT ACAGTCTTACGGTCACAAACAGGTACGCCCCCAGAGTC and Kan Rev—CGGTCTATGAACTTAGTGAGCGGATTTCTCGTGTA GCCGATGCAGTGAGGTAGCTTGCAG); spectinomycin resistance gene; and the pUC ori portion taken from pMO719 background (Keller et al., 2009). The products from the amplifications were transformed into *E. coli* α-select Silver Efficiency (Bioline), and successful transformants were isolated on lysogeny broth (LB)-agar medium containing 50 μg/mL kanamycin and 100 μg/mL spectinomycin. Correct isolates were identified by the expected PCR amplicons from the plasmid constructs and also by sequencing. The pMOIP01 plasmid was electroporated into *D. vulgaris* Hildenborough with the following parameters: 1500 V, 250 Ω, and 25 μF, from which the deletion strain for *hys*AB, IPAR01 (Δ*hys*AB), was obtained, by selection with MOYLS4 medium containing 400 μg/mL geneticin (Keller et al., 2011). The deletion of *hys*AB was confirmed by Southern blot using an upstream fragment of the *hys*AB gene as probe.

3.2 Expression vector and complemented strain for [NiFeSe] hydrogenase expression

To produce the recombinant soluble form of [NiFeSe] hydrogenase an expression vector was constructed encoding *hys*B followed by a Strep-tag and *hys*A, lacking the codons for the first 11 amino acid residues. To create this vector, two segments were amplified by PCR: the *hys*B gene (HysB Fw—AGGTTGGGAAGCCCTGCAATGCAGTCCCAGGAGGTACC ATATGAGTCTCACAAGGCGTGATTTCGTC and HysB Rev—TTTTTCGAACTGCGGGTGGCTCCACATGATAT CCTCCTGA AGCGACTGACGG) and the Strep-tag in the N-terminal of *hys*A gene (HysA Fw—TGGAGCCACCCGCAGTTCGAAAAAGGGGCCACCG GCAGGACGACCATC and HysA Rev—GATCGTGATCCCCTGC GCCATCAGATCCTTGGCTCGCGGCCCCTCCCCTTCATGAT); and then added into pMO9075 background shuttle plasmid (Keller et al., 2009) via SLIC, creating the pMOIP03 plasmid. The amplification products were transformed into *E. coli* α-select Silver Efficiency, and cells were plated on LB agar with 100 μg/mL spectinomycin. The correct plasmid construct was screened by colony PCR and later confirmed by sequencing. In the pMO9075 vector plasmid, the *hysA*B gene remains under the control of the Km^R gene-*aph(3′)-II* promoter, which is known to be constitutively expressed in *D. vulgaris* (Keller et al., 2009). The pMOIP03 vector was successfully introduced in the deletion strain IPAR01 by electroporation, with the following parameters: 1250 V, 250 Ω, 25 μF. The complemented strain for HysAB was grown in MOYLS4 medium containing 400 μg/mL geneticin and 100 μg/mL spectinomycin. The plasmid was confirmed by PCR amplification of the insert and also by sequencing. Western-Blot against the Strep-Tactin AP (IBA) antibody was performed to confirm expression.

3.3 Site-directed mutagenesis of the [NiFeSe] hydrogenase

For site-directed mutagenesis, the plasmid pMOIP03 carrying a *hys*A$_{Strep}$*hys*B fragment is used as a template. The amino acid exchange is carried out using the NZYMutagenesis kit (Nzytech), with mutagenic primers designed on the basis of the codon usage in *D. vulgaris* Hildenborough. For propagation, the mutagenic plasmid is transformed into *E. coli* Nzy5α (Nzytech), which is transferred to 1 mL LB medium for growth at 37°C, under aerobic conditions for 1 h. Cells are plated in LB-agar medium containing 100 μg/mL spectinomycin, and isolated colonies are grown on LB medium. The plasmid is extracted by Miniprep and sequenced to check for the correct amino acid exchange. The plasmid pMOIP03 carrying the correct mutation is then electroporated aerobically into the deletion IPAR01 strain using the Gene Pulser XCell (Bio-Rad), with the following parameters: 1250 V, 250 Ω, 25 μF. Electroporated cells are transferred to 1 mL of MOYLS4 medium and left to recover anaerobically overnight at 37°C. About 50–100 μL of these cells are plated, with molten MOYLS4 medium containing 400 μg/mL geneticin and 100 μg/mL spectinomycin poured over the cells and swirled. Once solidified, plates are inverted and placed in an airtight rectangular jar (Mitsubishi Gas Chemical Co., Inc., Japan) with an oxygen indicator strip (Biomérieux) and GenBox Anaer (Biomérieux) to absorb traces of O_2. The plates are incubated in a growth chamber at 37°C, between 2 and 5 days until *D. vulgaris* Hildenborough colonies appear (Fig. 4). The colonies are selected with a sterile toothpick and inserted into MOYLS4 liquid medium with 100 μg/mL spectinomycin. The first *D. vulgaris* Hildenborough [NiFeSe] hydrogenase variant was recently reported, where the Sec489 residue was exchanged for a Cys, transforming the active site of the [NiFeSe] hydrogenase into that of a [NiFe] hydrogenase (Marques et al., 2017).

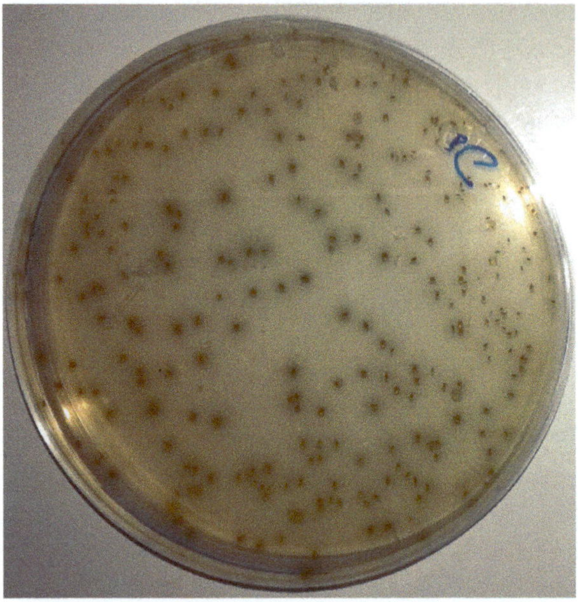

Fig. 4 Colonies of *D. vulgaris* Hildenborough plated in MOYLS4 medium.

3.4 Purification of the recombinant [NiFeSe] hydrogenase

Cells are suspended in 20 m*M* Tris–HCl pH 7.6 with DNase and disrupted by three passages through a French press at 6.9 MPa. The resulting extract is centrifuged at 12,000 *g* for 10 min at 4°C to remove cell debris. The supernatant is then centrifuged at 180,000 *g* for 90 min at 4°C to separate membranes from the soluble fraction. The soluble fraction is loaded onto a Q-Sepharose HP column (XK 26/10—GE Healthcare) equilibrated with 20 m*M* Tris–HCl pH 7.6, then a stepwise NaCl gradient is performed (0–350 m*M*, with 50 m*M* steps), whereupon fractions with $[NiFeSe]_r$ hydrogenase activity usually elute between 250 and 300 m*M* NaCl. The combined fractions are concentrated and the ionic strength adjusted using an Amicon stirred ultrafiltration cell with a 30 kDa cutoff membrane. Affinity chromatography is then performed by loading the fraction onto a gravity column containing Strep-Tactin resin (IBA Lifesciences) equilibrated with 100 m*M* Tris–HCl pH 8.0, 150 m*M* NaCl (buffer W). After five washing steps with buffer W, the recombinant protein was eluted with buffer W plus 2.5 m*M* desthiobiotin. This column yields pure [NiFeSe] hydrogenase as judged by SDS-PAGE. Protein concentration is determined based on $\varepsilon_{410} = 48{,}000\ M^{-1}\ \text{cm}^{-1}$, using the protein molecular mass of 88,000 Da.

4. Hydrogenase activities

The steady-state activity of the [NiFeSe] hydrogenase is routinely quantified by measuring the rates of hydrogen oxidation and proton reduction with artificial redox dyes. We use methyl viologen (MV) ($E = -449$ mV at all pH values) both as electron donor for H_2 production and as electron acceptor for H_2 oxidation. The natural physiological partner, Tp*Ic*$_3$, can also be used as electron acceptor or donor to assay hydrogenase activity in vitro (Valente et al., 2005). The H_2 oxidation activity can also be assessed qualitatively by activity-stained native gels.

The H/D isotope exchange activity measured by membrane-inlet mass spectrometry shows the reversibility of the [NiFeSe] hydrogenase catalytic function (Fig. 5). It is performed in the absence of a redox partner. Thus, no intermolecular electron transfer takes place; only H_2 cleavage/formation and proton exchange with the solution are involved during this catalytic process (Vignais, 2005).

Oriented immobilization of the [NiFeSe] hydrogenase on electrodes allows measuring its H_2 oxidation/production activities by electrochemical methods. Correct immobilization of the enzyme allows direct electron transfer with the electrode, thus avoiding the use of redox dyes as electron donors or acceptors (Rüdiger et al., 2010). The redox potential applied on the electrode drives the electrocatalytic process toward either proton reduction or H_2 oxidation, while the current is proportional to the enzymatic activity (Vincent, Parkin, & Armstrong, 2007). Benchmark H_2 oxidizing currents of $1.7\,\mathrm{mA\,cm^{-2}}$, with bioanodes based on the *D. vulgaris* Hildenborough [NiFeSe] hydrogenase immobilized in a viologen-modified hydrogel, were recently achieved (Ruff et al., 2017, 2018).

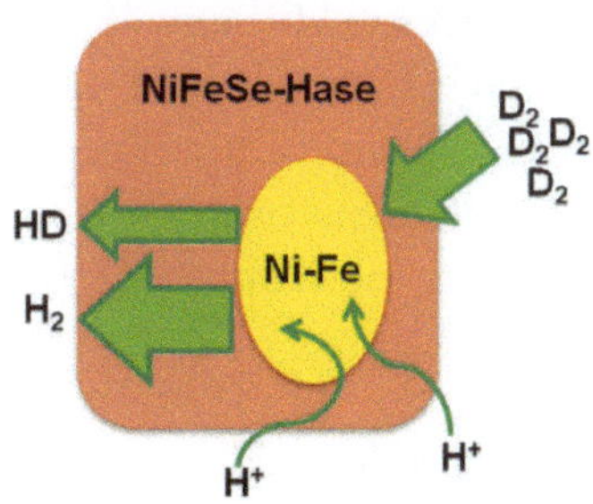

Fig. 5 Scheme depicting the H/D isotope exchange activity of [NiFeSe] hydrogenase in D_2/H_2O.

4.1 H_2 evolution and uptake with artificial mediators

4.1.1 H_2 evolution assay

The rate of H_2 evolution is obtained by assaying with gas chromatography (GC) in a Trace GC Ultra (Thermo Scientific) equipped with a thermal conductivity detector (TCD), using a MolSieve 5A 80/100 column (Althech) and N_2 as carrier gas. The assay is performed in a gas-tight screwcap vial containing 1 mL of 50 m*M* Tris–HCl buffer at pH 7 plus 1 m*M* MV, 15 m*M* sodium dithionite, 0.5 mg/mL bovine serum albumin, and 9 mL volume headspace filled with N_2. The reaction is initiated by adding to the vial 10 μL of approximately 9 n*M* enzyme with a gas-tight syringe. Dithionite-reduced MV is the electron donor for the [NiFeSe] hydrogenase. This reaction mixture is incubated in a shaker at 37°C for 10 min and a headspace sample is injected into the GC every 5–10 min. The GC measurements are calibrated by controlled injections of H_2 under identical liquid and gas volume headspaces. One unit of enzyme is defined as the amount of hydrogenase evolving 1 μmol of H_2/min. The purified $[NiFeSe]_r$ Hydrogenase exhibits a H_2-production activity of around 5650 ± 260 U/mg ($8280 \pm 380\,s^{-1}$).

4.1.2 H_2 uptake assay

The rate of H_2 uptake (oxidation) is measured spectrophotometrically in a Coy anaerobic chamber (95% N_2, 5% H_2), using a UV-1800 Shimadzu spectrophotometer. The enzyme is first activated with 0.5 bar H_2 at room temperature for approximately 30 min in an anaerobic vial containing 1 mL of 50 m*M* Tris–HCl buffer at pH 8 and 1 m*M* MV. Inside the anaerobic chamber, the reaction is initiated by adding 10 μL of approximately 25 n*M* enzyme to a cuvette containing 1 mL of H_2-saturated 50 m*M* Tris–HCl buffer at pH 8 and 2 m*M* MV, with constant stirring. The H_2 oxidation activity is measured from the reduction of MV, following the MV color change rate at 604 nm ($\varepsilon = 13.6\,mM^{-1}\,cm^{-1}$) at 30°C, according to the reaction: $H_2 + 2MV^{+2} \rightarrow 2H^+ + 2MV^+$. One unit of enzyme is defined as the amount of hydrogenase reducing 2 μmol of MV per minute, which is equivalent to 1 μmol H_2 oxidized per minute. The $[NiFeSe]_r$ hydrogenase exhibits an H_2 uptake activity of around 3310 ± 180 U/mg ($4450 \pm 220\,s^{-1}$).

4.1.3 Activity-stained native gels

For a qualitative assessment of the hydrogenase activity, either in crude cell extracts or in the purification fractions, activity staining is performed using

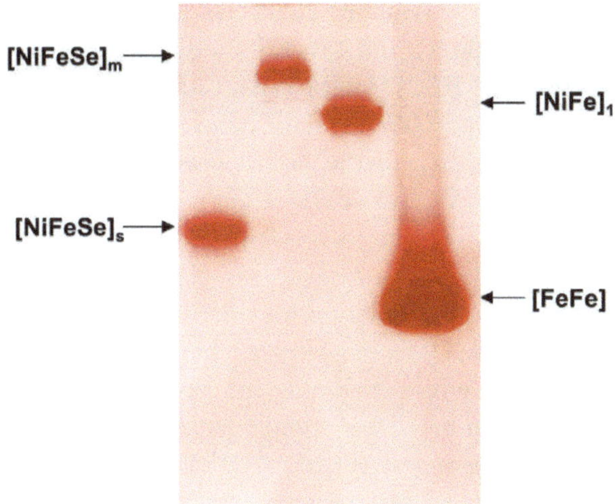

Fig. 6 Activity-stained native gel for hydrogenase activity of pure protein samples isolated from *D. vulgaris* Hildenborough. The $[NiFeSe]_r$ hydrogenase, not represented here, migrates on the gel with the same profile as the $[NiFeSe]_s$ hydrogenase.

native polyacrylamide gel electrophoresis (Fig. 6). The band intensity is related to the H_2 uptake activity, since the enzyme will oxidize H_2 and transfer the electrons to MV, giving rise to a blue band on the gel. Activity-stained gels are routinely used throughout the protein purification process, to identify fractions with hydrogenase activity as well as to screen rapidly for [NiFeSe] hydrogenase variants with significant differences in H_2 oxidation activity, based on the intensity of the bands.

Samples of 50–200 μg total protein from either soluble fractions or fractions taken from the ionic exchange chromatography are run in a 7.5% (v/v) polyacrylamide gel containing 0.1% (v/v) Triton X-100. Detergent is important for the membrane-associated hydrogenases to run properly since its absence results in band streaking. The gel is transferred to 50 mL of an N_2-saturated solution of 50 m*M* Tris–HCl pH 7.6 containing 0.5 m*M* MV and then incubated under 0.5 bar H_2. Once blue bands are well developed, they are fixed by adding 10 m*M* 2,3,5-triphenyltetrazolium chloride which is reduced by MV to give a permanent oxygen-insensitive red stain.

4.2 H/D isotope exchange activity

The H/D isotope exchange activity of the hydrogenase is measured in a thermostated, magnetically stirred 37°C reactor that is connected via a 14 μm Teflon membrane to a mass spectrometer (Pfeiffer Prisma)

(Gutiérrez-Sanz, Marques, Baltazar, et al., 2013). The output signal of the spectrometer for mass values 2 and 4 is calibrated by saturating buffered solutions with 100% H_2 and 20% D_2 in 80% Ar, respectively, which have been passed through an oxygen filter (Varian) before entering the reactor. The output signal is proportional to the partial pressure of the corresponding gas in the reaction vessel (Vignais, 2005). Calibration for the single exchange product of the enzymatic reaction (HD) is determined from the average values measured for H_2 and D_2.

Generally, the exchange activity is measured in 10 mL of buffered H_2O solution purged with 20% D_2 in Ar. After reaching a plateau value for mass 4, the reactor lid is closed while excluding the gas phase in the reactor. The enzymatic reaction is initiated by adding sequentially, through a rubber septum with gas-tight syringes, 100 μL of 1% DDM, 1 μL of 1 *M* sodium dithionite to eliminate residual oxygen, and typically 25 μL of 0.1 μ*M* hydrogenase. During the measurement, masses 2, 3, and 4 are scanned at a rate of 1 amu/s. The initial rates of H_2 and HD production are calculated from the difference in the slopes after and before enzyme addition (Gutiérrez-Sanz, Marques, Baltazar, et al., 2013). Alternatively, the isotope exchange activity can be measured by saturating a buffered D_2O solution with H_2. In this case HD and D_2 production rates are measured by monitoring the evolution of masses 3 and 4, respectively. In this case the hydrogenase sample needs to be preincubated overnight in the D_2O-buffered solution in order to convert exchangeable protons to deuterons (Gutiérrez-Sanz, Marques, Baltazar, et al., 2013).

4.3 Electrocatalytic H_2 production and oxidation

Covalent immobilization of the [NiFeSe] hydrogenase on a gold electrode modified with a 4-aminothiophenol (4-ATP) self-assembled monolayer (SAM) enables measuring its electrocatalytic H_2 oxidation/production activity to be measured by direct electron transfer. The 4-ATP SAM is formed by immersing a clean Au electrode in a 5 m*M* solution of 4-ATP in ethanol for 18 h at room temperature. After rinsing, the Au electrode surface is covered by 6 μL of 27 μ*M* hydrogenase in 10 m*M* MES, pH 6.5 for 30 min then, in a second step, 5.5 μL of 14 m*M* *N*-(3-dimethylaminopropyl)-*N'*-ethylcarbodiimide hydrochloride (EDC) and 4.5 μL of 21 m*M* *N*-hydroxysuccinimide, in the same buffer solution, are added in order to form amide bonds between the enzyme and the 4-ATP SAM on the electrode. The mixture is left to react for 90 min, then the modified Au surface is washed for 5 min in 0.1 *M* phosphate, pH 7.0, 0.25 *M* KCl buffer solution with stirring to remove noncovalently bound hydrogenase (Rüdiger et al., 2010). Electrocatalytic experiments are controlled by an Autolab potentiostat and measured in a three-electrode glass cell equipped with a platinum wire counter-electrode and a saturated calomel reference electrode separated from the main compartment in a sidearm connected by a Luggin capillary. The electrochemical cell is surrounded by a water jacket for controlling the temperature at 40°C. The gold working electrode is connected to an MSR electrode rotator (Pine Instruments). The electrocatalytic H_2 oxidation activity is measured by chronoamperometry at −0.25 V vs NHE, under 1 atm H_2 at an electrode rotation rate of 2500 rpm, whereas the H_2 production is measured at −0.45 V vs NHE under 1 atm N_2 (Rüdiger et al., 2010). The membrane form of the [NiFeSe] hydrogenase can be immobilized on Au gold electrodes modified with a phospholipid bilayer by using the enzyme's hydrophobic tail as an anchor to the supported biomimetic membrane. Two different orientations of the membrane-bound hydrogenase on the electrode are possible: (i) the enzyme is inserted via its hydrophobic tail onto a phospholipid bilayer supported on a 4-ATP-modified Au electrode with its 4Fe4S distal cluster facing the bulk solution; (ii) the enzyme is covalently bound to the 4-ATP-Au with its 4Fe4S distal cluster facing the electrode and a phospholipid bilayer is formed on top of the hydrogenase monolayer using the hydrophobic tail as scaffold (Gutiérrez-Sánchez et al., 2011). The first configuration is obtained by incubating the 4-ATP-Au electrode in a 0.6 mg/mL liposome suspension (300 nm diameter) from *E. coli* polar extract in 10 m*M* MES buffer at pH 5 for 45 min at room temperature. After rinsing carefully with MES buffer 10 m*M* at pH 7, the electrode is incubated in 27 μ*M* $[NiFeSe]_m$ hydrogenase in pH 6.0 MES buffer and 48 mg of dry CALBIOSORB adsorbent is added to the solution for 1.5 h to remove the detergent present in the hydrogenase solution (Gutiérrez-Sánchez et al., 2011). The second configuration is obtained in two steps by first immobilizing covalently on the 4-ATP-Au the $[NiFeSe]_m$ hydrogenase as described above but in the presence of 0.1% DDM. The second step involves incubating the hydrogenase-modified electrode in the liposomes suspension in the presence of CALBIOSORB adsorbent for 90 min (Gutiérrez-Sanz et al., 2013). It has been shown that the hydrogenase immobilized on electrodes in the latter configuration is able to create a proton gradient across the biomimetic membrane upon electrocatalytic H_2 oxidation (Gutiérrez-Sanz et al., 2015) and to induce ATP synthesis when an ATP synthase is incorporated in the phospholipid bilayer (Gutiérrez-Sanz et al., 2016).

4.4 Photocatalytic H_2 production

The [NiFeSe] hydrogenase can produce H_2 when combined with appropriate visible light-absorbing semiconductors such as In_2S_3 (Tapia et al., 2016). The In_2S_3 should be presented as a polycrystalline powder with a controlled pore size around 15–20 nm in diameter, suitable for hosting the enzyme and allowing electron transfer, and it is synthesized by a hydrothermal procedure as follows. A Pyrex beaker located inside a Teflon-lined steel high-pressure reactor is filled with 50 mL aqueous solution containing 148 m*M* $InCl_3$, 178 m*M* thiourea, and 80 μL HCl 37%, and the mixture is left to react at 162°C for 2 days (Lucena, Aguilera, Palacios, Wahnón, & Conesa, 2008). Afterward, the material is cleaned by three centrifugation-resuspension cycles in distilled water and another one in ethanol. Finally, the material is dried for 12 h at 333 K.

The [NiFeSe] hydrogenase is incubated for 6 h with the In_2S_3 nanoparticles to adsorb the enzyme on the surface. This process is performed at 4°C by mixing 0.26 pmol of hydrogenase with 22.1 μmol of In_2S_3 in 10 mL of a pH 7.0 solution containing 50 m*M* Tris–HCl and 0.2 *M* Na_2SO_3 (Tapia et al., 2016). The sample is then transferred to a glass reactor vessel connected to a mass spectrometer through a gas-permeable Teflon membrane, taking care to avoid a gas phase inside the reaction chamber. The [NiFeSe] hydrogenase is preactivated by bubbling a mixture of 1:4 H_2/Ar for 10 min followed by pure Ar to remove all the H_2 from the reactor. Upon irradiating the activated sample with light from a 450 W Xe lamp solar simulator, H_2 is produced, the rate being measured by mass spectrometry, monitoring mass = 2.

5. Structure determination of *D. vulgaris* Hildenborough [NiFeSe] hydrogenase by X-ray crystallography

5.1 Crystallization

The soluble forms of the [NiFeSe] hydrogenase (either native or recombinant) are crystallized aerobically at 20°C using the sitting drop vapor diffusion method, using microbridges in 24-well plates (Hampton Research). Crystallization drops are prepared by mixing 1 or 2 μL of pure protein having a concentration between 80 and 114 μ*M*, with an equal volume of reservoir solution containing ca. 20% PEG 1500 (w/v), 0.1 m*M* Tris–HCl, pH 7. The exact percentage of PEG 1500 varies somewhat depending on the protein batch, and some optimization is usually required to obtain the best quality crystals. Drops are equilibrated against 500 μL of

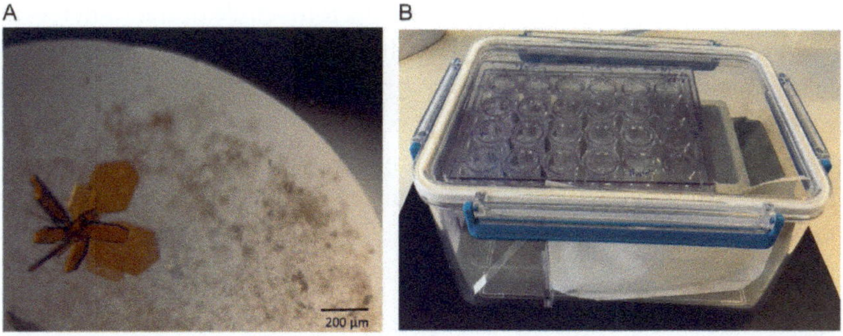

Fig. 7 (A) Crystals of the $[NiFeSe]_r$ hydrogenase. (B) Crystallization plate inside an air-tight jar to allow crystal growth under anaerobic conditions. An oxygen indicator strip and GenBox Anaer are placed inside the jar.

reservoir solution. Brown monoclinic blocks, with dimensions varying from 100 to 500 μm, grow over 4–6 days (Fig. 7A). Crystals are prepared for data collection with the reservoir solution adjusted to include 10% (v/v) glycerol and flash-cooled in liquid nitrogen or under a nitrogen gas stream at 100 K. Recombinant [NiFeSe] hydrogenase and variants usually crystallize in the monoclinic space group *C*2 with one molecule in the asymmetric unit, but the as-isolated native soluble enzyme was crystallized in monoclinic space group $P2_1$: crystallization occasionally occurs in other space groups.

For anaerobic crystallization, the procedure used is the same as described above except that it is carried out in a Coy anaerobic chamber (95% N_2, 5% H_2). All plastic materials and solutions are degassed and introduced to the chamber 1 day in advance. A 24-well crystallization plate with the crystallization drops is kept in an airtight rectangular jar (Mitsubishi Gas Chemical Co., Inc., Japan) with an oxygen indicator strip (Biomérieux) and GenBox Anaer (Biomérieux), outside the anaerobic chamber in a 20°C room (Fig. 7B). Harvesting crystals under anaerobic conditions is achieved by filling a container with argon, and the crystal plate and stereoscopic microscope are placed in this container.

The $[NiFeSe]_m$ hydrogenase has not been crystallized to date, despite screening of multiple crystallization conditions at the 100 nL scale using a robot.

5.2 Data collection and structure determination

Crystals of $[NiFeSe]_r$ hydrogenase usually display good diffraction quality on a home X-ray source (either a rotating anode X-ray generator or a microsource) coupled to a mirror system and a CCD detector, yielding diffraction data to a resolution of ca. 1.5 Å. The use of a four-circle diffractometer leads to very complete diffraction data with high multiplicity, obviating the cusp that can result from measuring a monoclinic crystal on a one-circle goniometer. Likewise, modern detectors and data collection software at synchrotron sources yield nearly 100% complete diffraction data to a resolution approaching 1 Å, with the added benefit of wavelength selection at some beamlines, which is very important to validate the structure of the active site, in particular the bimetallic NiFe and the position of the Se atom in the selenocysteine ligand.

The diffraction quality displayed by the first crystals of the as-isolated soluble form of [NiFeSe] hydrogenase was relatively poor. The first data set was obtained *in-house* to a resolution of 2.4 Å, and although a homologous structure from *Dm. baculatum* [NiFeSe] hydrogenase was already known (PDB 1CC1; Garcin et al., 1999) and a clear solution could be obtained by the molecular replacement method, the electron-density maps were not of sufficiently good quality to rebuild several regions in the structure (Marques et al., 2009).

Therefore, the crystal structure of $[NiFeSe]_s$ hydrogenase from *D. vulgaris* Hildenborough was determined at 2 Å resolution from a crystal of the as-isolated soluble form, using the Multiple Wavelength Anomalous Dispersion Method at the Fe K-edge (Marques et al., 2010) from data measured at Diamond Light Source beamline I04 (Didcot, UK) (Fig. 8). The peak and inflection point wavelengths were chosen to maximize the anomalous and dispersive signals with CHOOCH (Evans & Pettifer, 2001) from an X-ray fluorescence scan near the Fe K-edge, and the remote wavelength was chosen around 1 Å to minimize absorption effects and maximize dispersive differences during phasing. The diffraction images for the three data sets were processed with XDS (Kabsch, 1993).

The as-isolated soluble form of $[NiFeSe]_s$ hydrogenase from *D. vulgaris* Hildenborough crystallized in space group $P2_1$ with one heterodimer in the asymmetric unit; therefore, 14 Fe sites were expected (1 in the NiFe active site, 12 in the three 4Fe4S clusters, and 1 bound to the C-terminal region of the large subunit). Using the HKL2MAP graphical interface (Pape & Schneider, 2004) and the SHELXC/D/E program suite (Schneider & Sheldrick, 2002; Sheldrick, 2002, 2010) the data set was scaled and analyzed with SHELXC, the Fe-atom substructure was determined with SHELXD, and the phase problem solved with SHELXE. However, although a good solution for the Fe-atom substructure was obtained with SHELXD, a clear and unequivocal discrimination between the correct and inverted

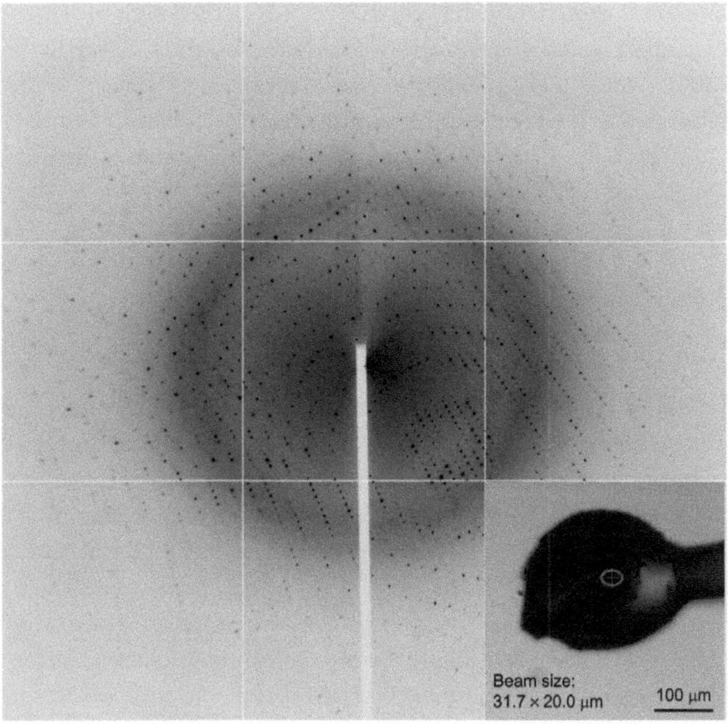

Fig. 8 Diffraction image from the data set measured at the Fe peak wavelength (1.7244 Å) used in the structure determination of *D. vulgaris* [NiFeSe] hydrogenase. The *inset* shows a typical cryocooled crystal used for diffraction data measurements at the Diamond Light Source.

substructures could not be obtained with SHELXE and the electron-density maps could not be interpreted. Improved phases were obtained using the maximum-likelihood heavy-atom parameter refinement in SHARP (Bricogne, Vonrhein, Flensburg, Schiltz, & Paciorek, 2003) and a subsequent optimizing density modification procedure using SOLOMON (Abrahams & Leslie, 1996) and DM (Cowtan, 1994). Automated model building with ARP/wARP (Perrakis, Morris, & Lamzin, 1999) produced a near-complete model for the protein chains of both subunits (except for a few disordered N-terminal residues). Prior to structure refinement, the 4Fe4S clusters and the NiFe cofactors were modeled with COOT (Emsley, Lohkamp, Scott, & Cowtan, 2010).

Subsequent crystal structures of the as-isolated soluble or recombinant forms of [NiFeSe] hydrogenase from *D. vulgaris* Hildenborough were determined by the Molecular Replacement Method with PHASER

(McCoy, 2006) using an earlier crystal structure as phasing model, with separate rotation/translation calculations for each subunit, and usually omitting the metal cofactors. High-resolution diffraction data for these structural studies have been measured at several European synchrotron sources: ESRF (Grenoble, France), SLS (Villigen, Switzerland), ALBA (Cerdanyola del Vallès, Spain), and DLS (Didcot, UK).

5.3 Structure refinement

The first crystal structure of the as-isolated soluble form of [NiFeSe] hydrogenase from *D. vulgaris* Hildenborough was refined against the 2.04 Å peak data measured at DLS using a maximum-likelihood procedure with REFMAC (Murshudov et al., 2011) as implemented in the CCP4i graphics user interface (Potterton, Briggs, Turkenburg, & Dodson, 2003). Prior to refinement, a random 5% sample of the reflection data was flagged for cross-validation using R-free (Brünger, 1992). Improved atomic displacement parameters at this resolution were obtained using a simple TLS (translation–libration–screw) rigid body model (Winn, Isupov, & Murshudov, 2001) where each of the two subunits (including its respective cofactors) was treated as a rigid body. During refinement, the structure was regularly inspected and corrected against σ_A-weighted $2|F_o|-|F_c|$ and $|F_o|-|F_c|$ electron-density maps using COOT. The map coefficients were calculated by REFMAC at the end of each refinement procedure. In this crystal structure, electron density for the aliphatic chains of four sulfobetaine 3–12 detergent molecules used in the purification was observed in two hydrophobic pockets open to the solvent and the visible atoms were included in the refinement. Solvent molecules were located with ARP/wARP and by inspection of σ_A-weighted $2|F_o|-|F_c|$ and $|F_o|-|F_c|$ electron-density maps, and included in the refinement.

Refinements of subsequent crystal structures of both as-isolated soluble and recombinant forms of [NiFeSe] hydrogenase from *D. vulgaris* Hildenborough were carried out using PHENIX (Adams et al., 2010). Although considerably slower than REFMAC, PHENIX allows for the refinement of occupation factors, which in the case of these structures is very important to elucidate the fine structural details of the active site. Since most datasets have been measured just above the K-absorption edge of selenium, the Se atom in selenocysteine displays significant anomalous dispersion properties ($\Delta f' \sim -4$ e$^-$ and $\Delta f'' \sim 3.5$ e$^-$) and anomalous refinement is used, increasing the number of experimental data and allowing the direct calculation of anomalous difference maps, which is very useful for validating structural changes in the active site and proximal 4Fe4S cluster. In all refinements, hydrogen atoms are included at calculated positions with REFMAC or the READYSET tool in PHENIX. At intermediate resolutions (up to ca. 1.5 Å) a TLS refinement of the atomic displacement parameters is followed by refinement of individual isotropic atomic displacement parameters for all nonhydrogen atoms. The TLS rigid bodies are chosen by submitting a previously refined structure with isotropic atomic displacement parameters to the TLS Motion Determination (TLSMD) server (Painter & Merritt, 2006) or by using the internal PHENIX tool. At higher resolutions, anisotropic refinement of atomic displacement parameters for all nonhydrogen atoms is undertaken. Solvent molecules can be added automatically during the refinement process in PHENIX or by inspection of the σ_A-weighted $2|F_o|-|F_c|$ and $|F_o|-|F_c|$ electron-density maps produced at the end of each refinement calculation.

6. *D. vulgaris* Hildenborough [NiFeSe] hydrogenase biophysical characterization

6.1 Infrared spectroscopy

The different redox states of the bimetallic active site of the hydrogenase can be studied by (Fourier-transform infrared spectroscopy) FTIR because its CO and CN^- ligands absorb infrared radiation and the vibrational frequencies of their bands are sensitive to changes in electron density at the active site (De Lacey, Fernández, Rousset, & Cammack, 2007). The FTIR-spectroelectrochemical experiments on [NiFeSe] hydrogenases are performed in a cell with an 8 μm path length described by Moss, Nabedryk, Breton, and Mantele (1990). For these measurements 10 μL of hydrogenase solution having a concentration of at least 0.8 m*M* in buffered solution with 0.2 *M* KCl is required. The sample must contain, in addition, a mixture of redox mediators, 0.05 m*M* in each case, the redox potentials of which span the range 0 to −450 mV vs NHE, in order to attain rapid equilibration with the electrode potential. The redox potential of the cell is controlled with a BAS CV-27 potentiostat and measured with a Fluke 77 multimeter. The temperature of the cell is controlled at 25°C with a thermostat. Spectra are recorded in an FTIR spectrometer with a Mercury Cadmium Telluride (MCT) detector and purged with a CO_2- and H_2O-free atmosphere (Fig. 9). The FTIR spectra are averaged from 1024 scans

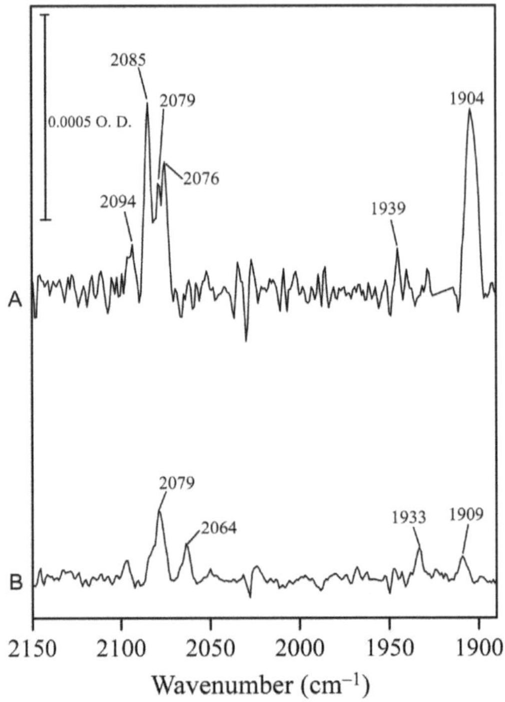

Fig. 9 FTIR spectra of *D. vulgaris* Hildenborough [NiFeSe] hydrogenase active site as isolated aerobically (A) and after reduction with 1 atm of H_2 (B).

and the spectral resolution is $2\,cm^{-1}$ (De Lacey, Gutiérrez-Sánchez, Fernández, Pacheco, & Pereira, 2008).

Surface-enhanced infrared absorption spectroscopy is a nonconventional infrared spectroscopy technique that is based on the surface enhancement effect of infrared absorption in rough (at the nanoscale) metals. It can be used for studying the [NiFeSe] hydrogenase immobilized on a 4-ATP-Au electrode. For this type of measurement, a trapezium-shaped silicon ATR-IR element covered with a nanostructured SEIRA Au film formed by electroless deposition is used. All spectra are recorded within a spectral window of $4000–1000\,cm^{-1}$ and with a resolution of $2\,cm^{-1}$ using an FTIR spectrometer with an MCT detector. For each spectrum, 620 scans are accumulated (Gutiérrez-Sanz, Marques, Pereira, et al., 2013).

6.2 Atomic force microscopy

The atomic force microscope (AFM) images the topography of modified surfaces describing the height and density of the proteins on the surface as well as the integrity and smoothness of a lipid bilayer. Measurements are performed using a flat substrate that consists of 200 nm gold evaporated over 1–4 nm chromium prepared on borosilicate glass (Metallhandel Schröder GMBH). Substrates are cleaned with "piranha" solution followed by exhaustive rinsing with Milli Q water and a final rinse with ethanol/water (2:1): they are then annealed to an orange glow for a few seconds in a propane flame (the heating is repeated five times). This treatment produces Au (111) grains of a few μm radius with atomically flat surfaces, suitable for AFM characterization. This atomically flat gold surface is then treated as described above (Section 4.3) to form the 4-ATP SAM and to covalently attach the enzyme for direct electron transfer. The protocol to form the bilayer on the gold substrate is also the same as that mentioned in Section 4.3. Imaging is carried out using either a Nanotec microscope operating in jumping mode or an Agilent Technologies (Santa Clara, CA) 5500 microscope used in tapping mode. Measurements were always made under liquid conditions in PBS buffer at room temperature using Olympus rectangular silicon nitride cantilevers (RC800PSA, $200 \times 20\,\mu m$) with a spring constant of 0.05–0.10 N/m, an estimated tip radius of 20 nm, and a resonance frequency in the liquid cell of approximately 27 kHz. Scanning rates are kept close to 1 Hz. Images contain 512×512 pixels and are first-order flattened using Picoimage software from Agilent or WSxM software from Nanotec Electronica (Gutiérrez-Sánchez et al., 2011; Rüdiger et al., 2010).

6.3 Quartz crystal microbalance

A quartz crystal microbalance (QCM) measures, in real time, the amount of material adsorbed on a surface. It is a useful instrument to aid in the sequential modification of electrodes described in Section 4.3 to measure electrocatalytic H_2 production and oxidation coupled to the creation of a proton gradient across a lipid membrane used to synthesize ATP. QCM measurements are used to verify the formation of a lipid bilayer containing ATPase synthase. Experiments are performed with 5 MHz SiO_2 quartz crystals from Q-Sense. Before each measurement SiO_2-coated crystals are cleaned in a 2% (w/v) sodium dodecyl sulfate (SDS) solution for 15 min, rinsed extensively with distilled water, dried under a stream of nitrogen gas, and exposed to a UV/ozone cleaner (Bioforce Nanoscience, model UV.TC:220) for at least 30 min. The UV light cleans the SiO_2 by eliminating organic contaminants and oxidizing the silicon surface, rendering it hydrophilic. After cleaning, crystals are immediately mounted in the

QCM chamber. All buffer solutions used for QCM measurements are degassed for 1 h using a 2510 Branson Sonicator with degas function (Branson Ultrasonic) immediately before use. The QCM measurements are performed using a QCM-Z500 resonator from KSV (Finland) with dissipation monitoring, equipped with a flow module. The sample is continuously delivered to the measurement chamber by a syringe pump (KD Scientific, model KDS120) at a flow rate of 50 μL/min. All measurements are performed at 23 ± 0.5°C, the temperature being controlled internally by a Peltier heat pump and externally by a refrigerated thermoblock (Nahita model 603/20). After obtaining a stable baseline, a temperature-equilibrated 0.1 mg/mL proteoliposome solution is injected into the QCM chamber. The excess of nonadsorbed proteoliposomes was removed from the chamber by rinsing with buffer (Gutiérrez-Sanz et al., 2016).

Acknowledgments

This work was financially supported by Fundação para a Ciência e Tecnologia (Portugal) through a fellowship SFRH/BD/100314/2014 (to S.Z.), grant PTDC/BBB-BEP/2885/2014 (to P.M.M. and I.A.C.P.), and R&D units UID/Multi/04551/2013 (Green-IT) and LISBOA-01-0145-FEDER-007660 (MostMicro) cofunded by FCT/MCTES and FEDER funds through COMPETE2020/POCI. Funding from the European Unions Horizon 2020 research and innovation program under grant agreement No 810856 and from MINECO/FEDER (projects CTQ2015-71290-R and FIS2015-70339-C2-2-R) is also acknowledged.

References

Abrahams, J. P., & Leslie, A. G. W. (1996). Methods used in the structure determination of bovine mitochondrial F1 ATPase. *Acta Crystallographica Section D: Biological Crystallography*, *52*(1), 30–42. https://doi.org/10.1107/S0907444995008754.

Adams, P. D., Afonine, P. V., Bunkóczi, G., Chen, V. B., Davis, I. W., Echols, N., et al. (2010). PHENIX: A comprehensive Python-based system for macromolecular structure solution. *Acta Crystallographica Section D: Biological Crystallography*, *66*(2), 213–221. https://doi.org/10.1107/S0907444909052925.

Baltazar, C. S. A., Marques, M. C., Soares, C. M., De Lacey, A. M., Pereira, I. A. C., & Matias, P. M. (2011). Nickel-iron-selenium hydrogenases—An overview. *European Journal of Inorganic Chemistry*, *2011*(7), 948–962. https://doi.org/10.1002/ejic.201001127.

Brandis, A., & Thauer, R. K. (1981). Growth of Desulfovibrio species on hydrogen and sulphate as sole energy source. *Microbiology*, *126*(1), 249–252. https://doi.org/10.1099/00221287-126-1-249.

Bricogne, G., Vonrhein, C., Flensburg, C., Schiltz, M., & Paciorek, W. (2003). Generation, representation and flow of phase information in structure determination: Recent developments in and around SHARP 2.0. *Acta Crystallographica Section D: Biological Crystallography*, *59*, 2023–2030. https://doi.org/10.1107/S0907444903017694.

Brünger, A. T. (1992). Free R value: A novel statistical quantity for assessing the accuracy of crystal-structures. *Nature*, *355*(6359), 472–475. https://doi.org/10.1038/355242a0.

Caffrey, S. M., Park, H.-S., Voordouw, J. K., He, Z., Zhou, J., & Voordouw, G. (2007). Function of periplasmic hydrogenases in the sulfate-reducing bacterium Desulfovibrio vulgaris Hildenborough. *Journal of Bacteriology*, *189*(17), 6159–6167. https://doi.org/10.1128/JB.00747-07.

Caputo, C. A., Gross, M. A., Lau, V. W., Cavazza, C., Lotsch, B. V., & Reisner, E. (2014). Photocatalytic hydrogen production using polymeric carbon nitride with a hydrogenase and a bioinspired synthetic Ni Catalyst. *Angewandte Chemie International Edition*, *126*(43), 11538–11542. https://doi.org/10.1002/anie.201406811.

Caputo, C. A., Wang, L., Beranek, R., & Reisner, E. (2015). Carbon nitride-TiO_2 hybrid modified with hydrogenase for visible light driven hydrogen production. *Chemical Science*, *6*(10), 5690–5694. https://doi.org/10.1039/C5SC02017D.

Cowtan, K. (1994). "Dm": An automated procedure for phase improvement by density modification. *Joint CCP4 and ESF-EACBM Newsletter on Protein Crystallography*, *31*, 34–38.

De Lacey, A. L., Fernández, V. M., Rousset, M., & Cammack, R. (2007). Activation and inactivation of hydrogenase function and the catalytic cycle: Spectroelectrochemical studies. *Chemical Reviews*, *107*(10), 4304–4330. https://doi.org/10.1021/cr0501947.

De Lacey, A. L., Gutiérrez-Sánchez, C., Fernández, V. M., Pacheco, I., & Pereira, I. A. C. (2008). FTIR spectroelectrochemical characterization of the Ni-Fe-Se hydrogenase from Desulfovibrio vulgaris Hildenborough. *Journal of Biological Inorganic Chemistry*, *13*(8), 1315–1320. https://doi.org/10.1007/s00775-008-0412-5.

Dong, A., Nam, H., Zhang, J., Andrei, V., Heidary, N., Wagner, A., et al. (2018). Solar water splitting with a hydrogenase integrated in photoelectrochemical tandem cells. *Angewandte Chemie International Edition*, *57*, 10595–10599. https://doi.org/10.1002/anie.201805027.

Emsley, P., Lohkamp, B., Scott, W. G., & Cowtan, K. (2010). Features and development of Coot. *Acta Crystallographica Section D: Biological Crystallography*, *66*(4), 486–501. https://doi.org/10.1107/S0907444910007493.

Evans, G., & Pettifer, R. F. (2001). CHOOCH: A program for deriving anomalous-scattering factors from X-ray fluorescence spectra. *Journal of Applied Crystallography*, *34*, 82–86.

Garcin, E., Vernede, X., Hatchikian, E. C., Volbeda, A., Frey, M., & Fontecilla-Camps, J. C. (1999). The crystal structure of a reduced [NiFeSe] hydrogenase provides an image of the activated catalytic center. *Structure (London, England: 1993)*, 7(5), 557–566. Available at: http://www.ncbi.nlm.nih.gov/pubmed/10378275.

Gutiérrez-Sánchez, C., Olea, D., Marques, M. C., Fernández, V. M., Pereira, I. A. C., Vélez, M., et al. (2011). Oriented immobilization of a membrane-bound hydrogenase onto an electrode for direct electron transfer. *Langmuir: The ACS Journal of Surfaces and Colloids*, *27*(10), 6449–6457. https://doi.org/10.1021/la200141t.

Gutiérrez-Sanz, O., Marques, M. C., Baltazar, C. S. A., Fernández, V. M., Soares, C. M., Pereira, I. A. C., et al. (2013). Influence of the protein structure surrounding the active site on the catalytic activity of [NiFeSe] hydrogenases. *Journal of Biological Inorganic Chemistry*, *18*(4), 419–427. https://doi.org/10.1007/s00775-013-0986-4.

Gutiérrez-Sanz, O., Marques, M. C., Pereira, I. A. C., De Lacey, A. L., Lubitz, W., & Rüdiger, O. (2013). Orientation and function of a membrane-bound enzyme monitored by electrochemical surface-enhanced infrared absorption spectroscopy. *Journal of Physical Chemistry Letters*, *4*(17), 2794–2798. https://doi.org/10.1021/jz4013678.

Gutiérrez-Sanz, O., Natale, P., Márquez, I., Marques, M. C., Zacarias, S., Pita, M., et al. (2016). H_2-fueled ATP synthesis on an electrode: Mimicking cellular respiration. *Angewandte Chemie International Edition*, *55*(21), 6216–6220. https://doi.org/10.1002/anie.201600752.

Gutiérrez-Sanz, O., Tapia, C., Marques, M. C., Zacarias, S., Vélez, M., Pereira, I. A. C., et al. (2015). Induction of a proton gradient across a gold-supported biomimetic membrane by electroenzymatic H_2 oxidation. *Angewandte Chemie International Edition, 54*, 2684–2687. https://doi.org/10.1002/anie.201411182.

Hedderich, R. (2004). Energy-converting [NiFe] hydrogenases from archaea and extremophiles: Ancestors of complex I. *Journal of Bioenergetics and Biomembranes, 36*(1), 65–75. https://doi.org/10.1023/B:JOBB.0000019599.43969.33.

Kabsch, W. (1993). Automatic processing of rotation diffraction data from crystals of initially unknown symmetry land cell constants. *Journal of Applied Crystallography, 26*(Pt. 6), 795–800. https://doi.org/10.1107/S0021889893005588.

Keller, K. L., Bender, K. S., & Wall, J. D. (2009). Development of a markerless genetic exchange system for Desulfovibrio vulgaris Hildenborough and its use in generating a strain with increased transformation efficiency. *Applied and Environmental Microbiology, 75*(24), 7682–7691. https://doi.org/10.1128/AEM.01839-09.

Keller, K. L., Wall, J. D., & Chhabra, S. (2011). Methods for engineering sulfate reducing bacteria of the genus *Desulfovibrio*. In *Methods in enzymology*. (1st ed.). Elsevier Inc. https://doi.org/10.1016/B978-0-12-385075-1.00022-6

Lee, C. Y., Park, H. S., Fontecilla-Camps, J. C., & Reisner, E. (2016). Photoelectrochemical H_2 evolution with a hydrogenase immobilized on a TiO_2-protected silicon electrode. *Angewandte Chemie International Edition, 55*(20), 5971–5974. https://doi.org/10.1002/anie.201511822.

Li, M. Z., & Elledge, S. J. (2007). Harnessing homologous recombination in vitro to generate recombinant DNA via SLIC. *Nature Methods, 4*(3), 251–256. https://doi.org/10.1038/NMETH1010.

Lucena, R., Aguilera, I., Palacios, P., Wahnón, P., & Conesa, J. C. (2008). Synthesis and spectral properties of nanocrystalline V-substituted In_2S_3, a novel material for more efficient use of solar radiation. *Chemistry of Materials, 20*(16), 5125–5127. https://doi.org/10.1021/cm801128b.

Maroney, M. J., & Hondal, R. J. (2018). Selenium versus sulfur: Reversibility of chemical reactions and resistance to permanent oxidation in proteins and nucleic acids. *Free Radical Biology and Medicine, 127*, 228–237. https://doi.org/10.1016/j.freeradbiomed.2018.03.035.

Marques, M. C., Coelho, R., De Lacey, A. L., Pereira, I. A. C., & Matias, P. M. (2010). The three-dimensional structure of [NiFeSe] hydrogenase from Desulfovibrio vulgaris Hildenborough: A hydrogenase without a bridging ligand in the active site in its oxidised, "as-isolated" state. *Journal of Molecular Biology, 396*(4), 893–907. https://doi.org/10.1016/j.jmb.2009.12.013.

Marques, M. C., Coelho, R., Pereira, I. A. C., & Matias, P. M. (2009). Purification, crystallization and preliminary crystallographic analysis of the [NiFeSe] hydrogenase from Desulfovibrio vulgaris Hildenborough. *Acta Crystallographica. Section F, Structural Biology and Crystallization Communications, 65*(Pt. 9), 920–922. https://doi.org/10.1107/S1744309109031261.

Marques, M. C., Coelho, R., Pereira, I. A. C., & Matias, P. M. (2013). Redox state-dependent changes in the crystal structure of [NiFeSe] hydrogenase from Desulfovibrio vulgaris Hildenborough. *International Journal of Hydrogen Energy, 38*(21), 8664–8682. https://doi.org/10.1016/j.ijhydene.2013.04.132.

Marques, M. C., Tapia, C., Gutiérrez-Sanz, O., Ramos, A. R., Keller, K. L., Wall, J. D., et al. (2017). The direct role of selenocysteine in [NiFeSe] hydrogenase maturation and catalysis. *Nature Chemical Biology, 13*(5), 544–550. https://doi.org/10.1038/nchembio.2335.

Matias, P. M., Soares, C. M., Saraiva, L. M., Coelho, R., Morais, J., Le Gall, J., et al. (2001). [NiFe] hydrogenase from Desulfovibrio desulfuricans ATCC 27774: Gene sequencing, three-dimensional structure determination and refinement at 1.8 Å and modelling studies of its interaction with the tetrahaem cytochrome c3. *Journal of Biological Inorganic Chemistry*, *6*(1), 63–81. https://doi.org/10.1007/s007750000167.

McCoy, A. J. (2006). Solving structures of protein complexes by molecular replacement with Phaser. *Acta Crystallographica Section D: Biological Crystallography*, *63*(1), 32–41. https://doi.org/10.1107/S0907444906045975.

Mersch, D., Lee, C., Zhang, J. Z., Brinkert, K., Fontecilla-camps, J. C., Rutherford, A. W., et al. (2015). Wiring of photosystem II to hydrogenase for photoelectrochemical water splitting. *Journal of the American Chemical Society*, *137*, 8541–8549. https://doi.org/10.1021/jacs.5b03737.

Moss, D., Nabedryk, E., Breton, J., & Mantele, W. (1990). Redox-linked conformational changes in proteins detected by a combination of infrared spectroscopy and protein electrochemistry: Evaluation of the technique with cytochrome c. *European Journal of Biochemistry*, *187*(3), 565–572. https://doi.org/10.1111/j.1432-1033.1990.tb15338.x.

Murshudov, G. N., Skubák, P., Lebedev, A. A., Pannu, N. S., Steiner, R. A., Nicholls, R. A., et al. (2011). REFMAC5 for the refinement of macromolecular crystal structures. *Acta Crystallographica Section D: Biological Crystallography*, *67*(4), 355–367. https://doi.org/10.1107/S0907444911001314.

Muth, E., Morschel, E., & Klein, A. (1987). Purification and characterization of an 8-hydroxy-5-deazaflavin-reducing hydrogenase from the archaebacterium Methanococcus voltae. *European Journal of Biochemistry*, *169*(3), 571–577. https://doi.org/10.1111/j.1432-1033.1987.tb13647.x.

Nonaka, K., Nguyen, N. T., Yoon, K.-S., & Ogo, S. (2013). Novel H_2-oxidizing [NiFeSe] hydrogenase from Desulfovibrio vulgaris Miyazaki F. *Journal of Bioscience and Bioengineering*, *115*(4), 366–371. https://doi.org/10.1016/j.jbiosc.2012.10.011.

Painter, J., & Merritt, E. A. (2006). Optimal description of a protein structure in terms of multiple groups undergoing TLS motion. *Acta Crystallographica Section D: Biological Crystallography*, *62*(4), 439–450. https://doi.org/10.1107/S0907444906005270.

Pape, T., & Schneider, T. R. (2004). HKL2MAP: A graphical user interface for macromolecular phasing with SHELX programs. *Journal of Applied Crystallography*, *37*(5), 843–844. https://doi.org/10.1107/S0021889804018047.

Papp, L. V., Lu, J., Holmgren, A., & Khanna, K. K. (2007). From selenium to selenoproteins: Synthesis, identity, and their role in human health. *Antioxidants & Redox Signaling*, *9*(7), 775–806. https://doi.org/10.1089/ars.2007.1528.

Parkin, A., Goldet, G., Cavazza, C., Fontecilla-Camps, J. C., & Armstrong, F. A. (2008). The difference a Se makes? Oxygen-tolerant hydrogen production by the [NiFeSe]-hydrogenase from Desulfomicrobium baculatum. *Journal of the American Chemical Society*, *130*(40), 13410–13416. https://doi.org/10.1021/ja803657d.

Pereira, I. A. C., Ramos, A. R., Grein, F., Marques, M., da Silva, S. M., & Venceslau, S. S. (2011). A comparative genomic analysis of energy metabolism in sulfate reducing bacteria and archaea. *Frontiers in Microbiology*, *2*, 69. https://doi.org/10.3389/fmicb.2011.00069.

Pereira, I. A. C., Romão, C. V., Xavier, A. V., Le Gall, J., & Teixeira, M. (1998). Electron transfer between hydrogenases and mono- and multiheme cytochromes in Desulfovibrio ssp. *Journal of Biological Inorganic Chemistry*, 494–498.

Perrakis, A., Morris, R., & Lamzin, V. S. (1999). Automated protein model building combined with iterative structure refinement. *Nature Structural Biology*, *6*(5), 458–463. https://doi.org/10.1038/8263.

Postgate, J. R. (1984). *Sulphate-reducing bacteria*. London, UK: Cambridge University Press. https://doi.org/10.1017/CBO9780511541490.

Potterton, E., Briggs, P., Turkenburg, M., & Dodson, E. (2003). A graphical user interface to the CCP4 program suite. *Acta Crystallographica, Section D: Biological Crystallography*, *59*(7), 1131–1137. https://doi.org/10.1107/S0907444903008126.

Reich, H. J., & Hondal, R. J. (2016). Why nature chose selenium. *ACS Chemical Biology*, *11*(4), 821–841. https://doi.org/10.1021/acschembio.6b00031.

Reisner, E., Fontecilla-camps, J. C., & Armstrong, F. A. (2009). Catalytic electrochemistry of a [NiFeSe]-hydrogenase on TiO_2 and demonstration of its suitability for visible-light driven H_2 production. *Chemical Communications*, (5), 550–552. https://doi.org/10.1039/b817371k.

Reisner, E., Powell, D. J., Cavazza, C., Fontecilla-Camps, J. C., & Armstrong, F. A. (2009). Visible light-driven H(2) production by hydrogenases attached to dye-sensitized TiO(2) nanoparticles. *Journal of the American Chemical Society*, *131*(51), 18457–18466. https://doi.org/10.1021/ja907923r.

Rodrigue, A., Chanal, A., Beck, K., Müller, M., Wu, L., & Mu, M. (1999). Co-translocation of a periplasmic enzyme complex by a hitchhiker mechanism through the bacterial tat pathway. *The Journal of Biological Chemistry*, *274*(19), 13223–13228. https://doi.org/10.1074/jbc.274.19.13223.

Rüdiger, O., Gutiérrez-Sánchez, C., Olea, D., Pereira, I. A. C., Vélez, M., Fernández, V. M., et al. (2010). Enzymatic anodes for hydrogen fuel cells based on covalent attachment of Ni-Fe hydrogenases and direct electron transfer to SAM-modified gold electrodes. *Electroanalysis*, *22*(7–8), 776–783. https://doi.org/10.1002/elan.200880002.

Ruff, A., Szczesny, J., Zacarias, S., Pereira, I. A. C., Plumeré, N., & Schuhmann, W. (2017). Protection and reactivation of the [NiFeSe] hydrogenase from Desulfovibrio vulgaris Hildenborough under oxidative conditions. *ACS Energy Letters*, *2*(5). https://doi.org/10.1021/acsenergylett.7b00167.

Ruff, A., Szczesny, J., Marković, N., Conzuelo, F., Zacarias, S., Pereira, I. A. C., et al. (2018). A fully protected hydrogenase/polymer-based bioanode for high-performance hydrogen/glucose biofuel cells. *Nature Communications*, *9*(1), 3675. https://doi.org/10.1038/s41467-018-06106-3.

Sakai, T., Mersch, D., & Reisner, E. (2013). Photocatalytic hydrogen evolution with a hydrogenase in a mediator-free system under high levels of oxygen. *Angewandte Chemie (International ed. in English)*, *52*(47), 12313–12316. https://doi.org/10.1002/anie.201306214.

Schneider, T. R., & Sheldrick, G. M. (2002). Substructure solution with SHELXD. *Acta Crystallographica Section D: Biological Crystallography*, *58*(10), 1772–1779. https://doi.org/10.1107/S0907444902011678.

Sheldrick, G. M. (2002). Macromolecular phasing with SHELXE. *Zeitschrift für Kristallographie*, *217*(12), 644–650.

Sheldrick, G. M. (2010). Experimental phasing with SHELXC/D/E: Combining chain tracing with density modification. *Acta Crystallographica Section D: Biological Crystallography*, *66*(4), 479–485. https://doi.org/10.1107/S0907444909038360.

Tapia, C., Zacarias, S., Pereira, I. A. C., Conesa, J. C., Pita, M., & De Lacey, A. L. (2016). In situ determination of photobioproduction of H_2 by In_2S_3-[NiFeSe] hydrogenase from Desulfovibrio vulgaris Hildenborough using only visible light. *ACS Catalysis*, *6*(9), 5691–5698. https://doi.org/10.1021/acscatal.6b01512.

Teixeira, M., Fauque, G., Moura, I., Lespinat, P. A., Berlier, Y., Prickril, B., Jr., et al. (1987). Nickel-[iron-sulfur]-selenium-containing hydrogenases from Desulfovibrio baculatus (DSM 1743). *European Journal of Biochemistry*, *58*, 47–58. https://doi.org/10.1111/j.1432-1033.1987.tb13302.

Teixeira, M., Moura, I., Fauque, G., Czechowski, M., Berlier, Y., Lespinat, P. A., et al. (1986). Redox properties and activity studies on a nickel-containing hydrogenase isolated from a halophilic sulfate reducer Desulfovibrio salexigens. *Biochimie*, *68*, 75–84. https://doi.org/10.1016/S0300-9084(86)81071-9.

Valente, F. M. A., Almeida, C., Pacheco, I., Saraiva, M., & Pereira, I. A. C. (2006). Selenium is involved in regulation of periplasmic hydrogenase gene expression in Desulfovibrio vulgaris Hildenborough. *Journal of Bacteriology, 188*(9), 3228–3235. https://doi.org/10.1128/JB.188.9.3228.

Valente, F. M. A., Oliveira, A. S. F., Gnadt, N., Pacheco, I., Coelho, A. V., Xavier, A. V., et al. (2005). Hydrogenases in Desulfovibrio vulgaris Hildenborough: Structural and physiologic characterisation of the membrane-bound [NiFeSe] hydrogenase. *Journal of Biological Inorganic Chemistry, 10*(6), 667–682. https://doi.org/10.1007/s00775-005-0022-4.

Valente, F. M. A., Pereira, P. M., Venceslau, S. S., Regalla, M., Coelho, A. V., & Pereira, I. A. C. (2007). The [NiFeSe] hydrogenase from Desulfovibrio vulgaris Hildenborough is a bacterial lipoprotein lacking a typical lipoprotein signal peptide. *FEBS Letters, 581*(18), 3341–3344. https://doi.org/10.1016/j.febslet.2007.06.020.

Vignais, P. M. (2005). H/D exchange reactions and mechanistic aspects of the hydrogenases. *Coordination Chemistry Reviews, 249*(15–16), 1677–1690. https://doi.org/10.1016/j.ccr.2005.01.026.

Vincent, K. A., Parkin, A., & Armstrong, F. A. (2007). Investigating and exploiting the electrocatalytic properties of hydrogenases. *Chemical Reviews, 107*(10), 4366–4413. https://doi.org/10.1021/cr050191u.

Winn, M. D., Isupov, M. N., & Murshudov, G. N. (2001). Use of TLS parameters to model anisotropic displacements in macromolecular refinement research papers use of TLS parameters to model anisotropic displacements in macromolecular refinement. *Acta Crystallographica, Section D: Biological Crystallography, 57*, 122–133.

Yamazaki, S. (1982). A selenium-containing hydrogenase from Methanococcus vannielii. Identification of the selenium moiety as a selenocysteine residue. *The Journal of Biological Chemistry, 257*, 7926–7929.

Zane, G. M., Yen, H. B., & Wall, J. D. (2010). Effect of the deletion of qmoABC and the promoter-distal gene encoding a hypothetical protein on sulfate reduction in Desulfovibrio vulgaris Hildenborough. *Applied and Environmental Microbiology, 76*(16), 5500–5509. https://doi.org/10.1128/AEM.00691-10.

CHAPTER EIGHT

[FeFe]-hydrogenases from green algae

Vera Engelbrecht, Thomas Happe*
AG Photobiotechnologie, Fakultät für Biologie und Biotechnologie, Ruhr-Universität Bochum, Universitätsstraße, Bochum, Germany
*Corresponding author: e-mail address: thomas.happe@rub.de

Contents

Abstract

Algal hydrogenases are among to the most efficient hydrogen (H_2) generating biocatalysts and use low-potential electrons from the photosynthetic light reactions. Thereby, photobiological H_2 evolution by eukaryotic microalgae represents a sustainable alternative to the energy intensive industrial production of H_2 based on fossil fuels. Novel algal hydrogenases are still being discovered and their biochemical and biophysical characterization has revealed unique features beneficial to overcome bottlenecks in photobiological H_2 production. Highly advanced techniques are available to study hydrogenases to atomic level accuracy. Yet, to benefit from these methods, one has to overcome the first and most fundamental obstacle, namely, obtaining sufficient amounts of active hydrogenase enzyme. The recombinant production of hydrogenases can be very challenging, so this chapter provides information and useful advice for the discovery, heterologous production, and analyses of hydrogenases from microalgae.

Methods in Enzymology, Volume 613
ISSN 0076-6879
https://doi.org/10.1016/bs.mie.2018.10.004

1. Introduction

In order to reduce consumption of limited fossil fuel resources, there is a rising interest in the environmentally friendly biotechnological production of molecular hydrogen (H_2). Among the organisms known to produce biohydrogen, algae are very interesting because they can split water into H_2 and oxygen (O_2) via photosynthesis. Special enzymes called hydrogenases are directly linked to the photosynthetic electron transfer pathway chain (Dubini & Ghirardi, 2015; Hemschemeier & Happe, 2011; Martin & Frymier, 2017; Oey, Sawyer, Ross, & Hankamer, 2016). Even though the ability of photosynthetic microorganisms to capture light energy and split water seems promising, commercial algal H_2 production is currently limited by several factors, including the O_2 sensitivity of hydrogenases and low reductant availability (Dubini & Ghirardi, 2015). However, the structural simplicity of most algal hydrogenases (40–50 kDa, monomeric) makes them ideal subjects of research. While the experimental toolbox to investigate the structural architecture of the hydrogenase polypeptide scaffold and catalytic cycle of its metal cofactor provides a broad array of biochemical and biophysical methods (electron paramagnetic resonance (EPR), Mössbauer and infrared (IR) spectroscopy, protein film electrochemistry (PPE), and 3D structure determination methods), the main subject focus of recent publications has been just one algal hydrogenase, HydA1 from *Chlamydomonas reinhardtii*. The natural variety of functions and structural differences of other algal hydrogenases has neither been fully considered nor exploited. A key step in the identification and characterization of novel algal hydrogenases is often the expression and purification of sufficient quantities of enzyme of high-enough quality. Therefore, this chapter focuses on approaches to express and purify algal [FeFe]-hydrogenases and encourages researchers to "think outside the box" and take advantage of the unique features of these often overlooked organisms.

2. Discovery and isolation of novel algal [FeFe]-hydrogenases

Classified by the composition of their active site metal cluster, algal hydrogenases belong to the [FeFe]-hydrogenase family, which is not phylogenetically related to the two other hydrogenase families, the [Fe]-hydrogenases and [NiFe]-hydrogenases (Lubitz, Ogata, Rüdiger, &

Reijerse, 2014; Peters et al., 2015; Winkler, Esselborn, & Happe, 2013). Prokaryotic [FeFe]-hydrogenases normally show a high degree of variation in regard to their additional N- and C-terminal iron sulfur (FeS) cluster-binding domains (F-cluster domains). However, most algal hydrogenase homologs consist only of the active site domain (H-cluster domain) and lack additional FeS clusters. Their active site is unique in that it comprises a standard [4Fe4S]-cluster with a cysteine linkage to an unusual 2Fe-subcluster with three CO and two CN^- ligands and an aza-dithiolato-bridge (Berggren et al., 2013; Cornish, Green, Gärtner, Mason, & Hegg, 2015; Florin, Tsokoglou, & Happe, 2001; Happe & Kaminski, 2002; Kamp et al., 2008; Peters et al., 2015; Ueno, Kurano, & Miyachi, 1999; Winkler, Heil, & Happe, 2002). Even though they lack additional F-clusters, which mediate electron transfer in prokaryotic [FeFe]-hydrogenases (Artz et al., 2017; Gauquelin et al., 2018; Peters, Lanzilotta, Lemon, & Seefeldt, 1998), small algal enzymes achieve very high turnover rates (Peters et al., 2015).

Green algal hydrogenases were discovered nearly 80 years ago (Gaffron, 1939, 1942; Gaffron & Rubin, 1942). However, apart from the well characterized hydrogenase HydA1 from *C. reinhardtii*, only a few other algal hydrogenases have been purified and characterized (Cornish et al., 2015; Engelbrecht et al., 2017; Florin et al., 2001; Kamp et al., 2008; King, Posewitz, Ghirardi, & Seibert, 2006; Ueno et al., 1999; Yan et al., 2011). Some of these other algal hydrogenases have interesting features such as a higher specific activity (Kamp et al., 2008) or different redox partner interaction mechanisms (Engelbrecht et al., 2017).

2.1 Step 1: Screening of H_2-producing microalgae

To identify new microalgal hydrogenases, the first requirement is a screening assay. Apart from using bioinformatic approaches, e.g., to search the increasing pool of completely sequenced algal genomes for potential new hydrogenase genes (Arriola et al., 2018; Blanc et al., 2010; Merchant et al., 2006; Nelson et al., 2017; Prochnik et al., 2010), environmental samples, or isolated algal strains can be examined by inducing H_2 metabolism. The presence of *HydA* transcripts or active enzymes can then be documented via quantitative real time or in vivo and in vitro activity assays, respectively (Hemschemeier, Melis, & Happe, 2009; Jacobs, Pudollek, Hemschemeier, & Happe, 2009; Timmins et al., 2009). Various assays have been developed to induce algal H_2 production; the most interesting are described below (Fig. 1).

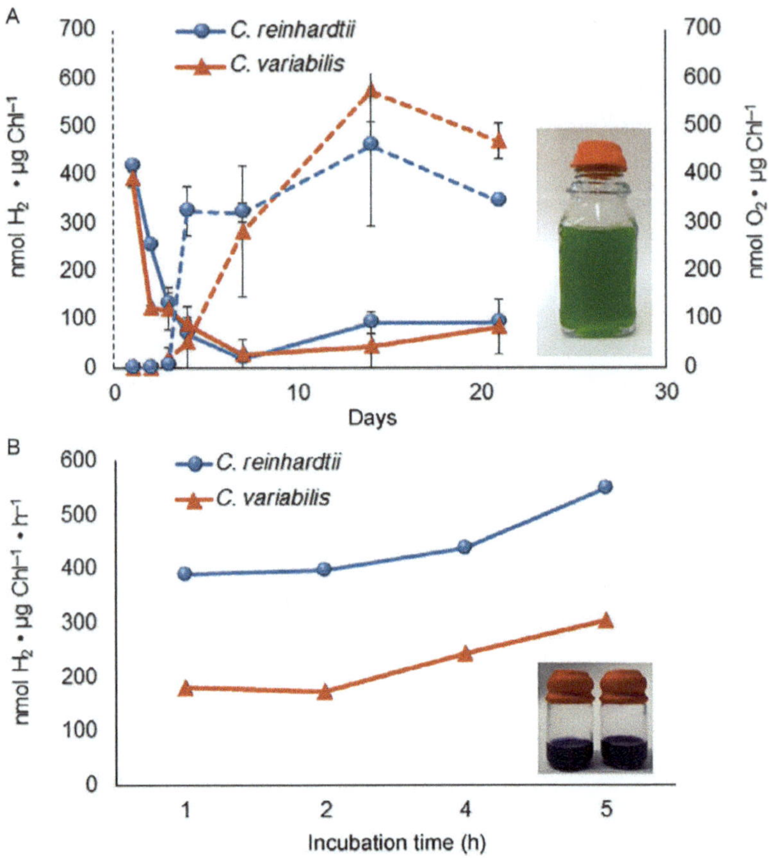

Fig. 1 Detection of the hydrogen production in various algal cultures. (A) Hydrogen (H_2, *dashed lines*) and oxygen (O_2, *continuous lines*) contents in the headspace of sulfur-deprived *C. reinhardtii* (*blue*) and *C. variabilis* (*red*) cultures. (B) In vitro hydrogenase activity of lysed *C. reinhardtii* (*blue*) and *C. variabilis* (*red*) cultures using sodium dithionite-reduced methyl viologen as an electron mediator. Detected amounts of H_2 were related to the chlorophyll (Chl) concentration of the cultures and to the duration of anaerobic incubation.

2.1.1 Anaerobic induction

During this artificial process, highly concentrated algal cultures are flushed with an inert gas such as argon (Ar) or nitrogen (N_2) in the dark.

Equipment

- Light bench with light intensity of about $80\,\mu E\,s^{-1}\,m^{-2}$
- 1000 mL vessels containing 200 mL of culture
- Centrifuge (e.g., Eppendorf 5810)
- N_2 gas supply system

- 50-mL Falcon tubes
- Gas chromatograph (GC) (e.g., Shimadzu GC2025)

Buffers and reagents

Tris acetate phosphate medium (TAP) (Gorman & Levine, 1965) or growth medium of choice.

Procedure

1. Algal cells are grown photoheterotrophically (in TAP medium) in the light with constant shaking, making sure that they are in the active growth phase ($\sim 1 \times 10^7$ cells mL^{-1}).
2. The culture is harvested by mild centrifugation (4 min 4000 × *g*, room temperature).
3. The pellet is resuspended immediately in fresh medium to a high chlorophyll concentration (0.5–5 mg mL^{-1}).
4. Exposure to light is prevented by wrapping aluminum foil around the culture tubes.
5. The culture is maintained at temperatures between 18 and 25°C.
6. The culture is flushed with Ar or N_2 by inserting a needle ($0.8 \times 120\,mm^2$) connected to a gas cylinder via a tube through a hole in the lid; avoid strong bubbling of the culture.
7. During the next 4–6 h of flushing, samples are taken every 30 min for in vitro measurements of H_2 production of the crude extract (Fig. 1B, for protocol see Section 2.4, in vitro activity of *Escherichia coli* crude extract, use 300–500 μL of algal culture (0.5–2 mg Chl mL^{-1}) with additional vortexing in the lysis step).

2.1.2 Sulfur deprivation

For some algal strains, anaerobic induction only leads to a very minimal H_2 production. In these cases, a sulfur deprivation assay (Ghirardi et al., 2000; Melis, Zhang, Forestier, Ghirardi, & Seibert, 2000) can be more advantageous. Sealed algal cultures grown in a sulfur-deprived growth medium show a decreased rate of photosynthesis (Wykoff, Davies, Melis, & Grossman, 1998) and effectively unchanged levels of respiration, leading to anaerobiosis and thus to *HydA* gene expression. For sulfur deprivation of algae, a detailed protocol and medium developed for *C. reinhardtii* is generally used (Hemschemeier et al., 2009; Timmins et al., 2009). However, most algal species need growth conditions tailored to their requirements and the success of this approach seems to depend on species-dependent medium optimization. For example, *Chlorella variabilis* NC64A needs additions of sucrose and glycine to normal or sulfur-deprived TAP medium in order to grow efficiently (Engelbrecht et al., 2017; Guillard & Ryther, 1962; Meuser et al., 2011). It should also be noted that for some algal strains, the induction of H_2 production can take longer compared to the quick induction of *C. reinhardtii*, where H_2 can be detected in the headspace as early as 2–3 days following sulfur deprivation (Fig. 1A).

Equipment

- Light bench with light intensity of about $80\,\mu E\,s^{-1}\,m^{-2}$
- 1000 mL vessels containing 200 mL of green algae culture (preculture)
- 100 mL square narrow neck glass bottles (e.g., Duran)
- Suba seals (no. 37, Sigma-Aldrich)
- Centrifuge (e.g., Eppendorf 5810)
- 50-mL falcon tubes
- 1-mL syringe, $0.45 \times 25\,mm^2$ needle

Buffers and reagents

TAP medium (or any other sufficient medium).

Procedure

1. Grow cells in standard medium in the light until a chlorophyll content of 15–23 $\mu g\,Chl\,mL^{-1}$ is reached.
2. Pellet cells by mild centrifugation (4 min $4000 \times g$, room temperature); discard supernatant.
3. Resuspend cell pellet in sulfur-deprived medium.
4. Repeat centrifugation (up to five washing steps).
5. Stir cells ($\sim 15\,\mu g\,Chl\,mL^{-1}$) in sealed square glass bottles in the light.

2.2 Step 2: Heterologous expression in bacterial systems

Once the hydrogenase of an interesting algal strain is identified, the next crucial step is choosing a host system for protein expression. The purification of *C. reinhardtii* HydA1 (*Cr*HydA1) has undergone a long optimization process since the protein was first purified to homogeneity from its native host, *C. reinhardtii* in 1993 (Happe & Naber, 1993). Major challenges for efficient hydrogenase expression are low protein yields and poor FeS-cluster maturation.

To express an active hydrogenase recombinantly, the [FeFe]-hydrogenase specific maturation system (maturase enzymes HydE, HydF, and HydG) is required to assemble the unique 2Fe-subcluster and transfer it to the preassembled standard [4Fe4S]-cluster (incorporated by the house keeping FeS-cluster assembly machinery of the host cell, Fig. 2) (Posewitz et al., 2004).

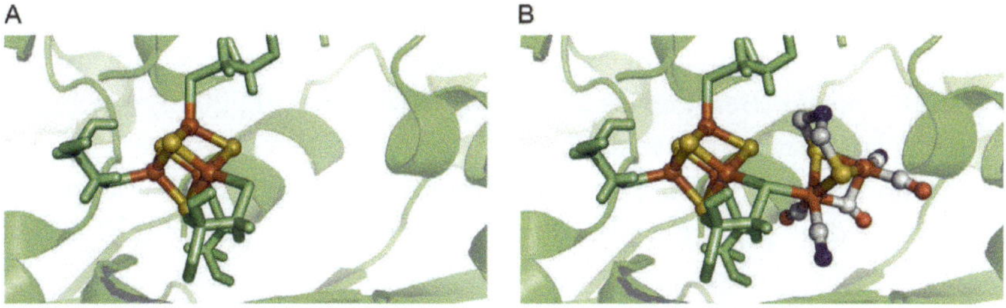

Fig. 2 Stick model representation of the H-cluster of *CpI* (hydrogenase from *Clostridium pasteurianum*, PDB ID: 4xdc) before (A) and after (B) integration of the [2Fe2S]-subcluster, colored according to element (C = *white*, O = *red*, N = *blue*, S = *yellow*, and F = *orange*), protein environment (cartoon model). H-cluster coordinating cysteines are depicted as *pale green sticks*.

In a clostridial expression host such as *Clostridium acetobutylicum*, the specific maturase enzymes are already present, since *C. acetobutylicum* itself possesses [FeFe]-hydrogenases (Gorwa, Croux, & Soucaille, 1996). Use of clostridial expression hosts resulted in hydrogenase yields about a 1000-fold higher than those obtained following homologous expression in *C. reinhardtii* (Abendroth et al., 2008; Girbal et al., 2005). However, heterologous expression of [FeFe]-hydrogenases in *E. coli* has several advantages, for example, facilitated growth conditions, capability for anaerobic respiration, fast doubling times, well-established heterologous expression techniques, low costs, and a lack of native [FeFe]-hydrogenases (Kuchenreuther et al., 2010; Terpe, 2006; Yacoby, Tegler, Pochekailov, Zhang, & King, 2012).

Despite these advantages, *E. coli* is devoid of the [FeFe]-hydrogenase specific maturation system and hydrogenase expression leads to the synthesis of inactive hydrogenases containing only the [4Fe4S]-cluster (Fig. 2) (Mulder et al., 2009). This problem was circumvented by the coexpression of maturation genes from *Shewanella oneidensis* or *C. acetobutylicum*, leading to further improvements in protein yields (King et al., 2006; Kuchenreuther et al., 2010). However, the high expression levels in the *E. coli* system led to lower Fe content and specific activities of the recombinant hydrogenases, most likely caused by the elevated demand for [FeS]-clusters (King et al., 2006).

To exploit the productivity of the *E. coli* expression but simultaneously increase maturation efficiency, an in vitro maturation approach can be used. Here, the [FeFe]-hydrogenase of choice is expressed in *E. coli* without maturase coexpression, and then chemically synthesized [2Fe2S]-subcluster mimics ($Fe_2[\mu\text{-}(SCH_2)_2NH](CN)_2(CO)4^{2-}$, Tard et al., 2005) are inserted into the purified "apo" form of the hydrogenase leading to a fully complemented H-cluster (Berggren et al., 2013; Esselborn et al., 2013) (see Fig. 2).

In the last decade, the total yield of active *Cr*HydA1 has increased from 1 μg protein per liter culture from the homologous expression system to up to 90 mg HydA1 per liter culture in *E. coli* (Happe & Naber, 1993; Noth et al., 2015; Yacoby et al., 2012).

Even though optimal expression and maturation conditions for *Cr*HydA1 have been established (Esselborn et al., 2013), yields of *Cr*HydA1 variants or other algal enzymes can be very low. In these cases, the expression conditions have to be customized. The most efficient screening method to check whether an alteration to the standard protocol is required is to determine whether there is active hydrogenase in the *E. coli* crude extract by performing an in vitro activity test with the cell lysate (see Section 2.4).

2.2.1 Promising alterations of expression protocol

The first attempt to express your hydrogenase of interest should be the adapted expression protocol of Kuchenreuther et al. (2010). To summarize, *E. coli* BL21 (DE3) ΔiscR, which has a deletion of the gene encoding the transcriptional negative regulator of the FeS-cluster assembly machinery (ISC), is used (Akhtar & Jones, 2008). Cells that have been transformed with the hydrogenase gene in the pET-21(b) expression vector (Novagene) are grown aerobically in LB (Luria Broth) medium supplemented with antibiotics (ampicillin 100 $mg L^{-1}$, kanamycin 40 $mg L^{-1}$), 0.5% (w/v) glucose, 2 m*M* ferric ammonium citrate, and 100 m*M* MOPS/NaOH (pH 7.4) at 37°C with constant stirring until optical densities (OD_{600nm}) of about 0.4 are reached. Cell cultures are then moved to an anaerobic glove box and stirred for at least 30 min. Prior to induction of protein expression with 0.5 mM IPTG (isopropyl β-D-1-thiogalactopyranoside), 2 m*M* cysteine and 25 m*M* sodium fumarate are added. The *E. coli* cells are stirred in an anaerobic atmosphere at room temperature (18–25°C) for about 17 h.

If the hydrogenase yield turns out to be low, there are different possible approaches to increase the outcome.

Protein toxicity

The most commonly used vectors for (hydrogenase) expression are T7 expression vectors such as pET21(b) (Novagen), which contain the T7 promotor upstream of the gene of interest. T7 expression hosts such as

E. coli DE3 strains contain a copy of the T7 RNA polymerase gene. After induction with lactose or its nonhydrolysable analog IPTG, T7 RNA polymerase is expressed and subsequently transcription can proceed (Rosano & Ceccarelli, 2014). Although protein expression with the pET system can be highly efficient—target proteins can account for up to 50% of the total cell protein (Graumann & Premstaller, 2006), the DE3 *lac*UV5 promotor, which controls T7 RNA polymerase expression, has the disadvantage of leaky expression in the absence of inducer, which is a problem if the target hydrogenase is toxic to the cells. In this case, the use of a low copy number plasmid with a tightly regulated promotor, for example, the pBAD vector, is recommended (Guzman, Belin, Carson, & Beckwith, 1995). It might also help to induce expression cultures at very high densities (Sivashanmugam et al., 2009) and to decrease the duration of expression (3–5 h, instead of overnight expression).

➔ Use a lower copy number plasmid with a tightly regulated promotor
➔ Induce at high densities
➔ Decrease induction time

Expression strain/codon usage

If protein expression in BL21 is low, differences in chaperones, codon usage bias, posttranslational modifications, or disulfide bridge formation in the algae compared to the *E. coli* expression strain could be the reason. In the case of codon usage bias, it might be most convenient to use a gene sequence that is already optimized for *E. coli*. Even though many methods optimize codon usage by enriching high-frequency host codons (Chin, Chung, & Lee, 2014; Grote et al., 2005; Liu, Deng, Wang, & Wang, 2014; Lorimer et al., 2009; Puigbò, Guzmán, Romeu, & Garcia-Vallvé, 2007; Villalobos, Ness, Gustafsson, Minshull, & Govindarajan, 2006), a recent study suggests using varying ratios of low-frequency to high-frequency codons in order to synchronize the translational speed of the protein according to its structural elements (Tian et al., 2017). Another strategy for overcoming the codon usage problem is to increase the availability of underrepresented tRNAs, which is possible with *E. coli* strains such as BL21-CodonPlus (Stratagene) or Rosetta (DE3) (Novagen). These strains contain plasmids with extra genes for rare tRNAs (Rosano & Ceccarelli, 2014). If disulfide bond formation is thought to be the critical step in proper folding of the recombinant hydrogenase in *E. coli*, expression in Origami (Novagen) or Shuffle (NEB) strains could be helpful. Both strains have mutations that lead to an oxidative cytoplasmic environment which favors disulfide bond formation (Derman, Prinz, Belin, & Beckwith, 1993). Another possibility is that the *E. coli* chaperone network might not be efficient enough to help with the recombinant protein folding. In that case, it has proven helpful to either induce the *E. coli* chaperone network by the addition of 10 m*M* benzyl alcohol 20 min before induction or coexpress plasmid-encoded molecular chaperones (de Marco, Vigh, Diamant, & Goloubinoff, 2005).

➔ Optimize codon usage
➔ Induce *E. coli* chaperone network by adding 10 m*M* benzyl alcohol before induction
➔ Use different *E. coli* strains such as Rosetta, BL21-CodonPlus, Origami, or Shuffle

New media formulation

For other FeS-cluster containing proteins, overexpression in *E. coli* has shown to be limited by the Fe and S content of the medium (Jaganaman, Pinto, Tarasev, & Ballou, 2007; Nakamura, Saeki, & Takahashi, 1999; Yano, Sled, Ohnishi, & Yagi, 1994). In our experience, for the expression of FeS proteins, it is preferable to vary Fe concentration and Fe and S additives (ferric ammonium citrate, ferrous sulfate, ferric citrate, and cysteine).

The total yield of protein may also depend on the density of the expression culture. It could be helpful to grow the expression culture to higher densities (for example, by adding carbon sources such as peptone or yeast extract and supplementing the LB with divalent cations) (Rosano & Ceccarelli, 2014). Other media recipes (TB (terrific broth), 2xYT, or Vogel–Bonner medium) may also prove useful (Lessard, 2013).

A very important parameter for increasing the protein yield is to have the *E. coli* starter culture in an early growth phase. This means that the starter culture should not grow overnight, or if it does, the temperature should be decreased to room temperature.

➔ Screen different Fe concentrations in growth medium
➔ Change growth medium
➔ Use preculture in early growth phase

Lower expression temperatures

To increase the amount of properly folded recombinant hydrogenase, it may be necessary to slow down the rate of protein synthesis. A small-scale screen with 100 mL of expression cultures at different expression temperatures

(4–25°C) can help to find conditions that minimize protein aggregation due to the temperature dependence of hydrophobic interactions and solubility of proteins (Kataeva et al., 2005). However, since reduced synthesis might result in lower protein yields and an effected chaperone network could additionally affect protein folding (Mendoza, Dulin, & Warren, 2000; Strocchi, Ferrer, Timmis, & Golyshin, 2006), lower temperatures may not lead to high quantities of protein (Rosano & Ceccarelli, 2014). To circumvent the negative influences on the chaperonin network, studies suggest the coexpression of cold-adapted chaperonins (ArticExpress, Stratagen) (Ferrer, Chernikova, Timmis, & Golyshin, 2004).

➔ Screen for optimal expression temperature
➔ Use ArticExpress strain to avoid folding problems at lower temperatures

Tuneable expression

In many cases, altered—mostly lower—inducer concentrations have led to improvements in protein expression (Guzman et al., 1995; Terpe, 2006). Nevertheless, in normal BL21 cells, concentration-dependent expression with IPTG is not possible due to the variable number of active Lac permeases in every cell (Rosano & Ceccarelli, 2014). To allow a uniform entry of IPTG, the Tuner (DE3) strain (Novagen) can be used to screen, efficiently, for an optimal inducer concentration.

➔ Lower inducer concentration
➔ For efficient screening, use Tuner (DE3) strain

2.3 Step 3: Purification

Most of the recently purified [FeFe]-hydrogenases are expressed with N- or C-terminal peptide tags (StrepII-tag, poly-His-tag) fused to the protein sequence with a short alanine-serine-linker (Cornish et al., 2015; Engelbrecht et al., 2017; King et al., 2006; Kuchenreuther et al., 2010; Noth et al., 2015). For *Cr*HydA1, the replacement of a Strep-tag II with a hexahistidine-tag led to an increased protein yield (Noth et al., 2015), but since some proteins are very sensitive to imidazole, even when rebuffered very quickly after elution, a C-terminal StrepII-tag might be the best choice to start with, when dealing with "new" hydrogenases.

For successful purifications of hydrogenases from *E. coli* crude extract, some general rules should be followed.

Since protein losses at every step of purification (lysis and ultracentrifugation, binding to column, wash, and concentration) are usually very high, it is advantageous to optimize the expression conditions and screen (in vitro activity assay with crude extract) for conditions in which the amount of active hydrogenase is the highest, before starting to purify.

Avoid any contamination with O_2

To avoid any contact of the hydrogenase of interest with O_2, in addition to working in a glove box under an inert atmosphere (N_2 with 1.5%–3% H_2 with maximal O_2 levels of 10 ppm), all buffers and lab materials should be rendered anaerobic (Fig. 3). For buffers, this means that they have to be sparged (vacuum pump or bubbling with N_2 or Ar for 1 h) and should be kept in the anaerobic chamber for at least 24 h before usage. Additionally, low amounts (1 m*M*) of reducing agent such as sodium dithionite (NaDT) should be added to all buffers. Laboratory materials, especially if made of polypropylene, should be stored in the anaerobic chamber for at least 1 day before usage. Anaerobic chambers should be equipped with palladium catalysts (Coy) to remove O_2 continuously.

Be sure to lyse cells properly

In our experience, low protein yields and even insolubility problems following ultracentrifugation have often turned out to be due to improper lysis of

Fig. 3 Anaerobic chamber (Coy). In order to avoid irreversible damage of the [FeFe]-hydrogenase by molecular oxygen, it is essential that expression, purification, and all following experiments are carried out in an anaerobic atmosphere.

the cells rather than a problem of inclusion body formation. For that reason, the choice of optimal lysis conditions can be of great importance. A combination of different lysis methods might be the safest: the cell pellet should be incubated in lysis buffer (0.1 *M* Tris–HCl pH 8, 0.15 *M* NaCl, 300 μg mL^{-1} lysozyme, 1%–2% Triton X-100, 10%–15% glycerol, if necessary, protease inhibitor) for 20 min, then sonicated in short bursts (three to six times at 30–45 s) while remaining immersed in an ice bath to avoid excessive heating.

Work fast and keep protein cold

Algal hydrogenases are often very temperature sensitive; this is why it is important not to leave the protein at temperatures above 10°C during the purification process. To avoid protein aggregation during purification, it is important to work as fast as possible and prechill buffers and columns before using.

Check if the enzyme is light sensitive

In some cases, the [2Fe2S]-subclusters of [FeFe]-hydrogenases have been shown to be completely destroyed under illumination (Albracht, Roseboom, & Hatchikian, 2006; Roseboom, De Lacey, Fernandez, Hatchikian, & Albracht, 2006). If the purified protein seems to be inactive, it is worth trying to carry out the purification in the dark.

Screen for appropriate storage conditions

For some proteins, it might be necessary to flash freeze them with 5%–25% glycerol in the storage buffer. To ensure that storage conditions do not harm the hydrogenase, it is advisable to maturate the protein (see Section 2.4) directly after purification, measure in vitro activity and screen for optimal storage temperatures (4°C, −20°C, −80°C with or without flash freezing) and buffers (glycerol concentration, NaCl concentration, pH). Highly concentrated protein stock solutions appear to have improved longevity.

2.4 Step 4: H-cluster assembly

In *E. coli*, nonhydrogenase-specific [FeS]-clusters are synthesized by the ISC or SUF system, which means that hydrogenases are expressed in the "apo-form," with the [4Fe4S] cubane cluster, but without the unique [2Fe2S]-subcluster (Fig. 2) (Mulder et al., 2009). Chemically synthesized [2Fe2S]-mimics either bound to the maturase HydF or even in the absence of any maturase protein can be used to maturate apo-hydrogenases in vitro and to build a functional H-cluster (Berggren et al., 2013; Esselborn et al., 2013). The aza-dithiolate [2Fe2S] active site mimic $Fe_2[\mu-(SCH_2)_2NH](CN)_2(CO)_4{}^{2-}$ is easily prepared by condensing $Fe_2(SH)_2(CO)_6$ (Seyferth & Henderson, 1981) with paraformaldehyde in the presence of primary amines (Li & Rauchfuss, 2002).

The so-called in vitro maturation, together with the possibility of chemical reconstitution of the [4Fe4S]-cluster component, opened up entirely new possibilities for hydrogenase research, since scientific curiosity concerning the interplay between protein scaffold and H-cluster would no longer be limited (Kertess, Wittkamp, et al., 2017b; Khanna, Esmieu, Meszaros, Lindblad, & Berggren, 2017; Noth et al., 2016; Siebel et al., 2015).

In vitro maturation leads to successful assembly of active H-cluster not only for *Cp*I and *Cr*HydA1 and various variants of these enzymes (Lampret et al., 2017; Mebs et al., 2018; Winkler et al., 2017) but also for several other [FeFe]-hydrogenases (Engelbrecht et al., 2017; Esselborn et al., 2013).

2.4.1 In vitro maturation

Equipment

- Anaerobic glove box (Coy, see Fig. 3)
- Size-exclusion chromatography column (NAP 5, GE Healthcare)

Buffers and reagents

- 100 m*M* K_2HPO_4/KH_2PO_4 buffer (pH 6.8) with 2 m*M* NaDT
- 100 m*M* Tris–HCl buffer (pH 8) with 2 m*M* NaDT
- [2Fe2S] complex $Fe_2[\mu-(SCH_2)_2NH](CN)_2(CO)_4{}^{2-}$ (Berggren et al., 2013; Esselborn et al., 2013; Li & Rauchfuss, 2002)

Procedure

1. Highly O_2-sensitive chemically synthesized [2Fe2S] active site mimic is incubated at room temperature with "apo"-hydrogenase (0.05–1 m*M*) with a 10-fold molar excess in 500 μL phosphate buffer for 1 h.
2. To remove excess unbound [2Fe2S] complex, size-exclusion chromatography is performed using a column equilibrated with Tris buffer.
3. Protein preparations can be concentrated using 30 kDa Amicon Ultra centrifugal filter units (Millipore).
4. To screen the success of maturation, an in vitro activity assay can be performed (see Section 3.2, Fig. 4).

Nevertheless, it is very common that the standard in vitro maturation protocol for *Cp*I and *Cr*HydA1 (see below) has to be adapted, even when

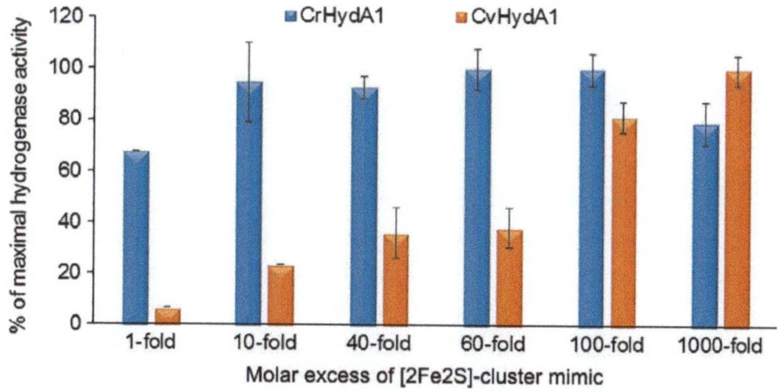

Fig. 4 Percentage of maximal H_2 production rates of *Chlorella variabilis* HydA1 (*Cv*HydA1) and *Chlamydomonas reinhardtii* HydA1 (*Cr*HydA1) maturated with different amounts of [2Fe2S]-subcluster mimic. Activity of *Cv*HydA1 and *Cr*HydA1 was measured with 10 m*M* methyl viologen and 100 m*M* sodium dithionite in 0.1 *M* K_2HPO_4/KH_2PO_4 buffer (pH 6.8) after in vitro maturation for 1 h at room temperature.

expressing single exchange variants of these enzymes but especially when with regard to maturating new [FeFe]-hydrogenases.

Possible adaptations

➔ Use different amounts of [2Fe2S] complex (1–1000-fold molar excess, Fig. 4).

➔ Vary incubation time (15 min to 17 h).

➔ Incubate at different temperatures (4°C, 15°C, room temperature).

➔ Measure *E. coli* crude extract activity: This is carried out by lysing 300 μL of *E. coli* culture in 0.1 *M* potassium phosphate buffer at pH 6.8 (final volume 500 μL) with 1% Triton X-100 (The Dow Chemical Company); incubating on ice for 5 min, agitating mildly, then proceeding as described earlier for purified protein.

3. Methods of characterization

Biophysical methods such as EPR, Mössbauer, Hyperfine sublevel correlation (HYSCORE), resonance Raman, IR, electron nuclear double resonance, nuclear magnetic resonance spectroscopy, and PFE have nicely shed light on the catalytic mechanisms and structures of [FeFe]-hydrogenases (Adamska-Venkatesh et al., 2014, 2015; Armstrong et al., 2016, 2018; Esselborn et al., 2016; Gilbert-Wilson et al., 2015; Kertess, Adamska-Venkatesh, et al., 2017a; Lampret et al., 2017; Meszaros,

Nemeth, Esmieu, Ceccaldi, & Berggren, 2018; Mulder, Guo, Ratzloff, & King, 2017; Mulder et al., 2013; Ratzloff et al., 2018; Rumpel et al., 2015; Rumpel, Sommer, Reijerse, Fares, & Lubitz, 2018).

3.1 IR spectroscopy

FTIR-spectroscopy has proven to be a powerful method in order to monitor intermediate states of the H-cluster of [FeFe]-hydrogenases. Equipping a standard IR-spectrometer (e.g., Bruker Tensor 27) with an attenuated total reflection (ATR) cell offers further possibilities to modify the protein sample to induce steady-state conditions. This method has recently contributed to the elucidation of molecular and mechanistic details during the H_2 conversion process at the H-cluster in numerous publications (Kertess, Adamska-Venkatesh, et al., 2017a; Lampret et al., 2017; Ratzloff et al., 2018; Senger et al., 2018; Winkler et al., 2017) and can further be applied to answer a broad range of general guiding principles in enzyme research.

FTIR-spectroscopy enables the detection of certain states having characteristic CO and CN^- ligand vibrations that are sensitive to the electron configuration of the iron atoms of the [2Fe2S]-subcluster to which they are bound (Lubitz, Reijerse, & van Gastel, 2007). This means that changes of the redox states as well as proton-coupled electron transfer steps can be monitored through specific shifts in the respective ligand vibrations (for wild-type *Cr*HydA1:CN^- stretching bands between 2020 and 2100 cm^{-1}, terminal CO stretching bands in 1850–2020 cm^{-1} range, bridging CO stretching bands at 1750–1850 cm^{-1}) (Adamska-Venkatesh et al., 2014; Lampret et al., 2017; Winkler et al., 2017). In order to accumulate hydrogenase samples in the oxidized (H_{ox}), reduced (H_{red}), super-reduced (H_{sred}), hydride (H_{hyd}), and carbon monoxide-inhibited ($H_{ox\text{-}CO}$, $H_{red\text{-}CO}$) states, samples can be treated before measurement by following the procedures below (Adamska et al., 2012; Chernev et al., 2014; Lambertz et al., 2011; Winkler et al., 2017):

H_{red} state: use buffer with 2 m*M* NaDT

H_{sred} state: add excess NaDT (20 m*M*) or flush for 15–20 min with moistured H_2 gas

H_{ox} state: add thionine (~ 2 m*M*) in small steps until no color change of the thionine (from blue (oxidized) to colorless (reduced)) can be observed

CO-inhibited states: flush for 60 min with moistured carbon monoxide gas

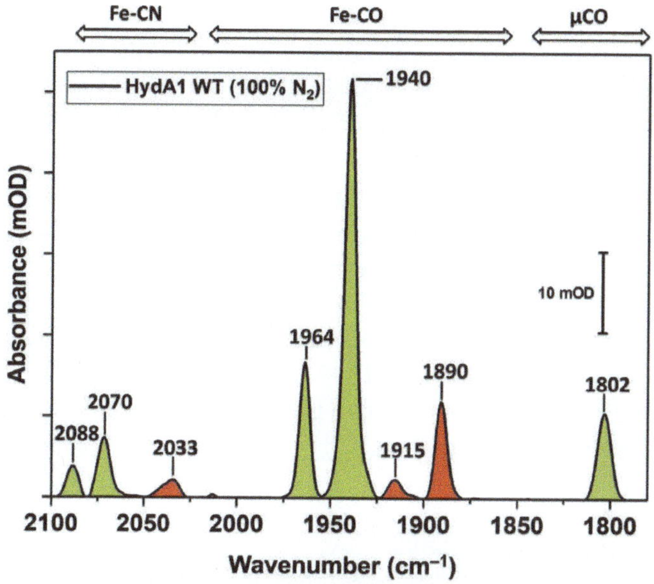

Fig. 5 FTIR-spectrum of *Cr*HydA1 (1 m*M*) under an aerosol of 100% N_2 (10 L min^{-1}) and Tris buffer (pH 8, supplemented with 2 m*M* NaDT) recorded at 30°C. Redox state labels are as follows: H_{ox}: *green*, H_{red}: *red*.

H_{hyd} state: increase substrate (H_2) and product (H^+) pressure during measurement by flushing with H_2 and using low pH buffer

However, if the spectrometer is equipped with an ATR cell, redox state adjustments can be undertaken by gassing the protein sample on the ATR crystal (usually a ZnSe crystal) with either dry gasses (H_2, N_2, O_2, and CO) or aerosols by humidifying a buffer before gassing the cell. The H_{ox} state accumulates upon constantly flushing with N_2 whereas the reduced states become dominant upon flushing with H_2. A schematic view of this setup is displayed in Fig. 5.

Equipment

Spectrometer (e.g., Tensor 27, Bruker GmbH, with mid-IR globar, silicon ATR crystal, and MCT (mercuric cadmium telluride) detector in an anaerobic glove box).

Buffers and reagents

N_2 (5.0), H_2, CO

Liquid N_2 (required to cool the MCT detector)

Procedure

1. The crystal surface is cleaned with isopropanol and MQ-H_2O.
2. The detector is precooled with liquid N_2 before usage.

3. The temperature is set to 30°C.
4. A background sample is acquired with the buffer that the sample is stored in.
5. Hydrogenase solution (15 μL of 0.5–1 m*M*, in 0.1 *M* TPW buffer with 2 m*M* NaDT) is applied on the crystal.
6. The cell is closed with a gas-tight lid.
7. The samples are dried with N_2 (the flow rate may range between 1 and 10 L min^{-1}) to obtain a stable protein film.
8. Formation of the protein film is checked by monitoring gradually decreasing water bands (3300 cm^{-1}) and increasing amide bands (1645 cm^{-1} for Amide I, c. 1540 cm^{-1} for Amide II).
9. As required, the gas and sample buffer compositions can be varied.
10. Data acquisition is initiated.

3.2 In vitro activity assays

Despite enzyme characterization often requiring costly equipment, low-cost in vitro assays are available which make the successful investigation of new [FeFe]-hydrogenases possible, even if the devices for the above-mentioned, highly expensive techniques are unavailable. To determine the kinetic parameters of enzymatic H_2 production (Winkler, Kuhlgert, Hippler, & Happe, 2009), O_2 sensitivity (Noth et al., 2015), and pH dependency of a novel hydrogenase (Happe & Naber, 1993) or to determine H_2 production activity with potential native electron partners (Engelbrecht et al., 2017) adapted protocols of an in vitro H_2 production assay with methyl viologen as an electron mediator and NaDT as electron donor can be employed.

3.2.1 In vitro activity assay H_2 production

Equipment

- Anaerobic glove box (Coy)
- 8 mL vessels with suba-seal septum stoppers (Sigma-Aldrich)
- Argon gas supply system
- Water bath
- GC with thermal conductivity detector and argon as carrier gas (e.g., Shimadzu, Fig. 5)

Buffers and reagents

Reaction buffer: 100 m*M* K_2HPO_4/KH_2PO_4 buffer (pH 6.8) with 100–200 m*M* NaDT and 10 m*M* methyl viologen.

Procedure

1. Stock solutions of 1 *M* NaDT and 1 *M* methyl viologen with anaerobic H_2O are prepared.
2. Samples (2 mL) of reaction buffer in 8 mL suba vessels are prepared.
3. Hydrogenase sample solution (8 n*M*) is added.
4. Vessels are sealed.
5. Headspaces are flushed with inert gas (Ar) for 5 min.
6. Samples are incubated for 20 min at 37°C with shaking.
7. The reaction is stopped by transferring the vessel to ice.
8. A headspace sample is injected into the GC (Fig. 6).
9. The H_2 production rate is determined.

Possible adaptations

➔ Use different buffers (200–100 m*M*, MES-NaOH for pH 5.5–6.7, MOPS-NaOH for pH 6.2–7.6, Tris–HCl for pH 7.4–8.8).

➔ Use different methyl viologen concentrations to determine Michaelis–Menten kinetics.

➔ To further test the capability of the hydrogenase to oxidize H_2, a colorimetric assay with benzyl viologen (ε_{600}: 10 mM^{-1} cm^{-1}) as electron acceptor in a sealed cuvette under an atmosphere of 1 bar H_2 with enzyme amounts between 1 ng and 1 μg can be performed.

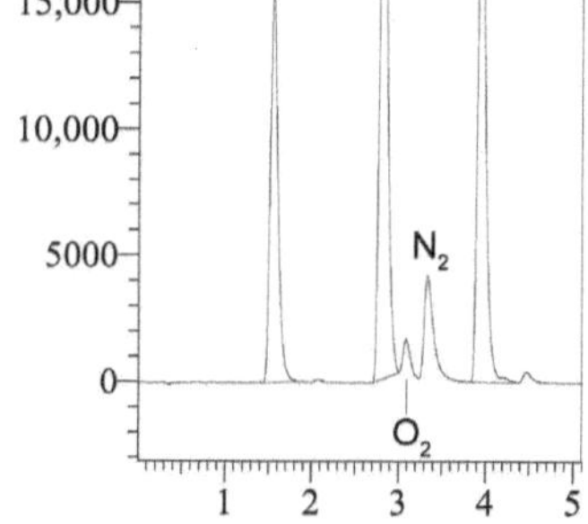

Fig. 6 Example output of a GC measurement (Shimadzu GC2025, Argon as carrier gas) of the headspace gas from suba vessels after an in vitro H_2 production assay. The *first peak* corresponds to H, in case of contamination with, e.g., room air during extraction or injection, O_2 (*second peak*) and N_2 (*third peak*) peaks can be detected.

Measurements with native interaction partners

In *C. reinhardtii*, and probably in most of the other known microalgae capable of photobiological hydrogen production, hydrogenases receive their electrons from photosystem I (PSI) via the plant-type ferredoxin PetF (Engelbrecht et al., 2017; Florin et al., 2001; Terauchi et al., 2009; Ueno et al., 1999). To determine, for example, Michaelis–Menten kinetics with the native electron partner, an adapted in vitro assay can be performed.

1. These assays use a previously described protocol (Jacobs et al., 2009) for the expression of algal ferredoxins in *E. coli*.
2. Protein concentration and [2Fe2S]-cluster properties of the oxidized ferredoxin are determined from the absorption spectrum (maxima at 276, 330, 420, and 460 nm expected; extinction coefficient of 8.38 mM^{-1} -cm^{-1} at 420 nm; Galván & Márquez, 1985).
3. A concentration range of 10–100 µM of purified ferredoxin in 100 mM K_2HPO_4/KH_2PO_4 buffer (pH 6.8) is used with 25 mM NaDT and 50–200 ng hydrogenase in 200 µL assays and 2 mL suba vessels.
4. The procedure as described earlier is followed for the standard in vitro assay, using prolonged incubation times (30–60 min) if necessary rather than higher amounts of hydrogenase.
5. Checks are made for background activity of the hydrogenase of interest, using NaDT.

4. Conclusion and outlook

The possibility for exploiting algal photobiological H_2 production for commercial use is of great interest regarding the current need for alternative fuels and sustainable energy. Apart from the research on the hydrogenases from the green algae *C. reinhardtii*, there is a wide range of other algal species which possess hydrogenases, and which might be better suited for whole-cell or molecular H_2 producing systems. The latter area, in particular, has been advancing fast, because apart from circumventing limitations of algal H_2 production with different genetic modification approaches (Dubini & Ghirardi, 2015), an alternative solution for sunlight-dependent proton reduction is to extract important molecular key players of algal H_2 production and to exploit them in photocatalytic systems. The insertion of O_2-insensitive metal catalysts into the reducing side of PSI is one strategy for a H_2 evolving in vitro system, but it relies on rare and therefore expensive noble metals (Evans, O'Neill, Hutchens, Bruce, & Greenbaum, 2004; Grimme, Lubner, & Golbeck, 2009; Iwuchukwu et al., 2010; Utschig, Soltau, & Tiede, 2015).

Hydrogenases can be used as catalysts by simply combining independently purified protein components (PSI, hydrogenase, and ferredoxin (wild-type enzymes or protein variants) from *C. reinhardtii*, Rumpel et al., 2014; Wiegand et al., 2018; Winkler et al., 2009), or, to avoid limitations that originate from diffusion, by fusing the necessary molecules. Hydrogen production rates comparable to the metal-based systems were achieved by linking cytochrome *c*6 and the [FeFe]-hydrogenase from *C. acetobutylicum* to PSI from *Synechococcus* sp. *PCC 7002* (Lubner et al., 2011).

To avoid the competition between hydrogenases and ferredoxin: $NADP^+$-oxidoreductase (FNR) a Fdx-HydA-fusion protein was developed to influence the bias of electron transfer from PSI in favor of hydrogenase, with more than 60% of the photosynthetic electrons being diverted to H_2 production in the presence of FNR (Eilenberg et al., 2016; Yacoby et al., 2011).

To overcome the lability of the large, and therefore costly protein complex PSI as photosensitizer, recent approaches used artificial photosensitizers such as proflavine or eosin in combination with *Cr*HydA1 to optimize hydrogenase-based in vitro systems with 40–70 times higher maximal turnover frequency numbers compared to the ones previously reported (Adam et al., 2017).

To improve these hydrogenase-based in vitro systems for biotechnological H_2 production, an attractive option is to expand the enzymatic toolbox by characterizing and screening for undiscovered features of novel algal hydrogenases. To achieve this aim, the research on yet-unknown algal hydrogenases has to be initiated by developing practical expression and purification protocols as described in this chapter.

Acknowledgments

The authors thank Alexander Günzel and Oliver Lampret for helpful discussions. The authors are also grateful to Dr. Anja Hemschemeier and Dr. Anne Sawyer for critical reading of the chapter. This work was supported by the Volkswagen Stiftung (Design of [FeS]-cluster containing Metallo-DNAzymes, Az 93412), the Deutsche Forschungsgemeinschaft (GRK 2341: Microbial Substrate Conversion), and the Studienstiftung des deutschen Volkes (V.E.).

References

Abendroth, G., Stripp, S., Silakov, A., Croux, C., Soucaille, P., Girbal, L., et al. (2008). Optimized over-expression of [FeFe] hydrogenases with high specific activity in Clostridium acetobutylicum. *International Journal of Hydrogen Energy*, *33*, 6076–6081.

Adam, D., Bösche, L., Castaneda-Losada, L., Winkler, M., Apfel, U. P., & Happe, T. (2017). Sunlight-dependent hydrogen production by photosensitizer/hydrogenase systems. *ChemSusChem*, *10*(5), 894–902.

Adamska, A., Silakov, A., Lambertz, C., Rüdiger, O., Happe, T., Reijerse, E., et al. (2012). Identification and characterization of the "super-reduced" state of the H-cluster in [FeFe] hydrogenase: A new building block for the catalytic cycle? *Angewandte Chemie (International Ed. in English)*, *51*(46), 11458–11462.

Adamska-Venkatesh, A., Krawietz, D., Siebel, J., Weber, K., Happe, T., Reijerse, E., et al. (2014). New redox states observed in [FeFe] hydrogenases reveal redox coupling within the H-cluster. *Journal of the American Chemical Society*, *136*(32), 11339–11346.

Adamska-Venkatesh, A., Simmons, T. R., Siebel, J. F., Artero, V., Fontecave, M., Reijerse, E., et al. (2015). Artificially maturated [FeFe] hydrogenase from *Chlamydomonas reinhardtii*: A HYSCORE and ENDOR study of a non-natural H-cluster. *Physical Chemistry Chemical Physics*, *17*(7), 5421–5430.

Akhtar, M. K., & Jones, P. R. (2008). Deletion of iscR stimulates recombinant clostridial Fe–Fe hydrogenase activity and H2-accumulation in *Escherichia coli* BL21(DE3). *Applied Microbiology and Biotechnology*, *78*(5), 853–862.

Albracht, S. P., Roseboom, W., & Hatchikian, E. C. (2006). The active site of the [FeFe]-hydrogenase from *Desulfovibrio desulfuricans*. I. Light sensitivity and magnetic hyperfine interactions as observed by electron paramagnetic resonance. *Journal of Biological Inorganic Chemistry*, *11*(1), 88–101.

Armstrong, F. A., Evans, R. M., Hexter, S. V., Murphy, B. J., Roessler, M. M., & Wulff, P. (2016). Guiding principles of hydrogenase catalysis instigated and clarified by protein film electrochemistry. *Accounts of Chemical Research*, *49*(5), 884–892.

Armstrong, F. A., Evans, R. M., & Megarity, C. F. (2018). Protein film electrochemistry of iron–sulfur enzymes. *Methods in Enzymology*, *599*, 387–407.

Arriola, M. B., Velmurugan, N., Zhang, Y., Plunkett, M. H., Hondzo, H., & Barney, B. M. (2018). Genome sequences of chlorella sorokiniana UTEX 1602 and micractinium conductrix SAG 241.80: Implications to maltose excretion by a green alga. *The Plant Journal*, *93*(3), 566–586.

Artz, J. H., Mulder, D. W., Ratzloff, M. W., Lubner, C. E., Zadvornyy, O. A., LeVan, A. X., et al. (2017). Reduction potentials of [FeFe]-hydrogenase accessory iron–sulfur clusters provide insights into the energetics of proton reduction catalysis. *Journal of the American Chemical Society*, *139*(28), 9544–9550.

Berggren, G., Adamska, A., Lambertz, C., Simmons, T. R., Esselborn, J., Atta, M., et al. (2013). Biomimetic assembly and activation of [FeFe]-hydrogenases. *Nature*, *499*(7456), 66–69.

Blanc, G., Duncan, G., Agarkova, I., Borodovsky, M., Gurnon, J., Kuo, A., et al. (2010). The *Chlorella variabilis* NC64A genome reveals adaptation to photosymbiosis, coevolution with viruses, and cryptic sex. *Plant Cell*, *22*(9), 2943–2955.

Chernev, P., Lambertz, C., Brunje, A., Leidel, N., Sigfridsson, K. G., Kositzki, R., et al. (2014). Hydride binding to the active site of [FeFe]-hydrogenase. *Inorganic Chemistry*, *53*(22), 12164–12177.

Chin, J. X., Chung, B. K., & Lee, D. Y. (2014). Codon Optimization OnLine (COOL): A web-based multi-objective optimization platform for synthetic gene design. *Bioinformatics*, *30*(15), 2210–2212.

Cornish, A. J., Green, R., Gärtner, K., Mason, S., & Hegg, E. L. (2015). Characterization of hydrogen metabolism in the multicellular green alga volvox carteri. *PLoS One*, *10*(4), e0125324.

de Marco, A., Vigh, L., Diamant, S., & Goloubinoff, P. (2005). Native folding of aggregation-prone recombinant proteins in *Escherichia coli* by osmolytes, plasmid- or benzyl alcohol-overexpressed molecular chaperones. *Cell Stress & Chaperones*, *10*(4), 329–339.

Derman, A. I., Prinz, W. A., Belin, D., & Beckwith, J. (1993). Mutations that allow disulfide bond formation in the cytoplasm of *Escherichia coli*. *Science*, *262*(5140), 1744–1747.

Dubini, A., & Ghirardi, M. L. (2015). Engineering photosynthetic organisms for the production of biohydrogen. *Photosynthesis Research*, *123*(3), 241–253.

Eilenberg, H., Weiner, I., Ben-Zvi, O., Pundak, C., Marmari, A., Liran, O., et al. (2016). The dual effect of a ferredoxin-hydrogenase fusion protein in vivo: successful divergence of the photosynthetic electron flux towards hydrogen production and elevated oxygen tolerance. *Biotechnology for Biofuels*, *9*, 182.

Engelbrecht, V., Rodríguez-Maciá, P., Esselborn, J., Sawyer, A., Hemschemeier, A., Rüdiger, O., et al. (2017). The structurally unique photosynthetic *Chlorella variabilis* NC64A hydrogenase does not interact with plant-type ferredoxins. *Biochimica et Biophysica Acta*, *1858*(9), 771–778.

Esselborn, J., Lambertz, C., Adamska-Venkatesh, A., Simmons, T., Berggren, G., Noth, J., et al. (2013). Spontaneous activation of [FeFe]-hydrogenases by an inorganic [2Fe] active site mimic. *Nature Chemical Biology*, *9*(10), 607–609.

Esselborn, J., Muraki, N., Klein, K., Engelbrecht, V., Metzler-Nolte, N., Apfel, U. P., et al. (2016). A structural view of synthetic cofactor integration into [FeFe]-hydrogenases. *Chemical Science*, *2*, 959–968.

Evans, B. R., O'Neill, H. M., Hutchens, S. A., Bruce, B. D., & Greenbaum, E. (2004). Enhanced photocatalytic hydrogen evolution by covalent attachment of plastocyanin to photosystem I. *Nano Letters*, *4*(10), 1815–1819.

Ferrer, M., Chernikova, T. N., Timmis, K. N., & Golyshin, P. N. (2004). Expression of a temperature-sensitive esterase in a novel chaperone-based Escherichia coli strain. *Applied and Environmental Microbiology*, *70*(8), 4499–4504.

Florin, L., Tsokoglou, A., & Happe, T. (2001). A novel type of iron hydrogenase in the green alga *Scenedesmus obliquus* is linked to the photosynthetic electron transport chain. *Journal of Biological Chemistry*, *276*(9), 6125–6132.

Gaffron, H. (1939). Reduction of carbon dioxide with molecular hydrogen in green algæ. *Nature*, *143*, 204.

Gaffron, H. (1942). The effect of specific poisons upon the photoreduction with hydrogen in green algae. *The Journal of General Physiology*, *26*(2), 195–217.

Gaffron, H., & Rubin, J. (1942). Fermentative and photochemical production of hydrogen in algae. *The Journal of General Physiology*, *26*(2), 219.

Galván, F., & Márquez, A. J. (1985). Physicochemical properties of ferredoxin from *Chlamydomonas reinhardii*. *Zeitschrift für Naturforschung*, *40*(c), 373–378.

Gauquelin, C., Baffert, C., Richaud, P., Kamionka, E., Etienne, E., Guieysse, D., et al. (2018). Roles of the F-domain in [FeFe] hydrogenase. *Biochimica et Biophysica Acta*, *1859*(2), 69–77.

Ghirardi, M. L., Zhang, L., Lee, J. W., Flynn, T., Seibert, M., Greenbaum, E., et al. (2000). Microalgae: A green source of renewable H(2). *Trends in Biotechnology*, *18*(12), 506–511.

Gilbert-Wilson, R., Siebel, J. F., Adamska-Venkatesh, A., Pham, C. C., Reijerse, E., Wang, H., et al. (2015). Spectroscopic investigations of [FeFe] hydrogenase maturated with [(57)Fe2(adt)(CN)2(CO)4](2-). *Journal of the American Chemical Society*, *137*(28), 8998–9005.

Girbal, L., von Abendroth, G., Winkler, M., Benton, P. M., Meynial-Salles, I., Croux, C., et al. (2005). Homologous and heterologous overexpression in *Clostridium acetobutylicum* and characterization of purified clostridial and algal Fe-only hydrogenases with high specific activities. *Applied and Environmental Microbiology, 71*(5), 2777–2781.

Gorman, D. S., & Levine, R. P. (1965). Cytochrome f and plastocyanin: Their sequence in the photosynthetic electron transport chain of Chlamydomonas reinhardi. *Proceedings of the National Academy of Sciences of the United States of America, 54*(6), 1665–1669.

Gorwa, M. F., Croux, C., & Soucaille, P. (1996). Molecular characterization and transcriptional analysis of the putative hydrogenase gene of *Clostridium acetobutylicum* ATCC 824. *Journal of Bacteriology, 178*(9), 2668–2675.

Graumann, K., & Premstaller, A. (2006). Manufacturing of recombinant therapeutic proteins in microbial systems. *Biotechnology Journal, 1*(2), 164–186.

Grimme, R. A., Lubner, C. E., & Golbeck, J. H. (2009). Maximizing H_2 production in photosystem I/dithiol molecular wire/platinum nanoparticle bioconjugates. *Dalton Transactions, 45*(45), 10106–10113.

Grote, A., Hiller, K., Scheer, M., Münch, R., Nörtemann, B., Hempel, D. C., et al. (2005). JCat: A novel tool to adapt codon usage of a target gene to its potential expression host. *Nucleic Acids Research, 33*(Web Server issue), W526–W531.

Guillard, R. R., & Ryther, J. H. (1962). Studies of marine planktonic diatoms. I. *Cyclotella nana* Hustedt, and *Detonula confervacea* (cleve) Gran. *Canadian Journal of Microbiology, 8*, 229–239.

Guzman, L. M., Belin, D., Carson, M. J., & Beckwith, J. (1995). Tight regulation, modulation, and high-level expression by vectors containing the arabinose PBAD promoter. *Journal of Bacteriology, 177*(14), 4121–4130.

Happe, T., & Kaminski, A. (2002). Differential regulation of the Fe-hydrogenase during anaerobic adaptation in the green alga *Chlamydomonas reinhardtii. European Journal of Biochemistry, 269*(3), 1022–1032.

Happe, T., & Naber, J. D. (1993). Isolation, characterization and N-terminal amino acid sequence of hydrogenase from the green alga *Chlamydomonas reinhardtii. European Journal of Biochemistry, 214*(2), 475–481.

Hemschemeier, A., & Happe, T. (2011). Alternative photosynthetic electron transport pathways during anaerobiosis in the green alga *Chlamydomonas reinhardtii. Biochimica et Biophysica Acta—Bioenergetics, 1807*(8), 919–926.

Hemschemeier, A., Melis, A., & Happe, T. (2009). Analytical approaches to photobiological hydrogen production in unicellular green algae. *Photosynthesis Research, 102*(2–3), 523–540.

Iwuchukwu, I. J., Vaughn, M., Myers, N., O'Neill, H., Frymier, P., & Bruce, B. D. (2010). Self-organized photosynthetic nanoparticle for cell-free hydrogen production. *Nature Nanotechnology, 5*(1), 73–79.

Jacobs, J., Pudollek, S., Hemschemeier, A., & Happe, T. (2009). A novel, anaerobically induced ferredoxin in *Chlamydomonas reinhardtii. FEBS Letters, 583*(2), 325–329.

Jaganaman, S., Pinto, A., Tarasev, M., & Ballou, D. P. (2007). High levels of expression of the iron–sulfur proteins phthalate dioxygenase and phthalate dioxygenase reductase in *Escherichia coli. Protein Expression and Purification, 52*(2), 273–279.

Kamp, C., Silakov, A., Winkler, M., Reijerse, E. J., Lubitz, W., & Happe, T. (2008). Isolation and first EPR characterization of the [FeFe]-hydrogenases from green algae. *Biochimica et Biophysica Acta, 1777*(5), 410–416.

Kataeva, I., Chang, J., Xu, H., Luan, C. H., Zhou, J., Uversky, V. N., et al. (2005). Improving solubility of *Shewanella oneidensis* MR-1 and *Clostridium thermocellum* JW-20 proteins expressed into *Esherichia coli. Journal of Proteome Research, 4*(6), 1942–1951.

Kertess, L., Adamska-Venkatesh, A., Rodriguez-Macia, P., Rudiger, O., Lubitz, W., & Happe, T. (2017a). Influence of the [4Fe-4S] cluster coordinating cysteines on active site maturation and catalytic properties of *C. reinhardtii* [FeFe]-hydrogenase. *Chemical Science*, *8*(12), 8127–8137.

Kertess, L., Wittkamp, F., Sommer, C., Esselborn, J., Rudiger, O., Reijerse, E. J., et al. (2017b). Chalcogenide substitution in the [2Fe] cluster of [FeFe]-hydrogenases conserves high enzymatic activity. *Dalton Transactions*, *46*(48), 16947–16958.

Khanna, N., Esmieu, C., Meszaros, L. S., Lindblad, P., & Berggren, G. (2017). In vivo activation of an [FeFe] hydrogenase using synthetic cofactors. *Energy & Environmental Science*, *10*(7), 1563–1567.

King, P. W., Posewitz, M. C., Ghirardi, M. L., & Seibert, M. (2006). Functional studies of [FeFe] hydrogenase maturation in an *Escherichia coli* biosynthetic system. *Journal of Bacteriology*, *188*(6), 2163–2172.

Kuchenreuther, J. M., Grady-Smith, C. S., Bingham, A. S., George, S. J., Cramer, S. P., & Swartz, J. R. (2010). High-yield expression of heterologous [FeFe] hydrogenases in *Escherichia coli*. *PLoS One*, *5*(11), e15491.

Lambertz, C., Leidel, N., Havelius, K. G., Noth, J., Chernev, P., Winkler, M., et al. (2011). O_2 reactions at the six-iron active site (H-cluster) in [FeFe]-hydrogenase. *Journal of Biological Chemistry*, *286*(47), 40614–40623.

Lampret, O., Adamska-Venkatesh, A., Konegger, H., Wittkamp, F., Apfel, U. P., Reijerse, E. J., et al. (2017). Interplay between CN(−) ligands and the secondary coordination sphere of the H-cluster in [FeFe]-hydrogenases. *Journal of the American Chemical Society*, *139*(50), 18222–18230.

Lessard, J. C. (2013). Chapter eleven—Growth media for *E. coli*. In J. Lorsch (Ed.), *Methods in enzymology*, 533 (pp. 181–189). Academic Press.

Li, H., & Rauchfuss, T. B. (2002). Iron carbonyl sulfides, formaldehyde, and amines condense to give the proposed azadithiolate cofactor of the Fe-only hydrogenases. *Journal of the American Chemical Society*, *124*(5), 726–727.

Liu, X., Deng, R., Wang, J., & Wang, X. (2014). COStar: A D-star lite-based dynamic search algorithm for codon optimization. *Journal of Theoretical Biology*, *344*, 19–30.

Lorimer, D., Raymond, A., Walchli, J., Mixon, M., Barrow, A., Wallace, E., et al. (2009). Gene composer: Database software for protein construct design, codon engineering, and gene synthesis. *BMC Biotechnology*, *9*, 36.

Lubitz, W., Ogata, H., Rüdiger, O., & Reijerse, E. (2014). Hydrogenases. *Chemical Reviews*, *114*(8), 4081–4148.

Lubitz, W., Reijerse, E., & van Gastel, M. (2007). [nife] and [fefe] hydrogenases studied by advanced magnetic resonance techniques. *Chemical Reviews*, *107*(10), 4331–4365.

Lubner, C. E., Applegate, A. M., Knorzer, P., Ganago, A., Bryant, D. A., Happe, T., et al. (2011). Solar hydrogen-producing bionanodevice outperforms natural photosynthesis. *Proceedings of the National Academy of Sciences of the United States of America*, *108*(52), 20988–20991.

Martin, B. A., & Frymier, P. D. (2017). A review of hydrogen production by photosynthetic organisms using whole-cell and cell-free systems. *Applied Biochemistry and Biotechnology*, *183*(2), 503–519.

Mebs, S., Kositzki, R., Duan, J., Kertess, L., Senger, M., Wittkamp, F., et al. (2018). Hydrogen and oxygen trapping at the H-cluster of [FeFe]-hydrogenase revealed by site-selective spectroscopy and QM/MM calculations. *Biochimica et Biophysica Acta*, *1859*(1), 28–41.

Melis, A., Zhang, L., Forestier, M., Ghirardi, M. L., & Seibert, M. (2000). Sustained photobiological hydrogen gas production upon reversible inactivation of oxygen evolution in the green alga *Chlamydomonas reinhardtii*. *Plant Physiology*, *122*(1), 127–136.

Mendoza, J. A., Dulin, P., & Warren, T. (2000). The lower hydrolysis of ATP by the stress protein GroEL is a major factor responsible for the diminished chaperonin activity at low temperature. *Cryobiology*, *41*(4), 319–323.

Merchant, S. S., Allen, M. D., Kropat, J., Moseley, J. L., Long, J. C., Tottey, S., et al. (2006). Between a rock and a hard place: Trace element nutrition in chlamydomonas. *Biochimica et Biophysica Acta, 1763*(7), 578–594.

Meszaros, L. S., Nemeth, B., Esmieu, C., Ceccaldi, P., & Berggren, G. (2018). In vivo EPR characterization of semi-synthetic [FeFe] hydrogenases. *Angewandte Chemie (International Ed. in English), 57*(10), 2596–2599.

Meuser, J. E., Boyd, E. S., Ananyev, G., Karns, D., Radakovits, R., Narayana Murthy, U. M., et al. (2011). Evolutionary significance of an algal gene encoding an [FeFe]-hydrogenase with F-domain homology and hydrogenase activity in *Chlorella variabilis* NC64A. *Planta, 234*(4), 829–843.

Mulder, D. W., Guo, Y., Ratzloff, M. W., & King, P. W. (2017). Identification of a catalytic iron-hydride at the H-cluster of [FeFe]-hydrogenase. *Journal of the American Chemical Society, 139*(1), 83–86.

Mulder, D. W., Ortillo, D. O., Gardenghi, D. J., Naumov, A. V., Ruebush, S. S., Szilagyi, R. K., et al. (2009). Activation of HydA(DeltaEFG) requires a preformed [4Fe-4S] cluster. *Biochemistry, 48*(26), 6240–6248.

Mulder, D. W., Ratzloff, M. W., Shepard, E. M., Byer, A. S., Noone, S. M., Peters, J. W., et al. (2013). EPR and FTIR analysis of the mechanism of H_2 activation by [FeFe]-hydrogenase HydA1 from *Chlamydomonas reinhardtii*. *Journal of the American Chemical Society, 135*(18), 6921–6929.

Nakamura, M., Saeki, K., & Takahashi, Y. (1999). Hyperproduction of recombinant ferredoxins in escherichia coli by coexpression of the ORF1-ORF2-iscS-iscU-iscA-hscB-hs cA-fdx-ORF3 gene cluster. *Journal of Biochemistry, 126*(1), 10–18.

Nelson, D. R., Khraiwesh, B., Fu, W., Alseekh, S., Jaiswal, A., Chaiboonchoe, A., et al. (2017). The genome and phenome of the green alga Chloroidium sp. UTEX 3007 reveal adaptive traits for desert acclimatization. *eLife, 6*, pii: e25783.

Noth, J., Esselborn, J., Guldenhaupt, J., Brunje, A., Sawyer, A., Apfel, U. P., et al. (2016). [FeFe]-hydrogenase with chalcogenide substitutions at the H-cluster maintains full H_2 evolution activity. *Angewandte Chemie (International Ed. in English), 55*(29), 8396–8400.

Noth, J., Kositzki, R., Klein, K., Winkler, M., Haumann, M., & Happe, T. (2015). Lyophilization protects [FeFe]-hydrogenases against O_2-induced H-cluster degradation. *Scientific Reports, 5*, 13978.

Oey, M., Sawyer, A. L., Ross, I. L., & Hankamer, B. (2016). Challenges and opportunities for hydrogen production from microalgae. *Plant Biotechnology Journal, 14*, 1487–1499.

Peters, J. W., Lanzilotta, W. N., Lemon, B. J., & Seefeldt, L. C. (1998). X-ray crystal structure of the Fe-only hydrogenase (*CpI*) from *Clostridium pasteurianum* to 1.8 angstrom resolution. *Science, 282*(5395), 1853–1858.

Peters, J. W., Schut, G. J., Boyd, E. S., Mulder, D. W., Shepard, E. M., Broderick, J. B., et al. (2015). [FeFe]- and [NiFe]-hydrogenase diversity, mechanism, and maturation. *Biochimica et Biophysica Acta, 1853*(6), 1350–1369.

Posewitz, M. C., King, P. W., Smolinski, S. L., Zhang, L., Seibert, M., & Ghirardi, M. L. (2004). Discovery of two novel radical S-adenosylmethionine proteins required for the assembly of an active [Fe] hydrogenase. *The Journal of Biological Chemistry, 279*(24), 25711–25720.

Prochnik, S. E., Umen, J., Nedelcu, A. M., Hallmann, A., Miller, S. M., Nishii, I., et al. (2010). Genomic analysis of organismal complexity in the multicellular green alga *Volvox carteri*. *Science, 329*(5988), 223–226.

Puigbò, P., Guzmán, E., Romeu, A., & Garcia-Vallvé, S. (2007). OPTIMIZER: A web server for optimizing the codon usage of DNA sequences. *Nucleic Acids Research, 35*(Web Server issue), W126–W131.

Ratzloff, M. W., Artz, J. H., Mulder, D. W., Collins, R. T., Furtak, T. E., & King, P. W. (2018). CO-bridged H-cluster intermediates in the catalytic mechanism of [FeFe]-hydrogenase CaI. *Journal of the American Chemical Society, 140*, 7623–7628.

Rosano, G. L., & Ceccarelli, E. A. (2014). Recombinant protein expression in *Escherichia coli*: Advances and challenges. *Frontiers in Microbiology*, *5*, 172.

Roseboom, W., De Lacey, A. L., Fernandez, V. M., Hatchikian, E. C., & Albracht, S. P. J. (2006). The active site of the [FeFe]-hydrogenase from *Desulfovibrio desulfuricans*. II. Redox properties, light sensitivity and CO-ligand exchange as observed by infrared spectroscopy. *Journal of Biological Inorganic Chemistry*, *11*(1), 102–118.

Rumpel, S., Siebel, J. F., Diallo, M., Fares, C., Reijerse, E. J., & Lubitz, W. (2015). Structural insight into the complex of ferredoxin and [FeFe] hydrogenase from *Chlamydomonas reinhardtii*. *ChemBioChem*, *16*(11), 1663–1669.

Rumpel, S., Siebel, J. F., Fares, C., Duan, J., Reijerse, E., Happe, T., et al. (2014). Enhancing hydrogen production of microalgae by redirecting electrons from photosystem I to hydrogenase. *Energy & Environmental Science*, 7(10), 3296–3301.

Rumpel, S., Sommer, C., Reijerse, E., Fares, C., & Lubitz, W. (2018). Direct detection of the terminal hydride intermediate in [FeFe] hydrogenase by NMR spectroscopy. *Journal of the American Chemical Society*, *140*(11), 3863–3866.

Senger, M., Mebs, S., Duan, J., Shulenina, O., Laun, K., Kertess, L., et al. (2018). Protonation/reduction dynamics at the [4Fe-4S] cluster of the hydrogen-forming cofactor in [FeFe]-hydrogenases. *Physical Chemistry Chemical Physics*, *20*(5), 3128–3140.

Seyferth, D., & Henderson, R. S. (1981). Di-μ-thiolbis(tricarbonyliron),(μ-HS)2 Fe2(CO)6: An inorganic mimic of organic thiols. *Journal of Organometallic Chemistry*, *218*(2), C34–C36.

Siebel, J. F., Adamska-Venkatesh, A., Weber, K., Rumpel, S., Reijerse, E., & Lubitz, W. (2015). Hybrid [FeFe]-hydrogenases with modified active sites show remarkable residual enzymatic activity. *Biochemistry*, *54*(7), 1474–1483.

Sivashanmugam, A., Murray, V., Cui, C., Zhang, Y., Wang, J., & Li, Q. (2009). Practical protocols for production of very high yields of recombinant proteins using *Escherichia coli*. *Protein Science*, *18*(5), 936–948.

Strocchi, M., Ferrer, M., Timmis, K. N., & Golyshin, P. N. (2006). Low temperature-induced systems failure in Escherichia coli: Insights from rescue by cold-adapted chaperones. *Proteomics*, *6*(1), 193–206.

Tard, C., Liu, X., Ibrahim, S. K., Bruschi, M., De Gioia, L., Davies, S. C., et al. (2005). Synthesis of the H-cluster framework of iron-only hydrogenase. *Nature*, *433*(7026), 610–613.

Terauchi, A. M., Lu, S. F., Zaffagnini, M., Tappa, S., Hirasawa, M., Tripathy, J. N., et al. (2009). Pattern of expression and substrate specificity of chloroplast ferredoxins from *Chlamydomonas reinhardtii*. *Journal of Biological Chemistry*, *284*(38), 25867–25878.

Terpe, K. (2006). Overview of bacterial expression systems for heterologous protein production: From molecular and biochemical fundamentals to commercial systems. *Applied Microbiology and Biotechnology*, 72(2), 211–222.

Tian, J., Yan, Y., Yue, Q., Liu, X., Chu, X., Wu, N., et al. (2017). Predicting synonymous codon usage and optimizing the heterologous gene for expression in *E. coli*. *Scientific Reports*, 7(1), 9926.

Timmins, M., Thomas-Hall, S. R., Darling, A., Zhang, E., Hankamer, B., Marx, U. C., et al. (2009). Phylogenetic and molecular analysis of hydrogen-producing green algae. *Journal of Experimental Botany*, *60*(6), 1691–1702.

Ueno, Y., Kurano, N., & Miyachi, S. (1999). Purification and characterization of hydrogenase from the marine green alga, *Chlorococcum littorale*. *FEBS Letters*, *443*(2), 144–148.

Utschig, L. M., Soltau, S. R., & Tiede, D. M. (2015). Light-driven hydrogen production from Photosystem I-catalyst hybrids. *Current Opinion in Chemical Biology*, *25*, 1–8.

Villalobos, A., Ness, J. E., Gustafsson, C., Minshull, J., & Govindarajan, S. (2006). Gene designer: A synthetic biology tool for constructing artificial DNA segments. *BMC Bioinformatics*, 7, 285.

Wiegand, K., Winkler, M., Rumpel, S., Kannchen, D., Rexroth, S., Hase, T., et al. (2018). Rational redesign of the ferredoxin-NADP(+)-oxido-reductase/ferredoxin-interaction for photosynthesis-dependent H_2-production. *Biochimica et Biophysica Acta*, *1859*(4), 253–262.

Winkler, M., Esselborn, J., & Happe, T. (2013). Molecular basis of [FeFe]-hydrogenase function an insight into the complex interplay between protein and catalytic cofactor. *Biochimica et Biophysica Acta-Bioenergetics*, *1827*(8–9), 974–985.

Winkler, M., Heil, B., & Happe, T. (2002). Isolation and molecular characterization of the [Fe]-hydrogenase from the unicellular green alga *Chlorella fusca*. *Biochimica et Biophysica Acta*, *1576*(3), 330–334.

Winkler, M., Kuhlgert, S., Hippler, M., & Happe, T. (2009). Characterization of the key step for light-driven hydrogen evolution in green algae. *Journal of Biological Chemistry*, *284*(52), 36620–36627.

Winkler, M., Senger, M., Duan, J., Esselborn, J., Wittkamp, F., Hofmann, E., et al. (2017). Accumulating the hydride state in the catalytic cycle of [FeFe]-hydrogenases. *Nature Communications*, *8*, 16115.

Wykoff, D. D., Davies, J. P., Melis, A., & Grossman, A. R. (1998). The regulation of photosynthetic electron transport during nutrient deprivation in *Chlamydomonas reinhardtii*. *Plant Physiology*, *117*(1), 129–139.

Yacoby, I., Pochekailov, S., Toporik, H., Ghirardi, M. L., King, P. W., & Zhang, S. G. (2011). Photosynthetic electron partitioning between [FeFe]-hydrogenase and ferredoxin:$NADP^+$-oxidoreductase (FNR) enzymes in vitro. *Proceedings of the National Academy of Sciences of the United States of America*, *108*(23), 9396–9401.

Yacoby, I., Tegler, L. T., Pochekailov, S., Zhang, S., & King, P. W. (2012). Optimized expression and purification for high-activity preparations of algal [FeFe]-hydrogenase. *PLoS One*, 7(4), e35886.

Yan, F., Chen, Z., Li, W., Cao, X., Xue, S., & Zhang, W. (2011). Purification and characterization of a hydrogenase from the marine green alga tetraselmis subcordiformis. *Process Biochemistry*, *46*(5), 1212–1215.

Yano, T., Sled, V. D., Ohnishi, T., & Yagi, T. (1994). Expression of the 25-kilodalton iron-sulfur subunit of the energy-transducing NADH-ubiquinone oxidoreductase of Paracoccus denitrificans. *Biochemistry*, *33*(2), 494–499.

CHAPTER NINE

Application of affinity purification methods for analysis of the nitrogenase system from *Azotobacter vinelandii*

Emilio Jiménez-Vicente[a], Julia Sanchez Martin Del Campo[a], Zhi-Yong Yang[b], Valerie L. Cash[a], Dennis R. Dean[a,*], Lance C. Seefeldt[b,*]

[a]Department of Biochemistry, Virginia Tech, Blacksburg, VA, United States
[b]Department of Chemistry and Biochemistry, Utah State University, Logan, UT, United States
*Corresponding authors: e-mail addresses: deandr@vt.edu; lance.seefeldt@usu.edu

Contents

Abstract

Nitrogenases are complex two-component metalloenzymes that catalyze biological nitrogen fixation. Three different nitrogenase types are found in the model nitrogen-fixing microbe *Azotobacter vinelandii*. In the case of the Mo-dependent enzyme, the two catalytic partners are referred to as the Fe protein and MoFe protein. In addition to genes encoding the catalytic components, there are a total of 68 other gene products known to be variously involved in producing, activating, protecting, sustaining, and

Methods in Enzymology, Volume 613
ISSN 0076-6879
https://doi.org/10.1016/bs.mie.2018.10.007

regulating formation of the Mo-dependent nitrogenase. In order to support experiments designed to gain insight into the catalytic mechanism and assembly of nitrogenase, four different affinity-based purification protocols have been developed. These include an improved Co^{2+}-based Immobilized Metal Affinity Chromatography (IMAC) method for the purification of MoFe protein, a newly developed StrepTactin Affinity Chromatography (STAC) method for the purification of MoFe protein and its assembly intermediates, a combined IMAC and STAC method for isolation of highly pure MoFe protein, and a STAC-based bait–prey method for isolation of complexes variously involved in the maturation process.

1. Introduction

Nitrogenases are enzymes that have the unique ability to catalyze the reduction of inert nitrogen gas (N_2) and are found only in a select and diverse group of bacteria and archaea called diazotrophs (Burris, 1991; Dixon & Kahn, 2004). Biological N_2 fixation accounts for approximately 50% of the fixed N input from N_2 into the global biogeochemical nitrogen cycle with the balance being provided by the industrial Haber–Bosch reaction (Erisman et al., 2015; Erisman, Sutton, Galloway, Klimont, & Winiwarter, 2008; Ladha et al., 2016). The Haber–Bosch reaction has proven very successful in supporting world-wide food production for the past century, yet the demands for H_2, heat, and pressure associated with the process currently consume up to 3% of the available fossil fuel supply (Chen et al., 2018). Furthermore, there are significant environmental, economic, and geopolitical penalties associated with the application of industrially produced nitrogenous fertilizers (Erisman et al., 2015). Consequently, there is interest in the development of a new generation of N_2 reducing catalysts that can utilize sustainably derived electrons and protons operating under benign conditions (Anderson, Rittle, & Peters, 2013; Brown et al., 2016; Paengnakorn et al., 2017). The rational design of such catalysts could be informed by a molecular level understanding of how nitrogenases accomplish the nucleotide-dependent reduction of N_2 at ambient temperature and pressure.

There are also ongoing efforts to endow food crops with an ability to fix N_2 without the aid of a microbial partner by transfer of the appropriate genetic determinants for biological nitrogen fixation from microbes to plants (Beatty & Good, 2011; Burén & Rubio, 2018; Curatti & Rubio, 2014; Vicente & Dean, 2017). These efforts rely on understanding how the nitrogenases, which are all complex two-protein component metalloenzymes, are assembled and maintained. Mechanistic features of nitrogenase, as well as the assembly of its associated metalloclusters, have been under intense investigation for more than 50 years. Significant advances in this area of research have often relied on the application of emergent genetic, biochemical and biophysical technologies. There is now a general understanding of many key mechanistic features related to nitrogenase catalysis (Hoffman, Lukoyanov, Yang, Dean, & Seefeldt, 2014; Seefeldt, Hoffman, & Dean, 2009) as well as a rudimentary understanding of the essential players involved in the assembly of an active nitrogenase (Jimenez-Vicente et al., 2018; Rubio & Ludden, 2008), yet many significant challenges remain. Many of these challenges are associated with the biochemical and genetic complexities of the system, as well as the extreme oxygen sensitivity of the catalytic components (Shah & Brill, 1973) and certain proteins involved in metallocluster assembly (Curatti, Ludden, & Rubio, 2006; Goodwin et al., 1998). This chapter outlines the general biochemical features and genetic organization of the Mo-dependent nitrogenase found in the model nitrogen-fixing microbe, *Azotobacter vinelandii*, and then describes affinity purification methods that can be applied to gain a deeper understanding of the system. Although the methods described here have only been applied to the Mo-dependent system so far, they are applicable to the other nitrogen-fixation systems present in *A. vinelandii*.

2. The nitrogenase systems in *A. vinelandii*

A schematic representation of the catalytic components of the Mo-dependent nitrogenase and the reaction it catalyzes is shown in Fig. 1. However, three different classes of nitrogenases are produced in *A. vinelandii* with a distinguishing feature among them being the identity of the heterometal (Mo, V, or Fe) contained in the active-site metal cluster that provides the site for substrate binding and reduction (Bishop et al., 1986; Chisnell, Premakumar, & Bishop, 1988). In the case of the Mo-dependent nitrogenase, this active-site metallocluster is called the "FeMo-cofactor." Although the key mechanistic features related to the activation of N_2 for reduction appear to be shared among all three systems (Harris, Lukoyanov, et al., 2018; Harris, Yang, Dean, Seefeldt, & Hoffman, 2018), the genes encoding their respective catalytic components are distinct (Bishop, Jarlenski, & Hetherington, 1980; Hamilton et al., 2011; Setubal et al., 2009). Most aspects of the assembly of metalloclusters necessary to activate all three systems also appear to be conserved, and, in some cases, the same protein,

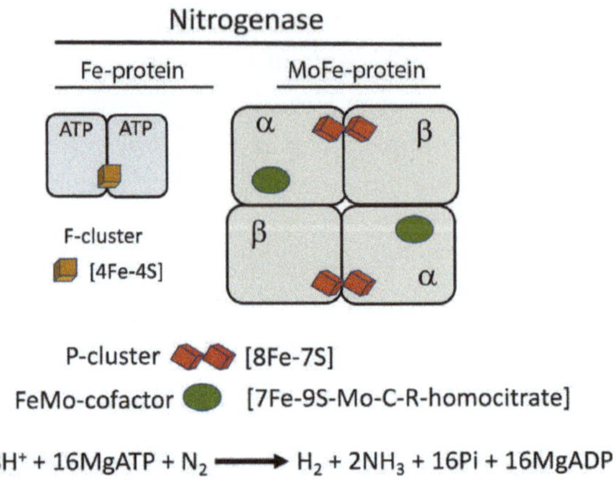

Fig. 1 Schematic representation of the catalytic components of Mo-dependent nitrogenase and composition and organization of the associated metal-sulfur-containing clusters.

encoded by a single unique gene, participates in metallocluster assembly for all three systems. For example, a protein encoded by the *nifV* gene is required for providing the organic constituent, homocitrate, for the different active site metalloclusters present in all three respective systems (Hoover, Imperial, Ludden, & Shah, 1989; Zheng, White, & Dean, 1997).

Mo-nitrogenase is composed of two component proteins designated the Fe protein and the MoFe protein. The Fe protein is a homodimer having two ATP-binding sites, one separately located on each subunit, and a single [4Fe-4S] cluster that bridges the two identical subunits (Fig. 1) (Georgiadis et al., 1992). Catalysis is initiated when the reduced, MgATP-bound form of the Fe protein docks with the MoFe protein, triggering intermolecular electron transfer and hydrolysis of the two ATP molecules. Various aspects of sequential events involving component protein association, intercomponent electron transfer, nucleotide hydrolysis, and complex dissociation have recently been reviewed (Harris, Yang, et al., 2018; Seefeldt, Hoffman, et al., 2018; Seefeldt, Peters, et al., 2018). The MoFe protein is an $\alpha_2\beta_2$ heterotetramer having two types of metal clusters. One of these, the [8Fe-7S] "P-cluster" is located at each $\alpha\beta$ interface in the MoFe protein near the Fe protein docking face where it facilitates electron transfer from the Fe protein to the active site [7Fe-9S-Mo-C-*R*-homocitrate] FeMo-cofactor, located entirely within the MoFe protein α-subunit (Kim & Rees, 1992). Binding of N_2 requires the prior accumulation of

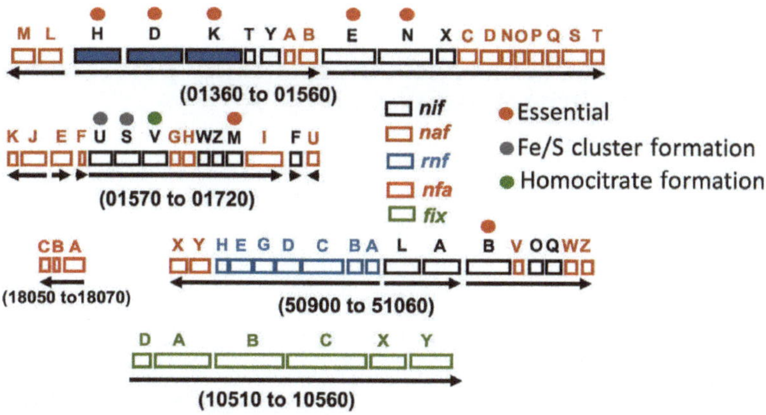

Fig. 2 Organization of genes whose products are associated with the Mo-dependent nitrogen fixation system from *A. vinelandii*. Numbers refer to gene designations used in the original annotation of the *A. vinelandii* genome (Setubal et al., 2009), and *arrows* indicate transcription units. Genes encoding the nitrogenase catalytic components (*nifH*, Fe protein; *nifDK*, MoFe protein subunits) are filled in *blue*. The seven genes whose products are strictly required to sustain Mo-dependent nitrogenase activity in *A. vinelandii* are indicated by *red dots*. Gene letters do not necessarily indicate analogous functions. For example, there is no structural or functional similarity between the products of *nifH* and *nafH*.

four electrons within the FeMo-cofactor in the form of bridging hydrides which are proposed to participate in a reductive elimination mechanism (Hoffman et al., 2014; Lukoyanov et al., 2016). The biochemical complexity of the Mo-dependent nitrogenase is reflected in the large number and genetic organization of genes (Fig. 2) whose products are involved in coupling cellular metabolism to the production of electrons needed to support N_2 reduction (Bennett, Jacobson, & Dean, 1988; Curatti, Brown, Ludden, & Rubio, 2005; Setubal et al., 2009), sequestration and activation of Fe, S, and Mo necessary for metallocluster assembly (Bosch, Rodríguez-Quiñones, & Imperial, 1997; Hernandez et al., 2008; Jacobson et al., 1989; Johnson, Dos Santos, & Dean, 2005; Kennedy & Dean, 1992), formation of homocitrate for FeMo-cofactor assembly (Zheng et al., 1997), protection of nitrogenase from oxygen damage (Moshiri, Crouse, Johnson, & Maier, 1995; Schlesier, Rohde, Gerhardt, & Einsle, 2016), and regulatory elements that control expression of the system under favorable anoxic conditions that demand N_2 fixation to support growth (Austin, Buck, Cannon, Eydmann, & Dixon, 1994; Bennett, Cannon, & Dean, 1988; Blanco, Drummond, Woodley, & Kennedy, 1993).

3. Strategies for application of affinity purification to the nitrogenase system

Various affinity purification-based methods have been developed for: (i) purification of nitrogenase components either in their native forms or those having amino acid substitutions (Chan, Wu, Dean, & Seefeldt, 2000; Christiansen, Goodwin, Lanzilotta, Seefeldt, & Dean, 1998; Danyal, Dean, Hoffman, & Seefeldt, 2011; Fay, Hu, Schmid, & Ribbe, 2007; Lukoyanov et al., 2011, 2016), (ii) purification of various recombinantly produced accessory proteins involved in nitrogenase maturation (Curatti et al., 2006; Goodwin et al., 1998; Hernandez et al., 2008, 2007; Hu, Fay, Lee, & Ribbe, 2007; Rubio, Singer, & Ludden, 2004; Smith et al., 2005), and (iii) purification of assembly intermediates captured at various stages in the maturation process as a consequence of having a gene involved in the assembly process inactivated (Christiansen et al., 1998; Jimenez-Vicente et al., 2018; Ribbe, Hu, Guo, Schmid, & Burgess, 2002). The affinity methods, depicted in Fig. 3A, rely on *A. vinelandii* strains that have an affinity-coding tag (either a "polyhistidine" tag or a "Strep" tag, see legend to Fig. 3) placed at either the N-terminus or C-terminus of *nifD* encoding the MoFe protein α-subunit. Affinity purification methods involving purification of His-tagged or Strep-tagged proteins are hereafter referred to as IMAC (Immobilized Metal Affinity Chromatography) or STAC (StrepTactin Affinity Chromatography). Strains indicated in Fig. 3A, which are generally available from our laboratories, were constructed using gene replacement methods (Dos Santos, 2011) so that the corresponding affinity tag-encoding region is incorporated directly into the

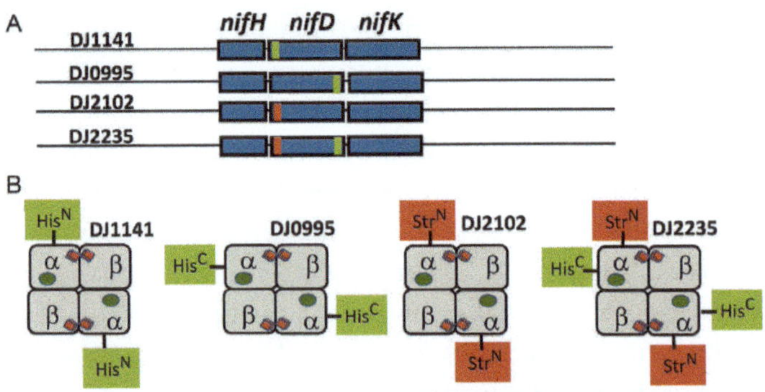

Fig. 3 Schematic representation of the location of the poly-histidine tag (Christiansen et al., 1998) or Strep-tag (Jimenez-Vicente et al., 2018) encoding regions incorporated into the *nifD*-coding sequence within the *A. vinelandii* genome (A) and schematic representation of correspondingly tagged MoFe proteins (B).

A. vinelandii genome and, therefore, no other function related to nitrogen fixation is disrupted. Importantly, integration of the affinity tags into the *A. vinelandii* genome has no effect on the capacity for diazotrophic growth in the corresponding strains. Furthermore, the catalytic and biophysical properties of MoFe proteins purified by either the IMAC or STAC procedures are essentially identical to MoFe protein purified using traditional chromatography methods (Fig. 4). The location of various affinity tags within MoFe protein α-subunits produced by the strains indicated in Fig. 3A is schematically depicted in Fig. 3B. A variation on the placement of a His-tag at the C-terminus of the MoFe protein α-subunit (produced by DJ0995) and the placement of a Strep-tag on the N-terminus of the MoFe protein α-subunit (produced by DJ2102) involves the combination of both tags, a Strep-tag at the N-terminus and a His-tag at the C-terminus of the MoFe protein α-subunit (produced by DJ2235). This latter construct enables the application of tandem STAC and IMAC procedures to obtain highly purified MoFe protein having no detectable contaminating proteins (Fig. 4B).

The availability of strains that produce affinity-tagged MoFe protein enabled isolation of a second generation of strains variously deleted for genes whose products are involved in formation of an active MoFe protein. For example, as shown in Fig. 4A (left panel), application of STAC for purification of a Strep-tagged MoFe protein produced by a *nifH* deletion strain or a Strep-tagged MoFe protein produced by a *nifB* deletion strain reveals copurification of different suites of associated assembly factors. MoFe protein produced by a *nifH* deletion strain is defective in both FeMo-cofactor and P-cluster and copurifies with the products of *nafH*, *nifW* and *nifZ*. In contrast, MoFe protein produced by a *nifB* deletion strain is defective only in FeMo-cofactor formation and copurifies with the products of *nifY* and *nafY*. Thus, STAC purification of MoFe protein produced in various genetic backgrounds represents an excellent tool for elucidation of individual steps in the MoFe protein maturation process (Jimenez-Vicente et al., 2018).

The observation that STAC purification of Strep-tagged MoFe proteins produced in different genetic backgrounds that are defective in aspects of metallocluster assembly results in the differential copurification of associated assembly factors led to development of a STAC-based bait–prey affinity protocol for purification of individual assembly complexes. In this procedure, various Strep-tagged versions of proteins associated with the Mo-dependent nitrogenase and heterologously produced in *Escherichia coli* are initially purified using a Strep-tag affinity purification protocol (Fig. 5). Proteins purified in this

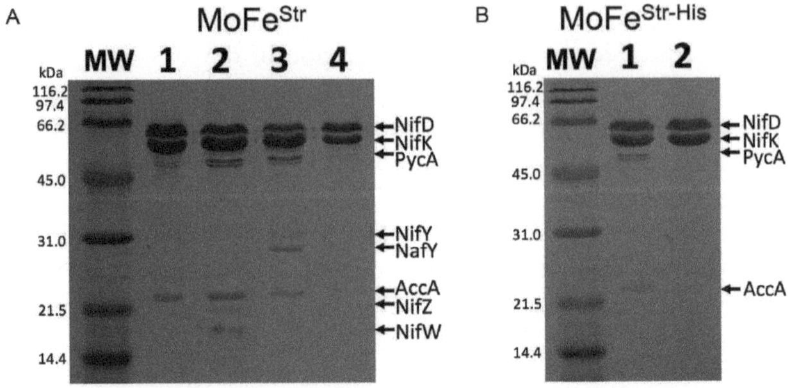

C

	Specific activity	Fe atoms per MoFe	Mo atoms per MoFe
DJ2102	1844.7 ± 81.3	30.1 ± 1.8	2.06 ± 0.13
DJ2235	1634 ± 89.2	28.9 ± 1.5	1.68 ± 0.24

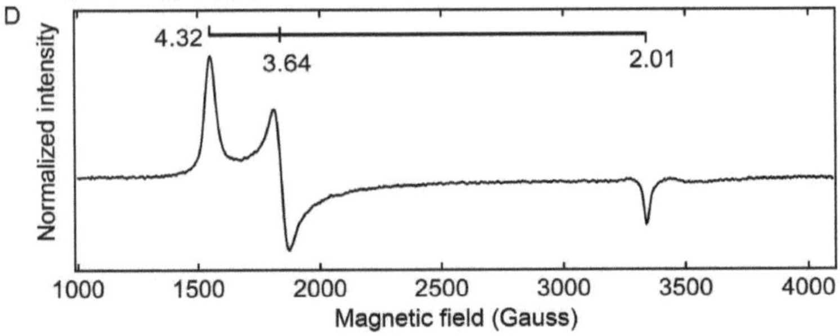

Fig. 4 Denaturing polyacrylamide gel electrophoresis (PAGE) of Strep-tagged MoFe proteins isolated by affinity purification. (A) Lane 1, STAC-purified MoFe protein from DJ2102 (see Fig. 3A); lane 2, STAC-purified Strep-tagged MoFe protein produced by a strain derived from DJ2102 deleted for *nifH*; lane 3, STAC-purified Strep-tagged MoFe protein produced by a strain derived from DJ2102 deleted for *nifB*; lane 4, STAC-purified Strep-tagged MoFe protein produced by DJ2102 that is further purified by an additional step to remove contaminating biotin-binding proteins. (B) Lane 1, STAC-purified Strep-tagged/His-tagged MoFe protein from DJ2235 (see Fig. 3A); lane 2, further purification of the sample shown in lane 1 of (B) by using IMAC. Note contaminating biotin-binding proteins in lanes 1, 2, and 3 of (A) and in lane 1 of (B) are not present in lane 4 of (A) nor in lane 2 of (B). Also note NifW and NifZ copurified with the Strep-tagged MoFe protein sample shown in lane 2 of (A). NafH also purifies with the Strep-tagged MoFe protein produced in a strain deleted for *NifH* but is not apparent at the sample concentration used for this analysis. NifY and NafY copurified with the Strep-tagged MoFe protein sample shown in lane 3 of (A). (C) Specific activities (nmoles of acetylene reduced per min per mg MoFe protein) and metal content for STAC-purified MoFe protein produced by DJ2102 (A, lane 4) and STAC/IMAC purified MoFe protein produced by DJ2235 (B, lane 2). (D) Electron paramagnetic resonance spectrum of STAC-purified Strep-tagged MoFe protein (A, lane 1) prepared from DJ2102. NifD and NifK correspond to the MoFe protein α- and β-subunits, respectively.

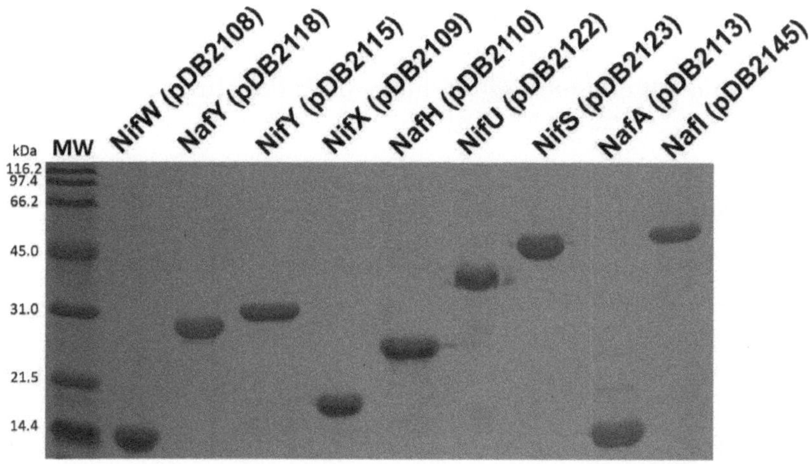

Fig. 5 STAC-purified samples of various nitrogen fixation-associated proteins from *A. vinelandii* produced in *E. coli* strain BL21 (DE3) using recombinant plasmids. Each plasmid directs the high level of expression of the indicated affinity tagged protein having the sequence MASWSHPQFEKH located at the N-terminus.

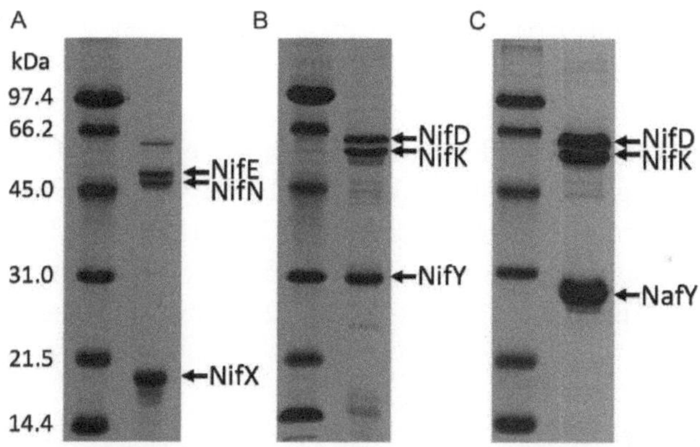

Fig. 6 Examples of affinity purification of nitrogenase-associated assembly complexes using recombinantly produced Strep-tagged proteins as bait. (A) Purified NifENX complex produced by a strain deleted for *nifDK* using NifX as bait; (B) Purified NifDKY complex produced by a strain deleted for *nifE* using NifY as bait; (C) Purified NifDKNafY complex produced by a strain deleted for *nifE* using NafY as bait. Note that excess bait proteins have not been removed from these samples. NifD and NifK, respectively, indicate the MoFe protein α- and β-subunits.

way can subsequently be reimmobilized on a Streptactin column and used as bait to capture their respective interacting partners by passing *A. vinelandii* crude extracts produced from various genetic backgrounds defective in the assembly process over the bait-charged column. After a washing step to remove unbound proteins, both bait and the captured interacting protein(s) can be eluted using a biotin-containing buffer. Fig. 6A–C shows the results of three examples of the application of the STAC-based bait–prey affinity purification procedure. In the following sections, details of the protocols used for application of affinity purification methods to the nitrogenase system are described.

4. Methods

4.1 Growth and cell-free extract preparation of *A. vinelandii*

4.1.1 Equipment

Glove box anaerobic chamber (Coy Laboratory Products) maintained using a 95% N_2, 5% H_2 atmosphere
Nano-DeBee homogenizer (B.E.E. International, Inc. South Easton, MA)
Two 500 mL Kimax side-arm flasks (Fisher scientific, Suwanee, GA)
Ultracentrifuge Optima LE-80K (Beckman Coulter, Brea, CA)
Type 45 Ti centrifuge rotor (Beckman Coulter, Brea, CA)
60-mL centrifuge tubes (Beckman Coulter, Brea, CA)
0.45 μ*M* polyethersulfone (PES) membrane vacuum filtration unit (VWR, Radnor, PA).

4.1.2 Buffers and reagents

Buffer A, 50 m*M* Tris–HCl (pH 7.9), 20% glycerol, 0.5 *M* NaCl, 1.6 m*M* $Na_2S_2O_4$, degassed and sparged with Ar before the addition of $Na_2S_2O_4$.
Buffer B, 50 m*M* Tris–HCl (pH 7.9), 20 m*M* EDTA, 0.2 m*M* phenylmethane sulfonyl fluoride (PMSF) (Sigma, St. Louis, MO), 2 μ*M* pepstatin (Sigma, St. Louis, MO), 2 μg/mL DNAse (Sigma, St. Louis, MO).

4.1.3 Procedure

1. The composition of media and general conditions used for the large-scale growth of *A. vinelandii* cells is described by Dos Santos (2011). In our laboratory, *A. vinelandii* cells are typically grown at 30°C in a 150-L custom-built fermenter (W.B. Moore, Inc. Easton, PA) using a modified Burk medium containing 10 μ*M* Na_2MoO_4 as the Mo source and 10 m*M* urea as a nitrogen source. When the 120-L culture reaches an approximate OD_{600} of 1.6, cells are harvested in a final volume of 5 L, to be resuspended in 120 L of fresh urea-free media. After 4 h of nitrogenase derepression, cells are harvested again, centrifuged at 5000 RPM at 4°C, and the pellets are frozen and kept refrigerated at −80°C until further use.

2. Frozen cells, 75 g, are resuspended in a 500-mL side-arm flask by addition of 75 mL of buffer A and gentle agitation. This step is performed in the anaerobic chamber.
3. The flask containing the resuspended cells is removed from the anaerobic chamber and the cells are disrupted using a Nano-DeBee homogenizer. The lines and reservoir of the Nano-DeBee homogenizer are flushed with buffer A prior to cell disruption and a stream of Ar is maintained over the reservoir during cell extract preparation.
4. Once the cell disruption is complete, 0.5 mL of 0.2 m*M* PMSF, 15 mg of DNAse, and 1.5 mg of pepstatin are added to the cell extract. The cell lysate is then dispensed into centrifuge tubes that have been prerinsed in buffer A, which are then tightly capped and centrifuged at 58,000 × *g* for 45 min at 4°C.
5. After centrifugation, the tubes containing cell extract are moved to the anaerobic chamber and carefully decanted onto a 0.45-μm PES membrane filter unit and vacuum filtered using a manual vacuum pump. The volume of crude extract recovered using this method is typically about 115 mL.

4.1.4 Notes

Cell-free extract preparation can be completed in 2 h. Cell-free extracts of *E. coli* cell-free extracts can be prepared following the same protocol with the exceptions that there is no need to maintain anoxic conditions and cells are resuspended in buffer B rather than buffer A. For preparation of *E. coli* extracts, 35 g of cells are resuspended in 140 mL of buffer B.

4.2 Affinity purification of Strep-tagged MoFe protein using STAC

4.2.1 Equipment

Glove box anaerobic chamber (Coy Laboratory Products) maintained using a 95% N_2, 5% H_2 atmosphere.

5 mL prepacked Strep-Tactin®XT Superflow® gravity flow column (IBA Lifescience, Göttingen, Germany).

4.2.2 Buffers and reagents

Buffer A, 50 m*M* Tris–HCl (pH 7.9), 20% glycerol, 0.5 *M* NaCl, and 1.6 m*M* $Na_2S_2O_4$

STAC elution buffer is prepared by the addition of solid biotin to buffer A to a final concentration of 50 m*M*. 8 μL of 4 *M* NaOH is added to each mL of STAC elution buffer followed by vigorous shaking to facilitate Biotin solubilization.

All buffers are degassed and sparged with Ar before the addition of $Na_2S_2O_4$.

Approximately 115 mL of cell-free extract is obtained from 75 g of DJ995 cells prepared according to Section 4.1.

4.2.3 Procedure

Samples for processing are continuously maintained under anoxic conditions inside a glove box.

1. A 5-mL prepacked Strep-Tactin®XT Superflow® gravity flow column is equilibrated by passing 50 mL of anoxic buffer A over the column.
2. Following equilibration the crude extract is manually applied to the column using gravity flow conditions. This step takes approximately 40 min.
3. The walls of the column reservoir above the loaded column matrix are rinsed with 25 mL of buffer A and the rinse is discarded.
4. The column reservoir is filled to the top with buffer A (25 mL) and the flow-through containing unbound proteins is discarded.
5. Elution of MoFe protein is initiated by the addition of 3 mL of STAC elution buffer to the column reservoir. After elution of the initial 3 mL, as MoFe protein is about to elute, an additional 5 mL of STAC elution buffer is added to the column reservoir. The dark brown fractions containing MoFe protein are collected.

4.2.4 Notes

A typical MoFe protein preparation from DJ2102 cells using this protocol yields approximately 30 mg of MoFe protein. The entire protocol can be completed within 60 min. Purification of immature forms of MoFe protein produced in strains defective in some aspect of the maturation process, for example, MoFe protein produced by a *nifB* deletion strain, is typically less abundant in crude extracts and, therefore, yields are generally somewhat lower for such samples. *A. vinelandii* cells produce high levels of biotin-binding proteins unrelated to nitrogenase, for example, acetate carboxylase (AccA) and pyruvate carboxylase (PycAB). These proteins are often detected as being nonspecifically copurified with MoFe protein when using this procedure. For most applications, such as EPR spectroscopy and nitrogenase assays, further purification is not necessary. However, copurifying contaminants can be removed using the protocol described in the following section. A typical PAGE analysis of Strep-tagged MoFe protein, Strep-tagged MoFe protein produced by a strain deleted for *nifH*, and Strep-tagged MoFe protein produced by a strain deleted for *nifB* is shown in Fig. 4A. The MoFe protein subunits, accessory proteins that differentially copurify with MoFe protein produced in strains deleted for *nifH* or *nifB*, and contaminating Biotin-binding proteins, are indicated in Fig. 4A.

4.3 Removal of biotin-binding proteins that copurify with Strep-tagged MoFe protein when using the STAC procedure

4.3.1 Equipment

Glove box anaerobic chamber (Coy Laboratory Products, Grass Lake, MI) operated with a 95% N_2, 5% H_2 atmosphere.
1-mL HiTrap Q HP column (GE Healthcare, Chicago, IL).
P-1 peristaltic pump (GE Healthcare, Chicago, IL).

4.3.2 Buffers and reagents

Buffer A, 50 m*M* Tris–HCl (pH 7.9), 1.6 m*M* $Na_2S_2O_4$
Buffer B, 50 m*M* Tris–HCl (pH 7.9), 175 m*M* NaCl, 1.6 m*M* $Na_2S_2O_4$
Buffer C, 50 m*M* Tris–HCl (pH 7.9), 300 m*M* NaCl, 1.6 m*M* $Na_2S_2O_4$
All buffers are degassed and sparged with Ar before the addition of $Na_2S_2O_4$

4.3.3 Procedure

Samples for processing are continuously maintained under anoxic conditions inside a glove box.

1. A peristaltic pump is used to pass 10 mL of buffer A over a 1-mL prepacked HiTrap Q HP column for column equilibration.
2. The sample prepared using STAC purification is diluted five times with buffer A.
3. A peristaltic pump is used to apply the diluted sample to the buffer A-equilibrated HiTrap Q HP column which is then washed with 5 mL of buffer A.
4. The loaded column is washed with 5 mL of buffer B.
5. MoFe protein is eluted with 5 mL of buffer C, the characteristic dark black/green color of the MoFe protein allowing visual monitoring.

4.3.4 Notes

The protocol can be completed in 60 min. A typical purification of MoFe protein using this protocol is shown in Fig. 4A, lane 4.

4.4 Purification of His-tagged MoFe protein using IMAC

4.4.1 Equipment

Glove box anaerobic chamber (Coy Laboratory Products, Grass Lake, MI) operated with a 95% N_2, 5% H_2 atmosphere.
2 in. diameter 140 mL DEAE resin column (GE Healthcare, Chicago, IL).
3 × 5 mL HiTrap IMAC FF five columns connected together in series (GE Healthcare, Chicago, IL).
5 mL HiTrap Q HP (GE Healthcare, Chicago, IL).
AKTA purifier (GE Healthcare, Chicago, IL).
Wheaton serum vials (38 mL) fitted with gray butyl rubber caps (Millville, NY).
Fisher scientific 1-L Kimax side-arm flasks (Suwanee, GA).

4.4.2 Buffers and reagents

Buffer A, 50 m*M* Tris–HCl (pH 7.9), 1.6 m*M* $Na_2S_2O_4$
Buffer B, 50 m*M* Tris–HCl (pH 7.9), 0.5 *M* NaCl, 1.6 m*M* $Na_2S_2O_4$
Buffer C, IMAC elution buffer, 50 m*M* Tris–HCl (pH 7.9), 150 m*M* NaCl, 150 m*M* imidazole, 1.6 m*M* $Na_2S_2O_4$
All buffers are degassed and sparged with Ar before the addition of $Na_2S_2O_4$.
4 *M* NaOH
Approximately 330 mL of cell-free extract is obtained from 150 g of DJ0995 cells prepared according to Section 4.1

4.4.3 Procedure

Samples and buffers are continuously maintained under anoxic conditions either in a glove box or in sealed vessels under a slightly pressurized Ar atmosphere on the bench.

1. The DEAE column is equilibrated overnight by slowly passing 2 L of buffer A through the column. Equilibration is complete when $Na_2S_2O_4$ can be detected at 280 nm in the eluate.
2. Approximately 330 mL of cell-free extract is loaded onto the column at a rate of 6 mL/min using a peristaltic pump. In the following steps, the application of wash and elution buffers is controlled using an AKTA purifier, unless otherwise indicated.

3. The loaded DEAE column is washed with a mixture of 80% buffer A and 20% buffer B which represents a final concentration of 100 m*M* NaCl. The flow rate is maintained at 6 mL/min, and the wash step is typically complete after application of approximately 300 mL.
4. A 1-L salt gradient having a flow rate of 5 mL/min and spanning concentrations of 100% buffer A plus 0% buffer B to 0% buffer A plus 100% buffer B is applied to the column for elution of the MoFe protein.
5. During the course of the salt gradient, when a conductivity of 15 mS is reached, 25 mL fractions are collected into serum bottles: these have been prerinsed with 5 mL buffer A, sealed with a rubber septum cap, and thoroughly flushed with Ar. During sample collection, positive pressure on the collection bottle is maintained using a 100% Ar line attached by a needle through the serum bottle septum. Also, an open-ended syringe needle is inserted into the serum bottle septum to relieve pressure during sample collection.
6. Collected fractions can be evaluated for peak concentrations of MoFe protein by the intensity of the black/green color associated with MoFe protein or by performing acetylene reduction activities on samples from individual fractions. Fractions having peak levels of MoFe protein are pooled and brought to a final NaCl concentration of approximately 500 m*M* using solid NaCl that has been stored under Ar.
7. Pooled fractions containing MoFe protein are applied to three tandemly connected prepacked 5 mL HiTrap IMAC HP columns charged with Co^{2+} at a rate of 3 mL/min. The HiTrap IMAC HP columns are charged with Co^{2+} according to the manufacturer's recommendation using a degassed and Ar-sparged 0.1 *M* $CoCl_2$ solution. The Co^{2+}-charged IMAC columns are equilibrated by passing 150 mL of buffer A over the columns prior to sample loading, which can be accomplished using either a peristaltic pump or an AKTA purifier.
8. The sample-loaded columns are first washed with buffer B until the effluent absorbance at 280 nm reaches approximately 0.
9. The columns are further washed using a 25 mL gradient starting with 100% buffer B plus 0% buffer C and ramping to 92% buffer B plus 8% buffer C so that the final imidazole concentration is 12 m*M* at the end of the gradient. Washing with the buffer B/C combination (yielding 12 m*M* imidazole) is continued until the effluent absorbance at 280 n*M* reaches approximately 0.
10. The columns are finally washed using a 25 mL gradient starting with 92% buffer B plus 8% buffer C and ramping to 82% buffer B plus

18% buffer C so that the final imidazole concentration is 27 m*M* at the end of the gradient. Washing with the buffer B/C combination (yielding 27 m*M* imidazole) is continued until the effluent absorbance at 280 reaches approximately 0.

11. Buffer C is used to elute MoFe protein which, after elution, is contained in buffer having an approximate imidazole concentration of 150 m*M* and contaminated by trace levels of Co^{2+} leached from the column. For most applications, both the contaminating Co^{2+} and imidazole must be removed.
12. To remove Co^{2+} and imidazole, the sample obtained in the previous step is diluted four times with buffer A and loaded using a peristaltic pump at a rate of 4 mL/min onto a 5-mL HiTrap Q HP column preequilibrated with 75 mL of buffer A. Buffer A (20 mL) and buffer B (5 mL) are combined to give a final NaCl concentration of 100 m*M*, and this is used to wash the loaded column.
13. The sample is eluted from the column using buffer B. Fractions containing MoFe protein have a deep black/green color.

4.4.4 Notes

The typical yield of His-tagged MoFe protein from strain DJ995 is approximately 120 mg. Starting at step 2, the purification protocol can be completed within approximately 10 h.

4.5 Purification of tandem Strep/His-tagged MoFe protein

4.5.1 Equipment

Glove box anaerobic chamber (Coy Laboratory Products, Grass Lake, MI) operated with a 95% N_2, 5% H_2 atmosphere.

5 mL prepacked Strep-Tactin®XT Superflow® gravity flow column (IBA Lifescience, Göttingen, Germany).

Prepacked HiTrap IMAC HP 1 mL column (GE Healthcare, Chicago, IL).

4.5.2 Buffers and reagents

Buffer A, 50 m*M* Tris–HCl (pH 7.9), 0.5 *M* NaCl, 1.6 m*M* $Na_2S_2O_4$.

4 *M* NaOH

STAC elution buffer, 50 m*M* Tris–HCl (pH 7.9), 0.5 *M* NaCl, 1.6 m*M* $Na_2S_2O_4$, 50 m*M* biotin

IMAC elution buffer, 50 m*M* Tris–HCl (pH 7.9), 150 m*M* NaCl, 25 m*M* imidazole, 1.6 m*M* $Na_2S_2O_4$

All buffers are degassed and sparged with Ar before the addition of $Na_2S_2O_4$

Approximately 115 mL of cell-free extract is obtained from 75 g of DJ2235 cells prepared according to Section 4.1

4.5.3 Procedure

The entire procedure is carried out under anoxic conditions inside a glove box 95% N_2, 5% H_2 atmosphere.

1. The procedure for STAC purification described in Section 4.2 is followed.
2. A Co^{2+}-charged 1-mL prepacked HiTrap IMAC column is equilibrated by passing 15 mL of buffer A over the column.
3. STAC-purified MoFe protein is loaded onto the column.
4. The column is washed with 10 mL of buffer A.
5. The MoFe protein is eluted using IMAC elution buffer.

4.5.4 Notes

A typical purification using extract prepared from 75 g DJ2235 cells yields approximately 30 mg of MoFe protein. The entire purification protocol, starting with crude extract, can be completed in 2 h. A PAGE analysis of MoFe protein prepared by the tandem STAC-IMAC method is shown in Fig. 4B. Note that all contaminating proteins are removed in the second, IMAC, step.

4.6 Purification of Strep-tagged Nif and Naf proteins recombinantly produced in *E. coli*

4.6.1 Equipment

50 mL Strep-Tactin®XT Superflow® resin (IBA Lifescience, Göttingen, Germany).

HiPrep 26/10 Desalting column (GE Healthcare, Chicago, IL).

AKTA purifier.

1 in. gravity chromatography column (Biorad, Hercules, CA).

2 in. gravity chromatography column (Biorad, Hercules, CA).

4.6.2 Buffers and reagents

Buffer A, 50 m*M* Tris–HCl (pH 8.0), 100 m*M* NaCl, 20 m*M* EDTA.

STAC elution buffer, buffer B, 50 m*M* Tris–HCl (pH 8.0), 100 m*M* NaCl, 20 m*M* EDTA.

50 m*M* biotin.

150 mL of cell-free extract obtained from 35 g of *E. coli* cells that produce Strep-tagged Nif or Naf proteins prepared according to Section 4.1.

4.6.3 Procedure

1. Buffer A (50 mL) is added to Strep-Tactin®XT Superflow® resin preequilibrated with buffer A and this is placed in a 500-mL screw-cap Nalgene bottle.
2. 150 mL of cell-free extract is added, then the bottle is capped and incubated for 2 h at 4°C under mild agitation to prevent resin settling.
3. After incubation, the resin-cell-free extract mixture is poured into a 2-in. diameter chromatography column permitting the resin to settle and the crude extract to partially elute.
4. Two washings with 100 mL of buffer A are carried out.
5. The resin is resuspended in another 100 mL of buffer A and transferred to a 1-in. diameter chromatography column. Once the resin has settled and excess buffer A has drained from the column, a disc of Whatman filter paper 3 is placed on top of the resin.
6. The column is connected to an AKTA purifier so the absorbance at 280 can be monitored.
7. Washing with buffer A is continued until the absorbance at 280 nm approaches 0.
8. Elution is carried out with buffer B and the peak fraction is collected (approximately 15 mL) as monitored by the absorbance at 280 nm.
9. Biotin is removed from the sample by buffer exchange, using a 53-mL HiPrep 26/10 desalting column, preequilibrated with buffer A.

4.6.4 Notes

The protocol can be completed in 5 h. The yield varies dependent on the particular protein recombinantly produced in *E. coli*. A typical yield for Strep-tagged NifW purified by this method is approximately 60 mg. Examples of recombinant Strep-tagged proteins isolated by this method are shown in Fig. 5.

4.7 Affinity purification of assembly complexes using Strep-tagged proteins as bait

4.7.1 Equipment

Glove box anaerobic chamber (Coy Laboratory Products, Grass Lake, MI) operated with a 95% N_2, 5% H_2 atmosphere.

P-1 peristaltic pump (GE Healthcare, Chicago, IL).

6×5 mL HiTrap Q sepharose FF 5 mL columns (GE Healthcare, Chicago, IL).
50 mL of Strep-Tactin XT Superflow resin, packed into a gravity column (IBA Lifescience, Göttingen, Germany).
3 mL of Strep-Tactin XT High capacity resin (IBA Life-science, Göttingen, Germany).
Pierce™ disposable plastic columns, 5 mL, 0.25 in. diameter (Thermo Scientific, Waltham, MA).

4.7.2 Buffers and reagents

Purified bait protein
Cell-free extract from the appropriate *A. vinelandii* strain (see Notes)
Buffer A, 50 m*M* Tris–HCl (pH 8.0), 100 m*M* NaCl, 20 m*M* EDTA
Buffer B, 50 m*M* Tris–HCl (pH 7.9), 1.6 m*M* $Na_2S_2O_4$
Buffer C, 50 m*M* Tris–HCl (pH 7.9), 100 m*M* NaCl, 1.6 m*M* $Na_2S_2O_4$
Buffer D, 50 m*M* Tris–HCl (pH 7.9), 0.5 *M* NaCl, 1.6 m*M* $Na_2S_2O_4$
Buffer E, 50 m*M* Tris–HCl (pH 7.9), 350 m*M* NaCl, 1.6 m*M* $Na_2S_2O_4$
Buffer F, 50 *mM* Tris–HCl (pH 7.9), 350 *mM* NaCl, 1.6 *mM* $Na_2S_2O_4$, 50 *mM* biotin
All buffers are degassed and sparged with Ar before addition of $Na_2S_2O_4$
Bradford reagent (Biorad, Hercules, CA)
4 *M* NaOH.

4.7.3 Procedure

1. Preparation of bait column:
   a. 60 mg of bait protein (prepared as described in Section 4.6) is mixed with 3 mL of Strep-Tactin XT® High Capacity resin for 1 h at room temperature. It is important that the concentration of the bait protein exceeds the binding capacity of biotin-binding protein to the resin. Saturation of available resin-binding sites by the bait protein can be evaluated by permitting the resin to settle and testing the buffer layer above the resin for the presence of bait protein using Bradford Reagent. Detection of protein in this layer means that the available binding sites are saturated. The resin is packed into a Pierce gravity flow column and excess bait protein unbound to the resin is removed by washing the column with 5 mL of buffer A. Buffer A is used to fill the rest of the column and both ends of the bait column are sealed (capped) and placed inside the glovebox. Sealing the bait column prior to its placement in the glove box is important in order to

prevent bubbles from forming in the column matrix during the vacuum evacuation steps that are required to equilibrate the glove box with a 95% N_2, 5% H_2 atmosphere.

b. Once the glove box is equilibrated with a 95% N_2, 5% H_2 atmosphere, both ends of the gravity flow column are unsealed and the column is washed using 7.5 mL of buffer B. The gravity flow bait column is now ready for use.

2. Cell-free extracts are processed prior to bait–prey affinity purification by the following steps. Six HiTrap Q FF columns are connected in series and equilibrated with buffer B by passing approximately 450 mL of buffer B over the columns using a flow rate of 5 mL/min controlled by a peristaltic pump. A peristaltic pump is then used to load approximately 115 mL of cell-free extract prepared from 75 g of cells onto the equilibrated columns, which are subsequently washed with 60 mL of buffer C at a flow rate of 5 mL/min.
3. Crude extract proteins bound to the columns are eluted using approximately 60 mL of buffer D at a rate of 5 mL/min with the effluent being directly loaded onto a 50-mL Strep-Tactin XT Superflow® gravity flow column equilibrated in buffer E. Biotin proteins such as acetate carboxylase and pyruvate carboxylase are captured on the column, whereas all other crude extract proteins are in the effluent.
4. The effluent from the previous step is directly loaded onto the gravity flow bait column prepared as described in step 1.
5. Once loaded with processed crude extract the gravity flow bait column is washed with 15 mL of buffer E and bait–prey protein complexes are eluted using 5 mL of buffer F.

4.7.4 Notes

The entire procedure can be completed within 5 h. Processed crude extracts produced in Section 4.7.3, step 3 will often saturate a single gravity bait column produced in Section 4.7.3, step 1, so it is useful to have more than one gravity bait column available for loading when performing this procedure. The results of three separate bait–prey affinity purification procedures are shown in Fig. 6. Fig. 6A shows the purification of a NifEN–NifX complex, involved in FeMo-cofactor formation, that is produced by a strain deleted for *nifHDK* (DJ1143). In this case NifX is the bait and the NifEN complex is the prey. Fig. 6B and C shows the purification of MoFe protein-NifY and MoFe protein-NafY complexes produced by a strain (DJ35) deleted for the *nifE* gene whose product is required for FeMo-cofactor formation, using respectively, NifY or NafY as the bait and the MoFe protein (NifDK) as the prey. There are no Strep-tagged proteins present in *A. vinelandii* crude extracts used in bait–prey complex purification strategies. Excess prey protein not present within the complex can be readily separated from the corresponding complex by standard size-exclusion chromatography methods.

5. Conclusions

A major technical breakthrough was achieved in the field when our laboratories developed strategies to take advantage of IMAC methods for the rapid purification of nitrogenase. This advance, together with our ability to rapidly construct strains that have one or more amino acid substitutions within nitrogenase components, was foundational to the identification and characterization of reduction intermediates in the catalytic process (Lukoyanov et al., 2011, 2016). The approach also made possible the isolation of apo-forms of the MoFe protein (Christiansen et al., 1998; Ribbe et al., 2002), as well as purification of the NifEN scaffold (Goodwin et al., 1998), upon which FeMo-cofactor assembly is completed. Since the development of the original IMAC-based protocol for purification of nitrogenase and various components involved in its maturation, there have been several improvements that are described in Section 4.4.

In spite of the enormous advantages of the IMAC method, there are also significant drawbacks, the first being the presence of, and, therefore, possible contamination by, the metal used to charge the affinity column. The second problem is the possible sequestration of nitrogenase-associated metals by the IMAC matrix. Finally, it is now known that the high concentration of imidazole required to elute most proteins immobilized on IMAC columns can cause dissociation of protein–protein complexes. To circumvent these issues, STAC-based methods were developed for the purification of nitrogenase, as well as assembly intermediates that are produced during the maturation process. These new methods have already revealed the sequential and differential interaction of a number of assembly factors with the MoFe protein during its maturation (Jimenez-Vicente et al., 2018). We believe that application of the STAC methods described in this chapter will ultimately lead to a comprehensive understanding of all the individual steps that occur during the complex maturation processes involving each of the three nitrogenase systems present in *A. vinelandii*. Because the approach can be scaled up or scaled down, with little or no effect on yield, sufficient amounts of material can be produced for the application of sophisticated biophysical analyses to explore various aspects of the nitrogenase mechanism, for example, the role of the central carbide in the FeMo-cofactor. Finally, it might now be possible to use a biotin-based system to tether Strep-tagged MoFe proteins to surfaces for electrochemical experiments.

References

Anderson, J. S., Rittle, J., & Peters, J. C. (2013). Catalytic conversion of nitrogen to ammonia by an iron model complex. *Nature*, *501*(7465), 84–87.

Austin, S., Buck, M., Cannon, W., Eydmann, T., & Dixon, R. (1994). Purification and in vitro activities of the native nitrogen fixation control proteins NifA and NifL. *Journal of Bacteriology*, *176*(12), 3460–3465.

Beatty, P. H., & Good, A. G. (2011). Plant science. Future prospects for cereals that fix nitrogen. *Science (New York, NY)*, *333*(6041), 416–417.

Bennett, L. T., Cannon, F., & Dean, D. R. (1988). Nucleotide sequence and mutagenesis of the *nifA* gene from *Azotobacter vinelandii*. *Molecular Microbiology*, *2*(3), 315–321.

Bennett, L. T., Jacobson, M. R., & Dean, D. R. (1988). Isolation, sequencing, and mutagenesis of the *nifF* gene encoding flavodoxin from *Azotobacter vinelandii*. *Journal of Biological Chemistry*, *263*(3), 1364–1369.

Bishop, P. E., Jarlenski, D. M., & Hetherington, D. R. (1980). Evidence for an alternative nitrogen fixation system in *Azotobacter vinelandii*. *Proceedings of the National Academy of Sciences of the United States of America*, 77(12), 7342–7346.

Bishop, P. E., Premakumar, R., Dean, D. R., Jacobson, M. R., Chisnell, J. R., Rizzo, T. M., et al. (1986). Nitrogen fixation by *Azotobacter vinelandii* strains having deletions in structural genes for nitrogenase. *Science (New York, N.Y.)*, *232*(4746), 92–94.

Blanco, G., Drummond, M., Woodley, P., & Kennedy, C. (1993). Sequence and molecular analysis of the *nifL* gene of *Azotobacter vinelandii*. *Molecular Microbiology*, *9*(4), 869–879.

Bosch, R., Rodríguez-Quiñones, F., & Imperial, J. (1997). Identification of gene products from the *Azotobacter vinelandii nifBfdxNnifOQ* operon. *Federation of European Microbiological Societies Microbiology Letters*, *157*(1), 19–25.

Brown, K. A., Harris, D. F., Wilker, M. B., Rasmussen, A., Khadka, N., Hamby, H., et al. (2016). Light-driven dinitrogen reduction catalyzed by a CdS:nitrogenase MoFe protein biohybrid. *Science (New York, NY)*, *352*(6284), 448–450.

Burén, S., & Rubio, L. M. (2018). State of the art in eukaryotic nitrogenase engineering. *Federation of European Microbiological Societies Microbiology Letters*, *365*(2), fnx274.

Burris, R. H. (1991). Nitrogenases. *Journal of Biological Chemistry*, *266*(15), 9339–9342.

Chan, J. M., Wu, W., Dean, D. R., & Seefeldt, L. C. (2000). Construction and characterization of a heterodimeric iron protein: Defining roles for adenosine triphosphate in nitrogenase catalysis. *Biochemistry*, *39*(24), 7221–7228.

Chen, J. G., Crooks, R. M., Seefeldt, L. C., Bren, K. L., Bullock, R. M., Darensbourg, M. Y., et al. (2018). Beyond fossil fuel-driven nitrogen transformations. *Science (New York, N.Y.)*, *360*(6391), eaar6611.

Chisnell, J. R., Premakumar, R., & Bishop, P. E. (1988). Purification of a second alternative nitrogenase from a *nifHDK* deletion strain of *Azotobacter vinelandii*. *Journal of Bacteriology*, *170*(1), 27–33.

Christiansen, J., Goodwin, P. J., Lanzilotta, W. N., Seefeldt, L. C., & Dean, D. R. (1998). Catalytic and biophysical properties of a nitrogenase Apo-MoFe protein produced by a nifB-deletion mutant of *Azotobacter vinelandii*. *Biochemistry*, *37*(36), 12611–12623.

Curatti, L., Brown, C. S., Ludden, P. W., & Rubio, L. M. (2005). Genes required for rapid expression of nitrogenase activity in *Azotobacter vinelandii*. *Proceedings of the National Academy of Sciences of the United States of America*, *102*(18), 6291–6296.

Curatti, L., Ludden, P. W., & Rubio, L. M. (2006). NifB-dependent *in vitro* synthesis of the iron–molybdenum cofactor of nitrogenase. *Proceedings of the National Academy of Sciences of the United States of America*, *103*(14), 5297–5301.

Curatti, L., & Rubio, L. M. (2014). Challenges to develop nitrogen-fixing cereals by direct nif-gene transfer. *Plant Science: An International Journal of Experimental Plant Biology*, *225*, 130–137.

Danyal, K., Dean, D. R., Hoffman, B. M., & Seefeldt, L. C. (2011). Electron transfer within nitrogenase: Evidence for a deficit-spending mechanism. *Biochemistry*, *50*(43), 9255–9263.

Dixon, R., & Kahn, D. (2004). Genetic regulation of biological nitrogen fixation. *Nature Reviews. Microbiology*, *2*(8), 621–631.

Dos Santos, P. C. (2011). Molecular biology and genetic engineering in nitrogen fixation. *Methods in Molecular Biology (Clifton, N.J.)*, *766*, 81–92.

Erisman, J. W., Galloway, J. N., Dise, N. B., Sutton, M. A., Bleeker, A., Grizzetti, B., et al. (2015). *Nitrogen: Too much of a vital resource*. Zeist, The Netherlands: WWF Netherlands. April.

Erisman, J. W., Sutton, M. A., Galloway, J., Klimont, Z., & Winiwarter, W. (2008). How a century of ammonia synthesis changed the world. *Nature Geoscience*, *1*, 636–639. September 28.

Fay, A. W., Hu, Y., Schmid, B., & Ribbe, M. W. (2007). Molecular insights into nitrogenase FeMoco insertion—The role of His 274 and His 451 of MoFe protein alpha subunit. *Journal of Inorganic Biochemistry*, *101*(11 − 12), 1630–1641.

Georgiadis, M. M., Komiya, H., Chakrabarti, P., Woo, D., Kornuc, J. J., & Rees, D. C. (1992). Crystallographic structure of the nitrogenase iron protein from *Azotobacter vinelandii*. *Science (New York, N.Y.)*, *257*(5077), 1653–1659.

Goodwin, P. J., Agar, J. N., Roll, J. T., Roberts, G. P., Johnson, M. K., & Dean, D. R. (1998). The *azotobacter vinelandii* NifEN complex contains two identical [4Fe-4S] clusters. *Biochemistry*, *37*(29), 10420–10428.

Hamilton, T. L., Ludwig, M., Dixon, R., Boyd, E. S., Dos Santos, P. C., Setubal, J. C., et al. (2011). Transcriptional profiling of nitrogen fixation in *Azotobacter vinelandii*. *Journal of Bacteriology*, *193*(17), 4477–4486.

Harris, D. F., Lukoyanov, D. A., Shaw, S., Compton, P., Tokmina-Lukaszewska, M., Bothner, B., et al. (2018). Mechanism of N_2 reduction catalyzed by Fe-nitrogenase involves reductive elimination of H_2. *Biochemistry*, *57*(5), 701–710.

Harris, D., Yang, Z., Dean, D. R., Seefeldt, L. C., & Hoffman, B. M. (2018). Kinetic understanding of N_2 reduction versus H_2 evolution at the E4(4H) janus state in the three nitrogenases. *Biochemistry*, *57*, 5706–5714.

Hernandez, J. A., Curatti, L., Aznar, C. P., Perova, Z., Britt, R. D., & Rubio, L. M. (2008). Metal trafficking for nitrogen fixation: NifQ donates molybdenum to NifEN/NifH for the biosynthesis of the nitrogenase FeMo-cofactor. *Proceedings of the National Academy of Sciences of the United States of America*, *105*(33), 11679–11684.

Hernandez, J. A., Igarashi, R. Y., Soboh, B., Curatti, L., Dean, D. R., Ludden, P. W., et al. (2007). NifX and NifEN exchange NifB cofactor and the VK-cluster, a newly isolated intermediate of the iron-molybdenum cofactor biosynthetic pathway. *Molecular Microbiology*, *63*(1), 177–192.

Hoffman, B. M., Lukoyanov, D., Yang, Z.-Y., Dean, D. R., & Seefeldt, L. C. (2014). Mechanism of nitrogen fixation by nitrogenase: The next stage. *Chemical Reviews*, *114*(8), 4041–4062.

Hoover, T. R., Imperial, J., Ludden, P. W., & Shah, V. K. (1989). Homocitrate is a component of the iron-molybdenum cofactor of nitrogenase. *Biochemistry*, *28*(7), 2768–2771.
Hu, Y., Fay, A. W., Lee, C. C., & Ribbe, M. W. (2007). P-cluster maturation on nitrogenase MoFe protein. *Proceedings of the National Academy of Sciences of the United States of America*, *104*(25), 10424–10429.
Jacobson, M. R., Cash, V. L., Weiss, M. C., Laird, N. F., Newton, W. E., & Dean, D. R. (1989). Biochemical and genetic analysis of the *nifUSVWZM* cluster from *Azotobacter vinelandii*. *Molecular & General Genetics: MGG*, *219*(1–2), 49–57.
Jimenez-Vicente, E., Yang, Z.-Y., Ray, W. K., Echavarri-Erasun, C., Cash, V. L., Rubio, L. M., et al. (2018). Sequential and differential interaction of assembly factors during nitrogenase MoFe protein maturation. *Journal of Biological Chemistry*, *293*(25), 9812–9823.
Johnson, D. C., Dos Santos, P. C., & Dean, D. R. (2005). NifU and NifS are required for the maturation of nitrogenase and cannot replace the function of *isc-gene* products in *Azotobacter vinelandii*. *Biochemical Society Transactions*, *33*, 90–93Pt. 1.
Kennedy, C., & Dean, D. (1992). The *nifU*, *nifS* and *nifV* gene products are required for activity of all three nitrogenases of *Azotobacter vinelandii*. *Molecular & General Genetics: MGG*, *231*(3), 494–498.
Kim, J., & Rees, D. C. (1992). Crystallographic structure and functional implications of the nitrogenase molybdenum-iron protein from *Azotobacter vinelandii*. *Nature*, *360*(6404), 553–560.
Ladha, J. K., Tirol-Padre, A., Reddy, C. K., Cassman, K. G., Verma, S., Powlson, D. S., et al. (2016). Global nitrogen budgets in cereals: A 50-year assessment for maize, rice, and wheat production systems. *Scientific Reports*, *6*, 19355.
Lukoyanov, D., Dikanov, S. A., Yang, Z.-Y., Barney, B. M., Samoilova, R. I., Narasimhulu, K. V., et al. (2011). ENDOR/HYSCORE studies of the common intermediate trapped during nitrogenase reduction of N2H2, CH3N2H, and N2H4 support an alternating reaction pathway for N2 reduction. *Journal of the American Chemical Society*, *133*(30), 11655–11664.
Lukoyanov, D., Khadka, N., Yang, Z.-Y., Dean, D. R., Seefeldt, L. C., & Hoffman, B. M. (2016). Reductive elimination of H_2 activates nitrogenase to reduce the N≡N triple bond: Characterization of the E4(4H) Janus intermediate in wild-type enzyme. *Journal of the American Chemical Society*, *138*(33), 10674–10683.
Moshiri, F., Crouse, B. R., Johnson, M. K., & Maier, R. J. (1995). The "nitrogenase-protective" FeSII protein of *Azotobacter vinelandii*: Overexpression, characterization, and crystallization. *Biochemistry*, *34*(40), 12973–12982.
Paengnakorn, P., Ash, P. A., Shaw, S., Danyal, K., Chen, T., Dean, D. R., et al. (2017). Infrared spectroscopy of the nitrogenase MoFe protein under electrochemical control: Potential-triggered CO binding. *Chemical Science*, *8*(2), 1500–1505.
Ribbe, M. W., Hu, Y., Guo, M., Schmid, B., & Burgess, B. K. (2002). The FeMoco-deficient MoFe protein produced by a *nifH* deletion strain of *Azotobacter vinelandii* shows unusual P-cluster features. *Journal of Biological Chemistry*, *277*(26), 23469–23476.
Rubio, L. M., & Ludden, P. W. (2008). Biosynthesis of the iron-molybdenum cofactor of nitrogenase. *Annual Review of Microbiology*, *62*, 93–111.
Rubio, L. M., Singer, S. W., & Ludden, P. W. (2004). Purification and characterization of NafY (apodinitrogenase γ subunit) from *Azotobacter vinelandii*. *Journal of Biological Chemistry*, *279*(19), 19739–19746.
Schlesier, J., Rohde, M., Gerhardt, S., & Einsle, O. (2016). A conformational switch triggers nitrogenase protection from oxygen damage by shethna protein II (FeSII). *Journal of the American Chemical Society*, *138*(1), 239–247.
Seefeldt, L. C., Hoffman, B. M., & Dean, D. R. (2009). Mechanism of Mo-dependent nitrogenase. *Annual Review of Biochemistry*, *78*, 701–722.

Seefeldt, L. C., Hoffman, B. M., Peters, J. W., Raugei, S., Beratan, D. N., Antony, E., et al. (2018). Energy transduction in nitrogenase. *Accounts of Chemical Research*, *51*(9), 2179–2186.

Seefeldt, L. C., Peters, J. W., Beratan, D. N., Bothner, B., Minteer, S. D., Raugei, S., et al. (2018). Control of electron transfer in nitrogenase. *Current Opinion in Chemical Biology*, *47*, 54–59.

Setubal, J. C., dos Santos, P., Goldman, B. S., Ertesvåg, H., Espin, G., Rubio, L. M., et al. (2009). Genome sequence of *Azotobacter vinelandii*, an obligate aerobe specialized to support diverse anaerobic metabolic processes. *Journal of Bacteriology*, *191*(14), 4534–4545.

Shah, V. K., & Brill, W. J. (1973). Nitrogenase. IV. Simple method of purification to homogeneity of nitrogenase components from *Azotobacter vinelandii*. *Biochimica et Biophysica Acta*, *305*(2), 445–454.

Smith, A. D., Jameson, G. N. L., Dos Santos, P. C., Agar, J. N., Naik, S., Krebs, C., et al. (2005). NifS-mediated assembly of [4Fe-4S] clusters in the N- and C-terminal domains of the NifU scaffold protein. *Biochemistry*, *44*(39), 12955–12969.

Vicente, E. J., & Dean, D. R. (2017). Keeping the nitrogen-fixation dream alive. *Proceedings of the National Academy of Sciences of the United States of America*, *114*(12), 3009–3011.

Zheng, L., White, R. H., & Dean, D. R. (1997). Purification of the *Azotobacter vinelandii nifV*-encoded homocitrate synthase. *Journal of Bacteriology*, *179*(18), 5963–5966.

CHAPTER TEN

Membrane-spanning electron transfer proteins from electrogenic bacteria: Production and investigation

Colin W.J. Lockwood, Jessica H. van Wonderen, Marcus J. Edwards, Samuel E.H. Piper, Gaye F. White, Simone Newton-Payne, David J. Richardson, Thomas A. Clarke, Julea N. Butt*

Centre for Molecular and Structural Biochemistry, School of Chemistry and School of Biological Sciences, University of East Anglia, Norwich Research Park, Norwich, United Kingdom

*Corresponding author: e-mail address: j.butt@uea.ac.uk

Contents

Abstract

Certain bacterial species have a natural ability to exchange electrons with extracellular redox partners. This behavior allows coupling of catalytic transformations inside bacteria to complementary redox transformations of catalysts and electrodes outside the cell. Electricity generation can be coupled to waste-water remediation. Industrially relevant oxidation reactions can proceed exclusively when electrons are released to anodes. Reduced products such as fuels can be generated when electrons are provided from

Methods in Enzymology, Volume 613
ISSN 0076-6879
https://doi.org/10.1016/bs.mie.2018.10.011

(photo)cathodes. Rational development of these opportunities and inspiration for novel technologies is underpinned by resolution at the molecular level of pathways supporting electron exchange across bacterial cell envelopes. This chapter describes methods for purification, engineering, and in vitro characterization of proteins providing the primary route for electron transport across the outer membrane lipid bilayer of *Shewanella oneidensis* MR-1, a well-described electrogenic bacterium and chassis organism for related biotechnologies.

1. Introduction

Redox enzymes provide attractive renewable electrocatalysts and often deliver performances that match, or exceed, those of their best abiotic counterparts. As a consequence these enzymes provide much inspiration for sustainable energy technologies. Electrocatalysis has been demonstrated for purified enzymes exchanging electrons with electrodes (Jenner & Butt, 2018; Masa & Schuhmann, 2016). Photocatalysis occurs when redox enzymes receive photoenergized electrons (holes) from molecular dyes, (nanostructured) semiconductors, and quantum dots (Bachmeier & Armstrong, 2015; Lee, Choi, Kuk, & Park, 2018). Nevertheless, there are significant bottlenecks to using such approaches for scalable synthesis of high-value chemicals, including fuels, not least because enzyme purification can be costly and time-consuming. Methods to overcome the limited stability of the purified materials may be necessary, for example, covalent attachment to solid supports or entrapment in O_2 removing polymers.

The need for protein purification is avoided when using cell-based systems, and here the opportunities for electrocatalyst self-repair and regeneration are also attractive (Lapinsonniere, Picot, & Barriere, 2012; Sakimoto, Wong, & Yang, 2016; Sakimoto, Kornienko, & Yang, 2017). Harnessing redox enzyme catalysis in such a context requires a flow of electrons into, or out from, the cell. For this purpose the outer membrane-associated porin:cytochrome complexes and extracellular cytochromes (Fig. 1A), recognized in a growing number of Gram-negative bacteria, are highly relevant (Shi et al., 2016; White et al., 2016). These proteins transfer electrons across an otherwise electrically insulating lipid bilayer in natural and engineered bacteria (Schuergers, Werlang, Ajo-Franklin, & Boghossian, 2017) in behavior supporting several energy-facing biotechnologies (Bursac, Gralnick, & Gescher, 2017; Fan, Dundas, Graham, Lynd, & Keitz, 2018; Flynn, Ross, Hunt, Bond, & Gralnick, 2010; Logan & Rabaey, 2012; Sasaki et al., 2018;

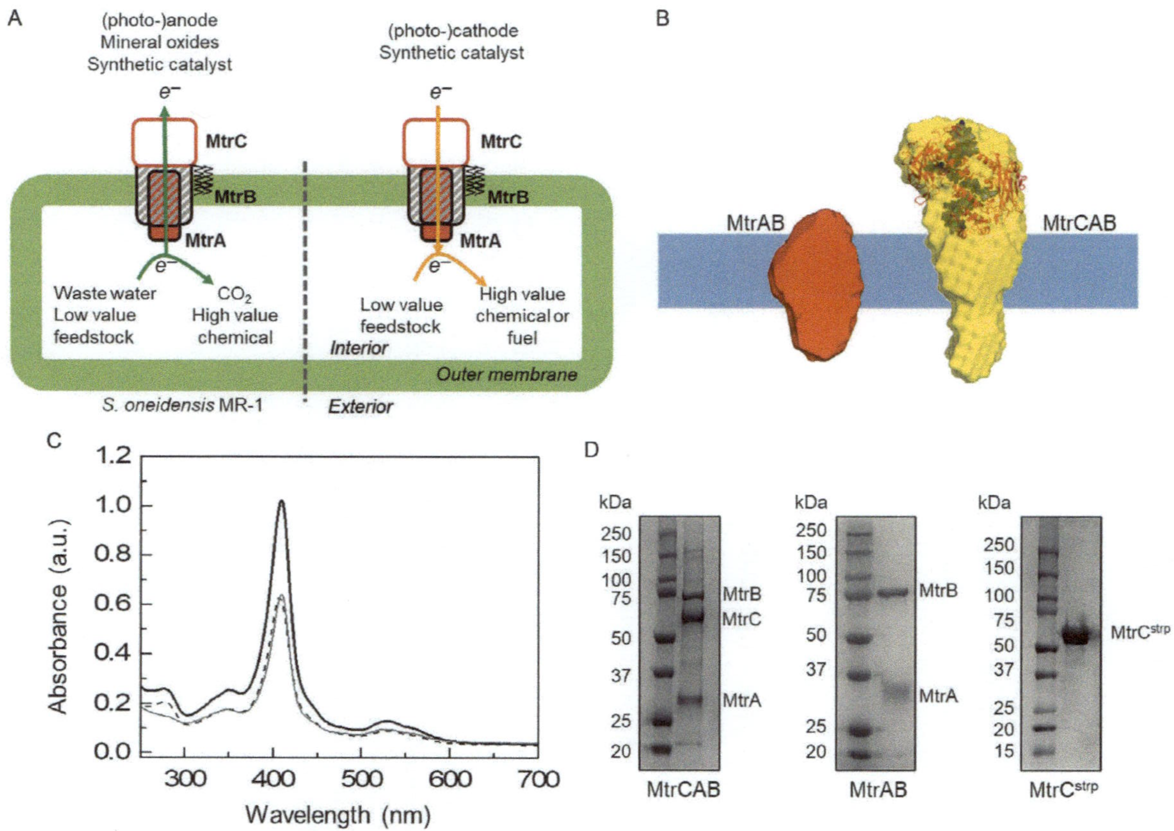

Fig. 1 Schematic of the MtrCAB complex (A) within the outer membrane (*green*) of MR-1, with the lipid modification of MtrC implied as a *black* "tail." *Arrows* represent the bidirectional flow of electrons across the membrane, which has the potential to be exploited for biotechnology (see text). Molecular envelopes (B) of MtrAB and MtrCAB resolved by SANS (Edwards et al., 2018) showing their proposed orientation within the outer membrane. The crystal structure of MtrC (PDB, 4LM8) positioned within (using SUPCOMB) the MtrCAB envelope: secondary protein structure is shown in *red* and the 10 *c*-type hemes are shown as *blue/gray spheres*. UV–visible absorption spectra (C) of air-oxidized 5 μ*M* MtrCAB (*black*), MtrAB (*broken line*), and MtrC (*gray*) in 50 m*M* HEPES, 50 m*M* NaCl, 5 m*M* LDAO, pH 7.8. Coomassie-stained SDS-PAGE gels (D) of MtrCAB, MtrAB, and $MtrC^{strp}$ as indicated with molecular weight markers on *left*.

Wang & Ren, 2013). Reductive transformations are driven by electrons entering cells from (photo)cathodes. Oxidized products form exclusively when fermentation delivers excess electrons to (photo)anodes. Furthermore, microbial fuel cells harness electricity produced by bacteria during oxidative waste-water remediation.

Shewanella oneidensis MR-1 (MR-1) has become a model organism for studying electron transfer across the outer membrane, and a chassis for synthetic biology demonstrating proof of principle in related biotechnologies, due to its genetic tractability, versatile metabolism, and phylogenetically widespread mechanism for electron exchange across the outer membrane (Shi et al., 2016; White et al., 2016). This chapter discusses approaches used by our laboratories to purify, clone, express, engineer, and characterize proteins associated with electron exchange across the MR-1 outer membrane.

2. Purification of the outer membrane-spanning MtrCAB complex from *S. oneidensis* MR-1

The primary route for electron exchange across the MR-1 outer membrane (Fig. 1A) is provided by the MtrCAB complex (Edwards et al., 2018; Hartshorne et al., 2009). MtrAB spans the lipid bilayer, Fig. 1B left: MtrA contains 10 covalently bound *c*-type hemes and MtrB is predicted to fold as a porin wrapped around MtrA. MtrC contains 10 *c*-type hemes and binds to the extracellular face of the MtrAB porin: cytochrome complex, Fig. 1B right. MtrCAB is purified as a stable, tightly bound heterotrimeric complex from MR-1 grown under (micro-)aerobic conditions in Lysogeny broth (LB) supplemented with lactate and Fe(III) citrate as described later. Pure MtrCAB is red in color and displays electronic absorbance typical of the His/His axially ligated, low-spin *c*-type hemes that it contains (Fig. 1C). When resolved by SDS-PAGE using 12% acrylamide gels and stained with Coomassie dye (Fig. 1D) three clear bands are revealed having apparent masses of approximately $\approx$75 kDa (MtrC), $\approx$65 kDa (MtrB), and $\approx$33 kDa (MtrA).

Triton X-100 is used to solubilize MtrCAB from MR-1 membrane fractions during purification. However, Triton X-100 masks the proteins' absorbance at UV wavelengths and these detergent micelles usually concentrate alongside MtrCAB producing viscous solutions. As a consequence, pure MtrCAB is exchanged into Fos-Choline-12 for Small Angle Neutron Scattering (SANS) and lauryldimethylamine *N*-oxide (LDAO) for incorporation into liposomes to study electron transfer across a lipid bilayer.

2.1 Growth and harvesting of *S. oneidensis* MR-1

Growth of all liquid cultures is at 30°C with shaking at 200 rpm. A single colony of MR-1 picked from an LB agar plate inoculates 10 mL LB that is cultured overnight. From the resulting culture, 1 mL inoculates 100 mL LB that is cultured overnight. The resulting culture inoculates 1 L of LB containing 50 m*M* sodium D, L-lactate and 20 m*M* Fe(III) citrate at pH 7.8. Note: the media pH is lowered by the presence of ferric citrate and aliquots of 10 *M* NaOH must be added to achieve the desired pH. A further overnight culture (≈18 h) is performed during which time iron precipitates can form and these are removed by low-speed centrifugation (500 × *g*, 4°C, 20 min) prior to harvesting the cells by more rapid centrifugation (6230 × *g*, 4°C, 20 min). The pelleted cells are resuspended and washed in 20 m*M* HEPES, 50 m*M* NaCl, pH 7.6 using approximately 15 mL per 1 L of culture. Cells are pelleted by a further round of centrifugation prior to flash-freezing in liquid nitrogen and storage at −20°C.

2.2 Membrane isolation

The cell pellet is thawed during gentle resuspension of cells from 1 L culture in 20 mL of 20 m*M* HEPES, pH 7.8 containing lysozyme and DNase. When thoroughly resuspended the cells are lysed by two passes through a French press (16,000 psi) then unbroken cells and cell debris are removed by centrifugation (15,000 × *g*, 4°C, 30 min). The membrane fraction (containing inner and outer membranes) is pelleted by ultracentrifugation (200,000 × *g*, 4°C, 100 min) and resuspended in 20 m*M* HEPES, 50 m*M* NaCl, pH 7.8 by gentle stirring overnight (4°C). Addition of 2% (w/v) sodium lauroyl sarcosinate (sarkosyl) over 45 min with gentle stirring (4°C) preferentially solubilizes the inner membrane. Ultracentrifugation (200,000 × *g*, 4°C, 100 min) produces a red-brown pellet containing predominantly outer membrane and its protein components including MtrCAB. That pellet is solubilized in 30 mL Triton X-100 5% (v/v unless stated otherwise), 20 m*M* HEPES, pH 7.8, by gentle stirring overnight (4°C). A final round of ultracentrifugation removes insoluble material and the red-brown supernatant containing MtrCAB is resolved by chromatography as described later.

2.3 Chromatography

The Triton X-100 solubilized MtrCAB-containing solution is loaded on to a Q-Sepharose column (60 mL) preequilibrated with 2% Triton X-100,

20 m*M* HEPES, 50 m*M* NaCl, pH 7.8 (Buffer A) using a flow rate of 2 mL min^{-1}. After washing the column with two-column volumes of Buffer A, it is developed with a linear gradient 0%–50% of Buffer B (Buffer A with 1 *M* NaCl) over 800 mL. MtrCAB elutes at ≈0.2 *M* NaCl and the corresponding fractions are identified by electronic absorbance and SDS-PAGE (Fig. 1C and D). MtrCAB-containing fractions are pooled, transferred to dialysis tubing (12 kDa MWCO), and dialyzed against 2% Triton X-100, 20 m*M* TRIS, 50 m*M* NaCl, pH 8.5 (Buffer C) overnight (2 × 5 L, 4°C). The dialyzed sample is loaded on to a DEAE column (130 mL) preequilibrated with Buffer C (2 mL min^{-1}). MtrCAB binds to the column, which is washed with Buffer C (two-column volumes) and eluted with a gradient of 0%–50% Buffer D (Buffer C + 1 *M* NaCl). The red fractions containing MtrCAB are pooled and concentrated to <5 mL. The latter is achieved using either a centrifugal concentrator, or by loading onto a HiTrap Q-Sepharose column (5 mL) equilibrated with Buffer A and eluting with a single step elution to 50% Buffer B; the column avoids concentration of Triton X-100 that occurs when using spin concentrators.

In the final stage of MtrCAB purification, the concentrated protein (<5 mL) is passed through a Superdex 200 26/60 column preequilibrated with 2% Triton X-100, 20 m*M* HEPES, 150 m*M* NaCl, pH 7.8 at a flow rate of 0.5 mL min^{-1}. MtrCAB forms a red band moving through the column and fractions containing the eluted complex should be pure when assessed by SDS-PAGE with Coomassie stain to visualize total protein. Purified protein is quantified by the electronic absorbance at 410 nm for the oxidized protein using $\varepsilon = 2200\,mM^{-1}\,cm^{-1}$ assuming each heme has $\varepsilon = 110\,mM^{-1}\,cm^{-1}$. A typical yield is 1 mg MtrCAB per 1 L of culture and samples containing pure MtrCAB are aliquoted and stored at −80°C after flash-freezing in liquid nitrogen.

2.4 Detergent exchange

MtrCAB as purified in Triton X-100 is diluted threefold with 20 m*M* HEPES, 50 m*M* NaCl, pH 7.8 containing the required detergent, i.e., 5 m*M* LDAO (Buffer E), ensuring the final detergent concentration is above the critical micelle concentration. The sample is loaded onto a HiTrap Q-Sepharose column (5 mL) equilibrated with Buffer E. The column is washed with four-column volumes (0.5 mL min^{-1}) of Buffer E, and the MtrCAB is eluted in a single step with Buffer F (Buffer E + 500 m*M* NaCl). The resulting sample can be concentrated with a Microcon-30 kDa centrifugal filter (Millipore) without the detergent (LDAO) concentration changing. Unlike Triton X-100, LDAO does not absorb at 280 nm revealing that MtrCAB has $A_{410nm}/A_{280nm} \approx 4.5$.

3. Purification of recombinant MtrAB and MtrC from *S. oneidensis* MR-1

The pBAD202 plasmid (Invitrogen) provides robust access to recombinant proteins in MR-1 (Shi, Lin, Markillie, Squier, & Hooker, 2005). Conferring resistance to kanamycin, this plasmid places the gene of interest under control of an arabinose-inducible promoter. Recombinant proteins are produced in MR-1 or a desired strain, for example, one deficient in the *mtr* operon encoding for MtrCAB, its paralog MtrFED and the extracellular decaheme cytochrome OmcA homologous to MtrC (MtrF) (Fig. 2A). We choose the host strain for ease of purification of the desired protein(s) or to facilitate in vivo analysis. Various strategies are reported for MR-1 transformation. In our hands the more efficient method is electroporation using a modification of the method of Myers and Myers (1997).

3.1 Electroporation of *S. oneidensis* MR-1

From a single colony of MR-1 on LB agar, a 100 mL LB culture having $OD_{600nm} = 0.4$ (it is essential that this value is below 0.5) is prepared by the method described above. Cells are pelleted by centrifugation (3220 × *g*, 4°C, 15 min) and the supernatant removed fully to minimize arching during electroporation. The cells are gently resuspended in 8 mL of an ice-cold, filtered solution of 1 *M* sorbitol, and pelleted by centrifugation (3220 × *g*, 4°C, 15 min). After removing the supernatant, the cells are gently resuspended in 2 mL of the ice-cold sorbitol solution. Electroporation is performed within 15 min and for this 390 μL of the cells are added to an ice-cold microcentrifuge tube (1.5 mL) and mixed gently with 0.1–0.5 μg of plasmid DNA added in a volume 1–2 μL; note the DNA should be prepared in sterile molecular biology grade water because arching can occur with too much salt in the sample. The mixture is transferred to an ice-cold 0.2 cm gap electroporation cuvette and treated with a single 1.10 kV pulse (Micropulser Electroporation Apparatus, BioRad) within 15 min of adding the DNA. *Immediately* after, 800 μL prewarmed (30°C) S.O.C. medium (Sigma-Aldrich) is added to the cuvette and the contents transferred to a 1.5-mL microcentrifuge tube where the cells are allowed to recover at 30°C for 1.5–2 h while

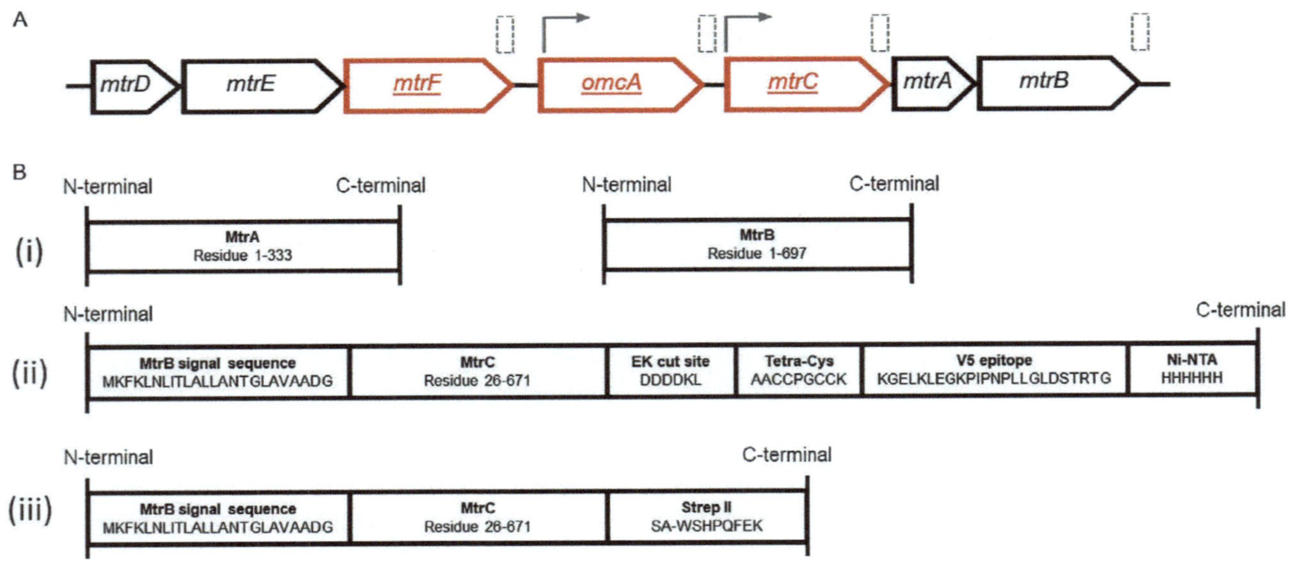

Fig. 2 Organization of MR-1 *mtr* gene locus (A) with genes for extracellular cytochromes (*red*) and porin:cytochrome components (*black*) with predicted promoters (*bent arrows*) and terminators (*dotted boxes*). Schematic (B) highlighting key features of recombinant (i) MtrAB, (ii) MtrC, (iii) $MtrC^{Strp}$, see text for details.

shaking at 200 rpm. The cells are pelleted in a microcentrifuge (4000 × *g*, room temperature, 5 min) resuspended in 100 μL fresh S.O.C. medium and plated on LB plates containing 20 μg mL^{-1} kanamycin. Red colonies characteristic of MR-1 typically become visible over 2 days incubation at 30°C.

3.2 Purification of recombinant MtrAB

Recombinant MtrAB proteins (Fig. 2Bi) are produced from a construct prepared by cloning *mtrA* (1002 bp), *mtrB* (2094 bp), and their intergenic region from the genomic DNA of MR-1, into pBAD202 (Edwards et al., 2018). The resulting plasmid (pCL001), when introduced into an *mtr*$^{-}$ MR-1 strain LS329, produces MR-1 strain *mtr*$^{-}$ *mtrAB*$^{+}$ which accumulates recombinant MtrAB in the outer membrane (Edwards et al., 2018). Purification of MtrAB is facilitated by the absence of other products from the *mtr* operon. SDS-PAGE of pure MtrAB stained by Coomassie dye reveals two bands (Fig. 1E) a tight band at ≈75 kDa and a diffuse band at ≈33 kDa corresponding to MtrB and MtrA, respectively.

MR-1 strain *mtr*$^{-}$ *mtrAB*$^{+}$ is grown in M72 media (Casein digest peptone 15 g L^{-1}, papaic digest of soybean meal 5 g L^{-1}, NaCl 5 g L^{-1}) supplemented with 20 m*M* sodium D, L-lactate, 20 m*M* fumarate, 20 m*M* HEPES, pH 7.8, and 30 μg mL^{-1} kanamycin. Cultures (1.5 L) are grown at 30°C in 2-L baffled flasks shaken at 200 rpm after inoculation (2%) with overnight cultures grown aerobically in LB containing 30 μg mL^{-1} kanamycin. In mid-exponential growth (OD_{590nm} ≈ 0.6) induction is by addition of 5 m*M* L-arabinose at which time shaking is stopped and the cultures left to grow microaerobically for a further 18 h. Cells are harvested by centrifugation (6230 × *g*, 4°C, 20 min), washed, and resuspended in 20 m*M* HEPES, 50 m*M* NaCl, pH 7.8 (Buffer F) using approximately 15 mL per 1 L of culture. Cell pellets are flash-frozen in liquid nitrogen and stored at −20°C until required.

Cell lysis and outer membrane isolation are as described above for purification of MtrCAB aside from preferential solubilization of the inner membrane using 1% (m/v) sarkosyl at 4°C for 45 min. Resolution of proteins in the outer membrane pellet by SDS-PAGE and stained by Coomassie dye shows MtrAB is largely free of contaminating proteins at this stage (Fig. 3A). The outer membrane pellet (Fig. 3B) is solubilized with 20 mL Buffer G (Buffer F + 5% Triton X-100). Resolution of the soluble material by SDS-PAGE and visualization by heme-dependent peroxidase activity reveals that MtrA is the only *c*-type cytochrome present in the sample.

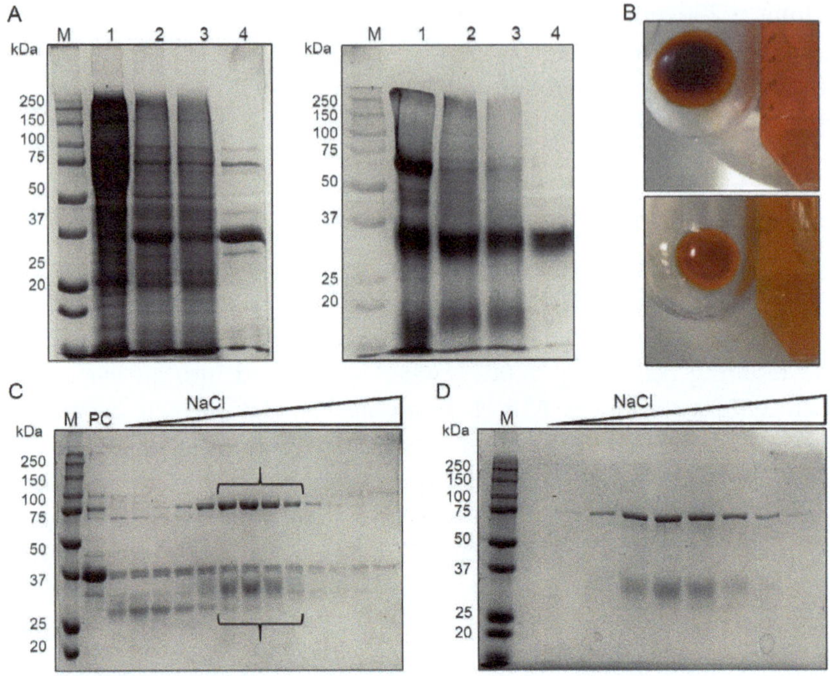

Fig. 3 SDS-PAGE gels (A) showing the purification of recombinant MtrAB. Proteins visualized by Coomassie stain (*left*) and heme-dependent peroxidase activity (*right*). Lanes (1) the soluble protein supernatant, (2) membrane pellet from first ultracentrifugation, (3) supernatant containing sarkosyl solubilized inner membrane, and (4) outer membrane pellet containing MtrAB from the second ultracentrifugation. Photograph (B) of the membrane pellets (*left*) and supernatant (*right*) for the first (*top*) and second (*bottom*) ultracentrifugation steps. Coomassie-stained SDS-PAGE gels resolve chromatographic purification of recombinant MtrAB by Q-Sepharose (C) followed by DEAE (D) columns. Lanes show Triton X-100-solubilized outer membrane pellet (PC) and fractions eluted with increasing NaCl concentration; those fractions carried forward from the Q-Sepharose column are highlighted.

Further purification of MtrAB follows chromatography as outlined above for purification of MtrCAB. MtrAB elutes from the Q-Sepharose column at $\approx 0.2\ M$ NaCl. The purest MtrAB-containing fractions as determined by SDS-PAGE (Fig. 3C) are pooled, dialyzed into Buffer C, and loaded on to the DEAE column. MtrAB typically elutes at $\approx 0.16\ M$ NaCl and again the fractions containing pure MtrAB as judged by SDS-PAGE are pooled and concentrated (Fig. 3D). At this stage, detergent exchange as described above can be carried out if required. This protocol typically yields 0.8 mg MtrAB per 1 L culture, the pure complex (in LDAO) typically has $A_{410nm}/A_{280nm} \approx 3.4$ and is quantified using $\varepsilon = 1100\,mM^{-1}\,cm^{-1}$ at 410 nm for the oxidized protein assuming each heme has $\varepsilon = 110\,mM^{-1}\,cm^{-1}$. Samples containing pure MtrAB are flash-frozen and stored at −80°C.

3.3 Purification of recombinant soluble forms of MtrC

Genomically encoded MtrC is a lipoprotein which is transported across the MR-1 outer membrane by the Type II Secretion System (Shi et al., 2008). Replacing the N-terminal acetylation site and native signal sequence of MtrC (MMNAQKSKIALLLAASAVTMALTGC) for the native signal sequence of MtrB allows for secretion of soluble forms of MtrC. These proteins are purified from spent media. Such a construct encoding for C-terminal extension including an enterokinase protease sequence and affinity tags (Fig. 2Bii) typically yields 15 mg MtrC per 1 L culture and enabled resolution of this protein's crystal structure by X-ray diffraction (Edwards et al., 2015). LC-MS of recently prepared protein revealed multiple sites of C-terminal cleavage, complicating other studies including protein film electrochemistry (Ainsworth et al., 2016; Hwang et al., 2017). We now routinely prepare soluble $MtrC^{Strp}$ from a construct *pJvW001* encoding for a 10 residue C-terminal extension where a two-residue spacer is followed by the Strep II tag (Fig. 2Biii). Following purification by affinity chromatography, as described below (Fig. 4A), LC-MS of pure

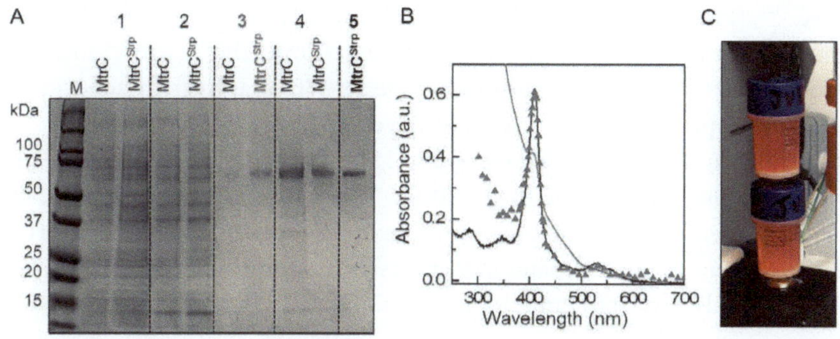

Fig. 4 Coomassie-stained SDS-PAGE gel (4%–20%) illustrating (A) production of MtrC soluble forms: $MtrC^{Strp}$ and MtrC, see text for details. Lanes (1) cells preinduction; (2) cells postinduction; spent media after pelleting cells; (3) 2× concentrated; and (4) 20× concentrated; (5) Pure $MtrC^{Strp}$ eluted from Strep-tactin XT column. Electronic absorbance spectra (B) of spent media concentrated 2× (*gray continuous line*, panel A, lane 3) and pure $MtrC^{Strp}$ (*black continuous line*, panel A, lane 5). The spectrum shown by *triangles* is that of $MtrC^{Strp}$ in spent media obtained by subtracting the spectrum of fresh media and normalizing to that of the pure protein at 550 nm. Photograph (C) of Strep-tactin XT column saturated with *red* $MtrC^{strp}$.

MtrC^{Strp} reveals a single mass as anticipated for the Strep II-tagged decaheme cytochrome. We note that spent media is loaded directly on to a Strep-tactin XT Superflow column, so care must be taken in the choice of media to minimize the possibility that its components interfere with protein binding.

MR-1 containing *pJvW001* is cultured as described above for recombinant MtrAB but with 1 L media in a 2-L baffled flask using 30 m*M* fumarate. Spent media containing MtrC^{Strp} is recovered as the supernatant after centrifugation (3600 × *g*, 4°C, 20 min). Electronic absorbance of the spent media supernatant reveals a small peak at 410 nm indicative of the cytochrome superimposed on the absorbance/scatter of media components and lysed cells (Fig. 4B, gray line and triangles). To each Liter of supernatant, 100 mL of 0.5 *M* NaH_2PO_4, 3 *M* NaCl, pH 8.0 containing 200 μL biotin blocking solution is added and the pH adjusted to 8. The resulting solution is applied to a Strep-tactin XT Superflow column (5 mL, IBA Solutions for Life Sciences) preequilibrated with 50 m*M* NaH_2PO_4, 300 m*M* NaCl, pH 8.0 (Buffer H, 1–2 mL min^{-1}). The column turns from white to deep red as MtrC^{Strp} binds (Fig. 4C). The loaded column is washed with 10-column volumes of Buffer H and MtrC^{Strp} and eluted with one- to two-column volumes of 20 m*M* biotin in Buffer H (final pH adjusted to 8.0). Columns can be stacked in series to increase capacity if needed. SDS-PAGE of the eluted material stained with Coomassie dye reveals pure MtrC^{Strp} (Fig. 4A, lane 5). The protocol typically yields 5–10 mg MtrC^{Strp} per 1 L culture, and the protein is exchanged into the desired biotin-free buffer-electrolyte solution prior to storage at −80°C as flash-frozen aliquots. Samples of pure MtrC^{Strp} typically have $A_{410nm}/A_{280nm} \approx 3.6$, (Fig. 4B, black line), and are quantified using $\varepsilon = 1260\ mM^{-1}\ cm^{-1}$ for oxidized protein at 410 nm (Hartshorne et al., 2007). The MtrC crystal structure shows that all 10 hemes have His/His ligation and as a consequence the electronic absorbance and electron paramagnetic resonance spectroscopies show evidence for only low-spin (and not high-spin) heme (Hartshorne et al., 2007).

4. Redox characterization of *S. oneidensis* MR-1 outer membrane cytochromes

Characteristic changes in the electronic absorbance spectra of the MR-1 outer membrane cytochromes accompany heme reduction, as illustrated for MtrC^{Strp} in Fig. 5A. The Soret band gains intensity, shifts to longer wavelength, and two sharp features appear in the Q-band with maxima at

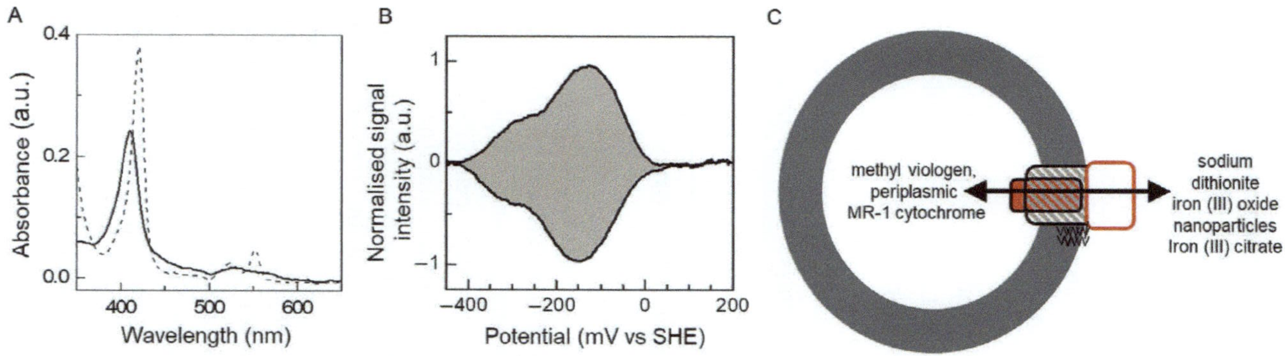

Fig. 5 Electronic absorbance spectra (A) of oxidized (*continuous line*) and dithionite-reduced (*broken line*) $MtrC^{Strp}$. Baseline-subtracted cyclic voltammogram of MtrCAB (B) adsorbed on a basal-plane graphite electrode, recorded at $30\,mV\,s^{-1}$, 0°C with 3000 rpm electrode rotation. Oxidative and reductive peaks are normalized to their respective peak currents. *Cartoon* (C) showing liposome with MtrCAB embedded; the *arrow* shows bidirectional electron transfer between the exterior and the interior compartment and relevant redox partners.

approximately 520 and 550 nm. These properties allow heme reduction potentials to be determined by potentiometric titration monitored by electronic absorbance at visible wavelengths (Hartshorne et al., 2007, 2009). The potential windows encompassing cytochrome redox activity can also be resolved by cyclic voltammetry (Fig. 5B) after adsorption of the proteins as electroactive (sub-)monolayer films on appropriate electrodes (Hartshorne et al., 2007, 2009; Reuillard et al., 2017). Such protein film electrochemistry provides a ready means to assess redox catalysis by these enzymes, for example, with iron (III) citrate (Hartshorne et al., 2007) and H_2O_2 (Reuillard et al., 2017) as substrates.

For MtrCAB we find that quantifying electron transfer across a lipid bilayer is readily achieved after inserting the complex into preformed liposomes (Edwards et al., 2018; Hartshorne et al., 2009; White et al., 2013). Homogeneous insertion occurs, placing MtrC on the external face of the liposome (Fig. 5C) allowing electron exchange between redox partners located on opposite sides of the bilayer to be quantified by electronic absorbance spectroscopy.

5. Protein–protein interactions

Analysis of the purified MR-1 outer membrane cytochromes by sedimentation equilibrium (SE) analytical ultracentrifugation is particularly valuable for complementing SDS-PAGE and LC-MS by directly measuring the average mass of protein and protein complexes in solution. Samples ($\approx$100 μL) are equilibrated in an eight-cell Ti50 rotor within a Beckman Optima XL-1 analytical ultracentrifuge, and data analysis is performed with Ultrascan II software. The buffer density and partial specific volume (buoyant mass) are determined using the utility software in Ultrascan II.

Analysis of MtrCAB preparations at a range of concentrations and rotation rates revealed that the complex behaved as a single homogeneous species with an apparent molecular mass of $\approx$210 kDa. This implies a 1:1:1 ratio of MtrC, MtrA, and MtrB in a heterotrimeric complex (Hartshorne et al., 2009) and is consistent with the SANS envelope of this complex (Edwards et al., 2018). Comparable studies (Fig. 6A) confirm that MtrAB, soluble MtrC and soluble OmcA (prepared as described in Edwards et al., 2014) are single homogeneous species with apparent molecular masses of $\approx$119, $\approx$77, and $\approx$80 kDa, respectively, in 0.1% Triton X-100, 50 m*M* NaH_2PO_4/Na_2HPO_4, 50 m*M* NaCl, pH 7.5. Triton X-100 micelles have an approximate micellular molecular weight of 80 kDa; however, the

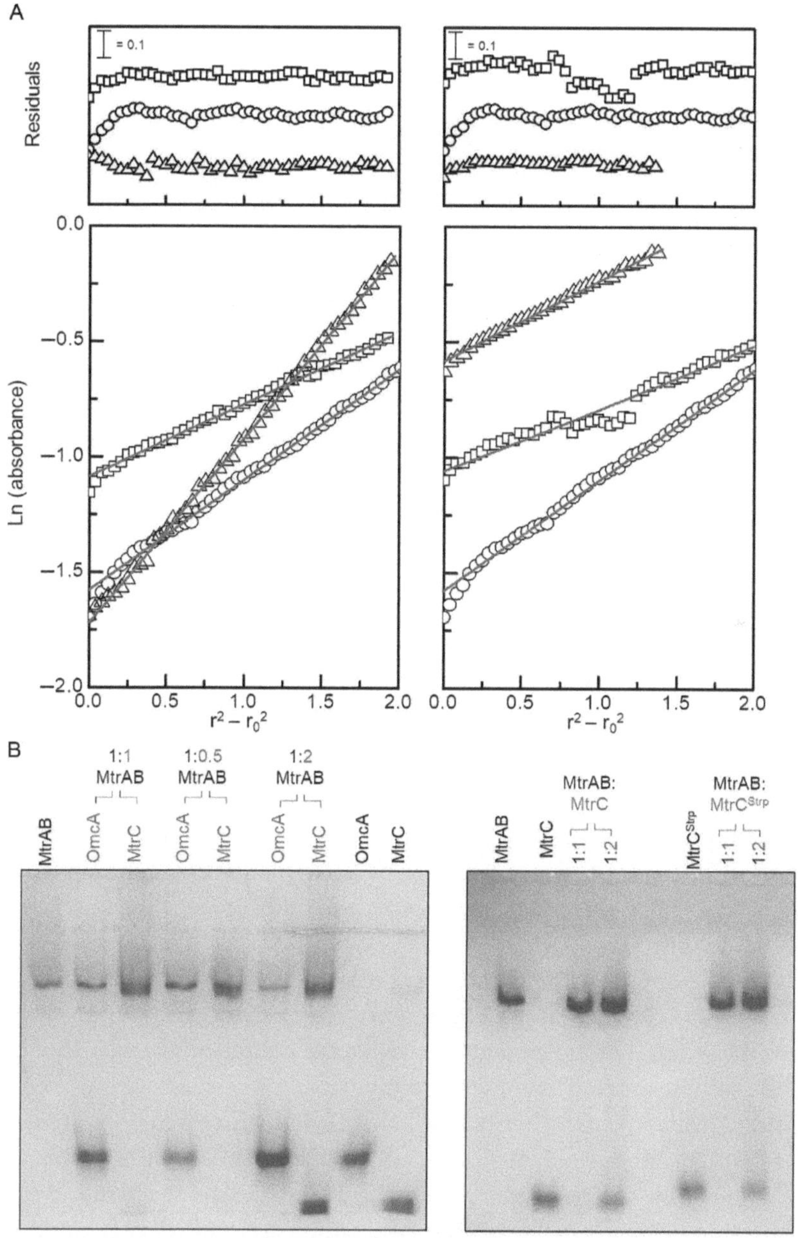

Fig. 6 Sedimentary equilibrium analysis (A) of MtrAB association with MtrC and OmcA at 8 krpm. Absorbance profiles of 0.5 μ*M* MtrC (*squares*) and 0.5 μ*M* MtrAB (*circles*) measured at 410 nm, 0.5 μ*M* MtrC + 0.5 μ*M* MtrAB (*triangles*) measured at 425 nm (*left*), and 0.5 μ*M* OmcA (*squares*), 0.5 μ*M* MtrAB (*circles*), and 0.5 μ*M* OmcA + 0.5 μ*M* MtrAB (*triangles*) measured at 410 nm. The data are fitted to single species (*lines*) with residual differences between the experimental data and fitted curves shown above and offset for clarity. Native-PAGE gels (B) resolving samples of MtrAB, OmcA, and MtrC (*left*) and MtrAB, MtrC, and MtrC^{Strp} (*right*) in isolation and after mixing, as indicated, protein visualized by Coomassie stain. Individual proteins at 5 μ*M* and mixtures contained 5 μ*M* MtrAB and either 2.5, 5, or 10 μ*M* OmcA, MtrC, or MtrC^{Strp}, as indicated by the molar ratios, in 20 m*M* HEPES, 50 m*M* NaCl, 5 m*M* LDAO, pH 7.2.

partial specific volume of Triton X-100 is 0.91 mL g^{-1}, which substantially decreases the micellular contribution to the overall mass of the protein complex.

A sample containing MtrAB (0.5 μ*M*) and an equal concentration of soluble MtrC behaved as a single homogeneous species with apparent molecular mass of ≈208 kDa. This mass is comparable to that of native MtrCAB demonstrating the ability of soluble MtrC to reconstitute with recombinant MtrAB to form a single homogeneous species with a 1:1:1 ratio of MtrC, MtrA, and MtrB. In contrast, SE of the mixture of equal concentrations (0.5 μ*M*) MtrAB and soluble OmcA failed to provide evidence for their association. Resolving the SE profile with a single-component analysis gave an apparent average molecular mass of ≈90 kDa, in between those of MtrAB and OmcA. Fitting the data to two noninteracting components gave apparent molecular masses of ≈80 and 119 kDa, in good agreement with the values resolved for OmcA and MtrAB individually.

Native PAGE (Fig. 6B) confirms and extends the description of protein: protein interactions resulting from SE. Native PAGE is performed with a stacking layer comprised of acrylamide (4%) in 0.2% Triton X-100, 370 m*M* Tris–HCl, pH 8.8, above a resolving layer comprised of acrylamide (5%) in 0.2% (v/v) Triton X-100, 370 m*M* Tris–HCl, pH 8.8. Samples are incubated on ice for 30 min prior to loading on the gel in a bromophenol blue-containing solution of 35% glycerol, 90 m*M* Tris–HCl, pH 6.8. Electrophoresis is performed at constant current (15 mA) with a running buffer of 25 m*M* Tris–HCl, 192 m*M* glycine, pH 8.4, and proteins are visualized by Coomassie stain.

MtrAB, MtrC, or the soluble form of OmcA (5 μ*M*) migrates as a single band with the distance of migration increasing as MtrAB < OmcA < MtrC. Thus, the MtrAB complex is stable under the conditions of electrophoresis and the soluble extracellular cytochromes have different charge/size characteristics consistent with their masses and crystal structures (Edwards et al., 2014, 2015). Mixtures containing 5 μ*M* MtrAB and OmcA at 2.5, 5, or 10 μ*M* give rise to two bands. One band migrates to the same extent as OmcA alone and the other band to the same extent as MtrAB alone. This behavior, together with the relative intensities of the bands for each sample, provides no evidence for interactions between MtrAB and soluble OmcA.

Quite different behavior is seen for samples containing MtrAB and soluble MtrC. By itself, MtrC migrates further than MtrAB. However, only one band is resolved for a 1:1 mixture of 5 μ*M* MtrAB and MtrC and the migration of this band, being slighter faster than MtrAB alone, is different to those of the starting materials. This new band is attributed to formation of an MtrAB–MtrC complex, an interpretation supported by SE, and the behavior of samples containing 5 μ*M* MtrAB and 2.5 or 10 μ*M* MtrC. Migration of the sample containing 1MtrAB:0.5MtrC revealed bands corresponding to MtrAB and the MtrAB–MtrC complex, and there was no indication of a band migrating as MtrC. However, for 1MtrAB:2MtrC, bands migrating as MtrC and the MtrAB–MtrC complex are resolved with no indication of MtrAB. Taken together, the data support spontaneous formation of a stable, high-affinity 1:1 complex between MtrAB and MtrC under the experimental conditions. The same pattern of behavior was seen using combinations of MtrAB and $MtrC^{Strp}$. Thus, formation of a high-affinity 1:1 complex of MtrAB and MtrC/$MtrC^{Strp}$ is independent of the nature of the residues in their C-terminal extensions and dependent on the residues present in the native protein.

6. Conclusion and prospects

Methods have been described for purification of native and recombinant proteins associated with the primary route for bidirectional electron exchange across the MR-1 outer membrane. Structural, spectroscopic, and biophysical studies of the purified proteins have allowed us to reveal their electron-transfer properties and structures, in some cases at the atomic level. The stage is now set for rational engineering of *Shewanella* and heterologous hosts to facilitate electronic coupling of internal biotransformations with external (photo-)electrode materials and/or synthetic catalysts. The opportunities to exploit such biohybrids for sustainable technologies contributing to global demands for chemicals, fuels, and electricity are wide-ranging.

Acknowledgments

We are grateful to the many colleagues, students, and researchers who have enriched our studies of proteins from *S. oneidensis* and related organisms with special thanks to Liang Shi, Jim Fredrickson, John Zachara, Lars Jeuken, and Erwin Reisner. Financial support has been provided by the UK Biotechnology and Biological Sciences Research Council (K009885, K009753, K010220, L020130, L022176, and L023733 and Doctoral Training Partnership PhD studentship to S.E.H.P.), Engineering and Physical Sciences Research Council (M001989) and a Royal Society Leverhulme Trust Senior Research Fellowship to J.N.B.

References

Ainsworth, E. V., Lockwood, C. W. J., White, G. F., Hwang, E. T., Sakai, T., Gross, M. A., et al. (2016). Photoreduction of *Shewanella oneidensis* extracellular cytochromes by organic chromophores and dye-sensitized TiO_2. *Chembiochem*, *17*(24), 2324–2333.

Bachmeier, A., & Armstrong, F. (2015). Solar-driven proton and carbon dioxide reduction to fuels—Lessons from metalloenzymes. *Current Opinion in Chemical Biology*, *25*, 141–151.

Bursac, T., Gralnick, J. A., & Gescher, J. (2017). Acetoin production via unbalanced fermentation in *Shewanella oneidensis*. *Biotechnology and Bioengineering*, *114*(6), 1283–1289.

Edwards, M. J., Baiden, N. A., Johs, A., Tomanicek, S. J., Liang, L. Y., Shi, L., et al. (2014). The X-ray crystal structure of *Shewanella oneidensis* OmcA reveals new insight at the microbe-mineral interface. *FEBS Letters*, *588*(10), 1886–1890.

Edwards, M. J., White, G. F., Lockwood, C. W., Lawes, M. C., Martel, A., Harris, G., et al. (2018). Structural modeling of an outer membrane electron conduit from a metal-reducing bacterium suggests electron transfer via periplasmic redox partners. *Journal of Biological Chemistry*, *293*(21), 8103–8112.

Edwards, M. J., White, G. F., Norman, M., Tome-Fernandez, A., Ainsworth, E., Shi, L., et al. (2015). Redox linked flavin sites in extracellular decaheme proteins involved in microbe-mineral electron transfer. *Scientific Reports*, *5*, 11677.

Fan, G., Dundas, C. M., Graham, A. J., Lynd, N. A., & Keitz, B. K. (2018). *Shewanella oneidensis* as a living electrode for controlled radical polymerization. *Proceedings of the National Academy of Sciences of the United States of America*, *115*(18), 4559–4564.

Flynn, J. M., Ross, D. E., Hunt, K. A., Bond, D. R., & Gralnick, J. A. (2010). Enabling unbalanced fermentations by using engineered electrode-interfaced bacteria. *MBio*, *1*(5). https://doi.org/10.1128/mBio.00190-10.

Hartshorne, R. S., Jepson, B. N., Clarke, T. A., Field, S. J., Fredrickson, J., Zachara, J., et al. (2007). Characterization of *Shewanella oneidensis* MtrC: A cell-surface decaheme cytochrome involved in respiratory electron transport to extracellular electron acceptors. *Journal of Biological Inorganic Chemistry*, *12*(7), 1083–1094.

Hartshorne, R. S., Reardon, C. L., Ross, D., Nuester, J., Clarke, T. A., Gates, A. J., et al. (2009). Characterization of an electron conduit between bacteria and the extracellular environment. *Proceedings of the National Academy of Sciences of the United States of America*, *106*(52), 22169–22174.

Hwang, E. T., Orchard, K. L., Hojo, D., Beton, J., Lockwood, C. W. J., Adschiri, T., et al. (2017). Exploring step-by-step assembly of nanoparticle: Cytochrome biohybrid photoanodes. *ChemElectroChem*, *4*(8), 1959–1968.

Jenner, L. P., & Butt, J. N. (2018). Electrochemistry of surface-confined enzymes: Inspiration, insight and opportunity for sustainable biotechnology. *Current Opinion in Electrochemistry*, *8*, 81–88.

Lapinsonniere, L., Picot, M., & Barriere, F. (2012). Enzymatic versus microbial bio-catalyzed electrodes in bio-electrochemical systems. *ChemSusChem*, *5*(6), 995–1005.

Lee, S. H., Choi, D. S., Kuk, S. K., & Park, C. B. (2018). Photobiocatalysis: Activating redox enzymes by direct or indirect transfer of photoinduced electrons. *Angewandte Chemie (International Ed. in English)*, *57*(27), 7958–7985.

Logan, B. E., & Rabaey, K. (2012). Conversion of wastes into bioelectricity and chemicals by using microbial electrochemical technologies. *Science*, *337*(6095), 686–690.

Masa, J., & Schuhmann, W. (2016). Electrocatalysis and bioelectrocatalysis—Distinction without a difference. *Nano Energy*, *29*, 466–475.

Myers, C. R., & Myers, J. M. (1997). Replication of plasmids with the p15A origin in *Shewanella putrefaciens* MR-1. *Letters in Applied Microbiology*, *24*(3), 221–225.

Reuillard, B., Ly, K. H., Hildebrandt, P., Jeuken, L. J. C., Butt, J. N., & Reisner, E. (2017). High performance reduction of H_2O_2 with an electron transport decaheme cytochrome on a porous ITO electrode. *Journal of the American Chemical Society*, *139*(9), 3324–3327.

Sakimoto, K. K., Wong, A. B., & Yang, P. D. (2016). Self-photosensitization of non-photosynthetic bacteria for solar-to-chemical production. *Science*, *351*(6268), 74–77.

Sakimoto, K. K., Kornienko, N., & Yang, P. D. (2017). Cyborgian material design for solar fuel production: The emerging photosynthetic biohybrid systems. *Accounts of Chemical Research*, *50*(3), 476–481.

Sasaki, K., Sasaki, D., Kamiya, K., Nakanishi, S., Kondo, A., & Kato, S. (2018). Electrochemical biotechnologies minimizing the required electrode assemblies. *Current Opinion in Biotechnology*, *50*, 182–188.

Schuergers, N., Werlang, C., Ajo-Franklin, C. M., & Boghossian, A. A. (2017). A synthetic biology approach to engineering living photovoltaics. *Energy & Environmental Science*, *10*(5), 1102–1115.

Shi, L., Deng, S., Marshall, M. J., Wang, Z. M., Kennedy, D. W., Dohnalkova, A. C., et al. (2008). Direct involvement of type II secretion system in extracellular translocation of *Shewanella oneidensis* outer membrane cytochromes MtrC and OmcA. *Journal of Bacteriology*, *190*(15), 5512–5516.

Shi, L., Dong, H. L., Reguera, G., Beyenal, H., Lu, A. H., Liu, J., et al. (2016). Extracellular electron transfer mechanisms between microorganisms and minerals. *Nature Reviews. Microbiology*, *14*(10), 651–662.

Shi, L., Lin, J. T., Markillie, L. M., Squier, T. C., & Hooker, B. S. (2005). Overexpression of multi-heme C-type cytochromes. *BioTechniques*, *38*(2), 297–299.

Wang, H. M., & Ren, Z. Y. J. (2013). A comprehensive review of microbial electrochemical systems as a platform technology. *Biotechnology Advances*, *31*(8), 1796–1807.

White, G. F., Edwards, M. J., Gomez-Perez, L., Richardson, D. J., Butt, J. N., & Clarke, T. A. (2016). Mechanisms of bacterial extracellular electron exchange. *Advances in Microbial Physiology*, *68*, 87–138.

White, G. F., Shi, Z., Shi, L., Wang, Z. M., Dohnalkova, A. C., Marshall, M. J., et al. (2013). Rapid electron exchange between surface-exposed bacterial cytochromes and Fe(III) minerals. *Proceedings of the National Academy of Sciences of the United States of America*, *110*(16), 6346–6351.

CHAPTER ELEVEN

Reductive activation of CO_2 by formate dehydrogenases

Dimitri Niks, Russ Hille*
Department of Biochemistry, University of California, Riverside, CA, United States
*Corresponding author: e-mail address: russ.hille@ucr.edu

Contents

Abstract

Two factors, climate change brought on by rising atmospheric CO_2 levels and the accelerating shift toward renewable energy sources, have together worked to heighten interest in understanding how biological catalysts so effectively bring about the reduction of CO_2 to formate, with potential applications for both bioremediation and energy storage. Most metal-dependent formate dehydrogenases, containing either molybdenum or tungsten in their active sites, function physiologically in the direction of formate oxidation to CO_2, but it has become clear that many, if not all, are also effective in catalyzing the reverse reaction. In this chapter, we describe methods for isolating and characterizing these enzymes.

1. Introduction

The oxidation of formate to CO_2 is a widespread metabolic process catalyzed by two quite different classes of enzymes: simple metal-independent enzymes that are NAD^+-dependent and act via a ternary E·formate·NAD^+ complex with direct hydride transfer between the two substrates, and molybdenum- or tungsten-dependent enzymes (invariably from bacterial or archaeal sources, and often components of quite large assemblies) that

Methods in Enzymology, Volume 613
ISSN 0076-6879
https://doi.org/10.1016/bs.mie.2018.10.013

operate via a ping-pong mechanism in which the metal center (in the Mo^{VI} or W^{VI} oxidation state) reacts with formate to become reduced, with the reducing equivalents thus obtained subsequently transferred to other redox-active centers (and eventually on to an oxidizing substrate) to regenerate the oxidized metal center. This chapter focuses on these latter enzymes. While most metal-containing formate dehydrogenases function physiologically in the direction of formate oxidation to CO_2, it has become clear that many, if not all, are also effective catalysts in the opposite direction as CO_2 reductases; indeed, at least one functions physiologically in this direction, using reducing equivalents from a source such as H_2 to reduce CO_2 to formate in a reaction of considerable interest for bioremediation of atmospheric CO_2 and the energy economy (Schuchmann & Müller, 2013).

Molybdenum is highly bioavailable owing to the high solubility of molybdate in water, and it is hardly surprising that living organisms have taken advantage of its chemical versatility. Furthermore, there is accumulating evidence that the Last Universal Common Ancestor (LUCA) to all extant life forms made extensive use of molybdenum and the closely related tungsten, which under the highly reducing, anaerobic and sulfur-rich environment in which LUCA existed (Weiss et al., 2016) would also have been readily bioavailable. The molybdenum-containing formate dehydrogenases are members of the DMSO reductase family of molybdenum enzymes, and the tungsten-containing enzymes are members of the eponymous formate dehydrogenase family. The two groups are closely related structurally, in each case having the metal coordinated by two equivalents of a pyranopterin cofactor via an enedithiolate side chain; in both Mo- and W-containing formate dehydrogenases, the pyranopterins are elaborated as the dinucleotide of guanine (typically referred to in the literature as MGD, for molybdopterin— or metallopterin—guanine dinucleotide). The remainder of the metal coordination sphere consists of a cysteine or selenocysteine residue and, as has become generally recognized, a catalytically essential Mo=S or W=S group (Schrapers et al., 2015).

Mo- and W-containing formate dehydrogenases can be found in all three major compartments of the bacterial cell (for recent reviews, see Hartmann, Schwanhold, & Leimkühler, 2015; Maia, Moura, & Moura, 2017). Some, like the *Desulfovibrio gigas* FdhAB enzyme, are periplasmically localized (Raaijmakers et al., 2002), while others (e.g., the *Escherichia coli* FdhGHI and FdoGHI enzymes; Jormakka, Törnroth, Byrne, & Iwata, 2002) are membrane integral and, in conjunction with membrane-integral nitrate reductases, are important respiratory systems. Finally, there are the cytosolically oriented FdhF subunits of formate:hydrogen lyases (which can frequently be isolated on their own, separate from the complex; Axley, Grahame, & Stadtman, 1990) and the cytosolic, NAD^+-dependent FdsABG enzymes from organisms such as *Cupriavidus necator* (also known as *Ralstonia eutropha*) (Oh & Bowien, 1998). The architectural complexity of these enzymes is also highly varied, ranging from the monomeric FdhFs from *E. coli* and *Pectobacterium atrosepticum*, enzymes that contain just a single [4Fe–4S] cluster in addition to their molybdenum centers, to the $(\alpha\beta\gamma)_3$ FdhGHI, with a Mo- and [4Fe–4S]-containing α subunit similar to the FdhF proteins, as well as a 4 × [4Fe–4S]-containing β subunit and a di-heme-containing membrane-integral γ subunit. The FdsABG enzymes are comparably complex, being organized as $(\alpha\beta\gamma)_2$ dimers of trimers. They possess an α subunit with a C-terminal domain, again similar to the FdhF proteins but with an N-terminal region that possesses an additional three [4Fe–4S] clusters and one [2Fe–2S] cluster, a β subunit with FMN (where NAD^+ is reduced) and a [4Fe–4S] cluster, and a γ subunit with a [2Fe–2S] cluster. These three subunits bear considerably homology to the Nqo3, Nqo1, and Nqo2 subunits, respectively, of the *Thermus thermophilus* NADH dehydrogenase (the Nqo3 subunit of which does not possess a molybdenum center, although the domain structure of the portion of the protein homologous to FdhF is preserved). Like NADH dehydrogenases, the FdsABG formate dehydrogenase exhibits high levels of diaphorase ($NADH:O_2$ oxidoreductase) activity, as discussed further later.

Both Mo- and W-containing enzymes are represented among the periplasmic and cytosolic proteins described earlier, but there are to date no known tungsten-containing enzymes homologous to the membrane-integral FdnGHI enzymes. The known periplasmic and membrane-integral enzymes all contain selenocysteine, with the cysteine-containing enzymes confined to the cytosol. The selenocysteine-containing FdhF component of formate:hydrogen lyases is also cytosolic. Hyperthermophilic organisms (all strict anaerobes) utilize tungsten, while thermophiles utilize either tungsten or molybdenum, and mesophiles (including facultative anaerobes such as *E. coli* and *Rhodobacter capsulatus*, as well as aerobes such as *C. necator*) utilize molybdenum exclusively. Selenocysteine is found only in enzymes from strict anaerobes or facultative anaerobes grown under anaerobic conditions, although at least one cysteine-containing enzyme, the FdhF from *P. atrosepticum*, is expressed under anaerobic growth conditions as well. The enzymes from aerobes and facultative anaerobes grown under aerobic conditions contain cysteine.

A general spectrophotometric assay for the oxidation of formate to CO_2 follows the reduction of benzyl or methyl viologen in the presence of formate at pH 7–7.5 under strictly anaerobic conditions. An assay system for the *E. coli* FdhF consists of 2 m*M* benzyl viologen, 20 m*M* formate in 50 m*M* Tris–HCl, pH 7.5, with the reaction monitored at 578 nm ($\varepsilon_{578} = 11{,}000\ M^{-1}\ cm^{-1}$; Axley et al., 1990). The presence of stabilizing agents (which are more often than not competitive inhibitors of formate, see later) in the enzyme stock solutions is mitigated by dilution of the enzyme 100-fold into the assay mix. Assays can be performed in an anaerobic chamber (small footprint, inexpensive plate readers can be used for this purpose) or outside the anaerobic chamber in septa-sealed quartz cuvettes. In the authors' laboratory, an inexpensive Ocean Optics spectrophotometer is used inside the anaerobic chamber for routine measurements.

In addition to the above general activity assay, the NAD^+-dependent FdsABG activity can be followed conveniently and specifically using the absorbance change associated with reduction of NAD^+($\varepsilon_{340} = 6220\ M^{-1}\ cm^{-1}$); this assay can be performed aerobically and is routinely performed at 30°C in 75 m*M* K-PO_4, pH 7.7 with 2 m*M* NAD^+ and 40 m*M* sodium formate (Niks, Duvvuru, Escalona, & Hille, 2016). An assay for the reverse reaction, i.e., the reduction of CO_2 to formate, uses a solution saturated with CO_2 (at 29.5 m*M*) and NADH (where applicable) or dithionite-reduced benzyl/methyl viologen as the source of reducing equivalents. For FdsABG, the reaction is routinely performed in a septum-sealed quartz cuvette which contains 200 μ*M* NADH in 100 m*M* K-PO_4, adjusted to pH 7.0 after the solution is saturated with CO_2; gas bubbling is maintained as the pH is adjusted. The reaction is initiated by addition of enzyme via an argon-purged gas-tight syringe (Yu, Niks, Mulchandani, & Hille, 2017). It is important to recognize that significant amounts of bicarbonate and carbonate accumulate in CO_2-saturated solutions above pH 7.0, which dramatically increases the ionic strength of the assay mix. Enzymes such as FdsABG are quite sensitive to ionic strength effects, and this must be taken into consideration in evaluating the reverse reaction at higher pH values.

2. Air sensitivity

Selenocysteine- and/or tungsten-containing enzymes tend to be extremely air-sensitive, whereas the FdsABG enzymes are (at least in the presence of nitrate, which is routinely added at a concentration of 10 m*M* as a protectant; Friedebold & Bowien, 1993) reasonably stable in air. Anaerobic growth and suitable protocols for isolating and handling these enzymes are therefore often essential. Anaerobic cell growth is best carried out outside the anaerobic chamber using flasks that are small enough to fit into the antechamber. In the authors' laboratory, 2.8-L Fernbach flasks are fitted with rubber stoppers pierced with 2 × 18-gauge needles for use as ports (one for gas entry and the second for venting). A long 21-gauge needle, connected to a N_2 or Ar source, is inserted into one of the ports and used to sparge the flask. The flask is stirred during growth, and the temperature is maintained as necessary using a stir/heat plate. Alternatively, a 3-L stirred flask/bioreactor with a heating blanket and temperature control unit can be utilized.

Cells are typically induced for no more than 24 h, with final yields of 2–10 g cells/L. Upon completion of growth, the cultures are transferred to an anaerobic chamber, placed into centrifuge tubes, sealed with gas-tight caps, removed from the anaerobic chamber, and pelleted by centrifugation. The centrifuge tubes are then returned to the anaerobic chamber, where the supernatant is decanted off and cells resuspended in an appropriate buffer. Cell lysis is accomplished using a cell homogenizer or sonicator inside the anaerobic chamber. The lysed cells can then once again be loaded into centrifuge tubes inside the anaerobic chamber and then centrifuged to remove cell debris. For recombinant systems with, for example, His-tagged proteins, purification can be readily accomplished by batch methods. For native proteins, more complicated procedures can be performed outside the anaerobic chamber with sparged buffers. When using FPLC, for example, outlet tubing can be inserted through anaerobic chamber ports and a fraction collector placed inside the anaerobic chamber. Purified enzymes can be concentrated inside the anaerobic chamber using centrifugal concentrators. The authors favor ammonium sulfate precipitations as a method for quickly concentrating protein prior to buffer exchange/desalting for systems amenable to such treatment. Concentrated protein samples aliquoted for storage can be easily flash-frozen in liquid nitrogen and transferred to liquid nitrogen storage dewars. The reader is referred to a recent review of various anaerobic procedures for the growth and expression of metalloproteins for a more detailed description of the methodologies involved (Tsai & Tainer, 2018).

Specialized anaerobic glassware used in conjunction with an anaerobic train (Schlenk line) makes it possible to perform many types of experiments in the open laboratory; this is far more convenient than using an anaerobic chamber, especially with the ready availability and affordability of Ar or N_2 gas with a built-in-purifier that reduces O_2 to less than 10 ppb (Airgas AR ISP300 or NI ISP300, respectively). Either Ar or N_2 can be used for anaerobiosis, although the heavier Ar is more efficient. Buffer solutions are conveniently made anaerobic by bubbling for 10–15 min in a serum-stoppered vessel fitted with an exit needle. Enzyme solutions, on the other hand, must be placed in an appropriate gas-tight vessel, which is then made anaerobic by alternately evacuating and flushing with N_2 or Ar on the anaerobic train. Six to eight cycles over the course of 60–90 min are typically sufficient. An alternative procedure that does not require a vacuum pump involves flowing gas over a stirred enzyme solution inside a gas-tight vessel for 60–90 min with a small gauge (typically 25–26 gauge) exit needle that creates sufficient back pressure for efficient gas exchange. When working at room temperature, the incoming gas should be bubbled through water to minimize evaporation of the sample, but when the sample is on ice the incoming gas must be dry to avoid condensation inside the vessel. These procedures are far superior to, for example, simply placing a protein solution in an anaerobic chamber overnight, even with stirring.

Specialized anaerobic glassware includes apparatus such as anaerobic cuvettes with which spectrophotometric experiments can be performed, or tonometers (sometimes equipped with a detachable sidearm cuvette) that can be mounted on instrumentation such as a stopped-flow or freeze-quench apparatus. Other appropriate vessels include septum-sealed vials, frequently used in conjunction with gas-tight syringes for anaerobic transfer of solutions previously made anaerobic. In addition, long 20–24 gauge steel cannulae can be used to transfer anaerobic enzyme solutions directly from septum-sealed vials (where enzyme was made anaerobic) to septum-sealed EPR tubes. The reader is referred to references for detailed descriptions of various types of apparatus that have been developed over the years (Beinert, Orme-Johnson, & Palmer, 1978; Foust, Burleigh, Mayhew, Williams, & Massey, 1969).

A sample of anaerobic glassware used in the authors' laboratory is depicted in Fig. 1. The tonometer shown here includes a Luer ground glass joint connected to a double oblique stopcock with 1–2 mm inner bore; the Luer tip is used to mount the tonometer on to a stopped-flow spectrophotometer. The Luer end of the tonometer can be connected to an anaerobic train via gas-impermeable tubing, and the contents of the tonometer made anaerobic by alternately evacuating and flushing with either Ar or N_2. The tonometer has a sidearm (14/20 standard taper) that can be fitted with an anaerobic cuvette, enabling absorption spectra to be recorded, or a cap that may contain a solution (e.g., formate) to be tipped into the enzyme solution in the body of the tonometer subsequent to being made anaerobic. The sidearm cuvette is constructed from a Quartz-to-Pyrex graded seal cuvette

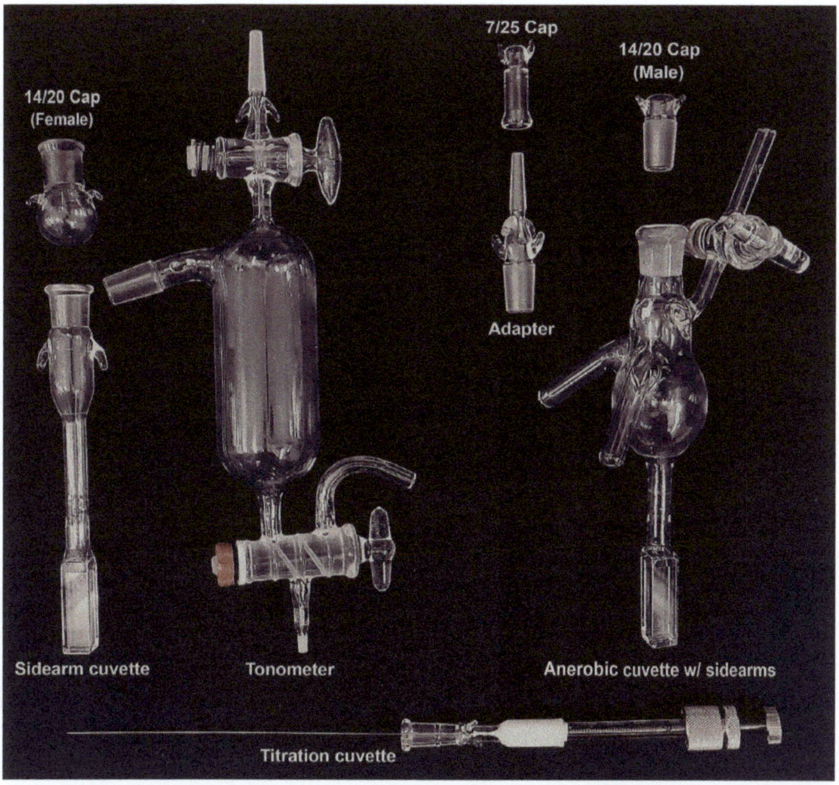

Fig. 1 Glassware used in the authors' laboratory for a variety of anaerobic experiments.

with two or five polished windows (see Starna 9F-Q-10/GS). The end of the tonometer opposite the Luer tip is fitted with a stopcock having a 7/25 standard taper joint to which an anaerobic titration syringe can be fitted once anaerobiosis has been achieved. The syringe is fitted by opening the top stopcock while the tonometer is under continuous positive pressure, and quickly positioning the syringe. The titration syringe is a Hamilton threaded plunger syringe (see Model 1750 LT) modified so as to be able to mount on to either the anaerobic cuvette or the tonometer via an adapter depicted in Fig. 1. All ground joints are sealed with Apiezon N vacuum grease and the various parts are secured with rubber bands via glass hooks. The anaerobic cuvette shown in Fig. 1 is also made from a cuvette with a graded glass seal and is fitted with sidearms for solutions that may be added in succession subsequent to the enzyme being made anaerobic (by alternate evacuation and flushing on an anaerobic train). The cuvette has a 14/20 standard taper top that can be sealed with a cap (or rubber septum if rigorous anaerobiosis is not required), or alternatively fitted with an adapter to accommodate the titration syringe for spectrophotometric titrations. An example of the use of the anaerobic cuvette is described below.

3. Reversibility of the reaction: CO_2 reduction

As indicated earlier, there are documented systems that function physiologically as CO_2 reductases (Schuchmann & Müller, 2013), but the literature presents a rather confusing picture as to whether one or another of the enzymes that function physiologically to oxidize formate is capable of catalyzing the reverse reaction, the reduction of CO_2 to formate. In particular, there is no apparent correlation with reported reactivity in the reverse direction as regards molybdenum vs tungsten, or whether cysteine or selenocysteine is coordinated to the metal. Given the importance of the reaction, both from the standpoint of clearing a greenhouse gas from the atmosphere and also from the standpoint of energy storage, the situation is particularly problematic. The *Desulfovibrio desulfuricans* formate dehydrogenase has been clearly demonstrated to catalyze CO_2 reduction using a viologen as reductant, and the reaction has recently been comprehensively examined, with k_{cat} in the reverse direction being determined to be $47\,s^{-1}$ (as compared to the k_{cat} for formate oxidation in the forward direction of $543\,s^{-1}$) (Maia, Fonseca, Moura, & Moura, 2016). Similarly with the *C. necator* FdsABG enzyme, which had been reported previously to be unable to catalyze CO_2 reduction (Friedebold & Bowien, 1993), the enzyme is found to be effective in utilizing NADH to drive CO_2 reduction, with a k_{cat} of $11\,s^{-1}$, K_m^{NADH} of $46\,\mu M$, and $K_m^{CO_2}$ of $2.7\,mM$ (Yu et al., 2017). These parameters can be compared with k_{cat} of $211\,s^{-1}$, K_m^{NAD+} of $130\,\mu M$, and $K_m^{formate}$ of $310\,\mu M$ in the forward direction. Importantly, it is found that the observed steady-state kinetic parameters for the reaction in the forward and reverse directions comply with the expected Haldane relationships and the known reduction potentials for the NAD^+/NADH and CO_2/formate couples of −320 and −420 mV vs SHE at pH 7.0, respectively. Two important points have emerged: first, the enzyme's considerable diaphorase activity makes it essential to conduct assays anaerobically so as to avoid NADH consumption not associated with CO_2 reduction. Second, it is critical to use CO_2-saturated buffers in preparing substrate solutions for the reverse reaction, as bicarbonate is found to be a poor surrogate for CO_2. According to Henry's Law, an aqueous solution equilibrated under 1 atm. of CO_2 has a CO_2

concentration of 29.5 m*M*. The bicarbonate (and carbonate) concentrations increase dramatically with pH, but the concentration of CO_2 itself remains constant. CO_2-equilibrated solutions are required in these assays because the CO_2/bicarbonate/carbonate equilibria are quite sluggish on the timescale of a typical enzyme assay. These points likely account for the many false-negative results that have been reported in the literature, and it has been concluded that all the Mo- and W-containing formate dehydrogenases are in fact likely to be able to catalyze the reduction of CO_2 (Yu et al., 2017).

4. Native enzyme preparations

Isolation of formate dehydrogenases in substantial quantities from native organisms is often quite feasible, as growth conditions can usually be identified under which the source organism is heavily dependent on formate dehydrogenase activity and makes a great deal of one or another forms of the enzyme depending on the specific growth conditions. Among the factors to consider, particularly when using minimal media, is ensuring that the molybdenum or tungsten levels are sufficiently high to support expression levels of the desired molybdenum or tungsten enzyme.

Enzyme preparations themselves continue to be improved upon with, for example, modern ceramic hydroxyapatite FPLC chromatography resins replacing the notoriously poorly performing gravity columns used in earlier protocols. Even with these newer methodologies, however, identifying the optimal sequence of chromatographic steps is often nontrivial, particularly for enzymes that are membrane associated or even membrane integral. A major challenge is always ensuring that the enzyme remains stable and active throughout the purification process, and activity must be carefully monitored in the course of developing new purification protocols. Some enzymes rapidly lose activity on certain column materials (for example, in the authors' laboratory the FdsABG formate dehydrogenase from *C. necator* has been found to lose a substantial amount of its activity on GE MonoQ columns, and such materials are to be avoided). Working as expeditiously as possible and maintaining high protein concentrations throughout purification (by use of filter concentrators, for example) generally improves yields.

A common issue in the purification of molybdenum- and tungsten-containing enzymes has to do with the degree of functionality of the isolated material. Most preparations of enzymes from native sources (as distinct from what is typically seen with recombinant systems, see later) are replete in the molybdenum or tungsten center (or nearly so) and such other redox-active centers as may be found in the enzyme, but may still have lower than maximal specific activity. In many cases, lower specific activity is due to loss of (or failure to incorporate) the catalytically essential sulfido ligand to the metal. The sulfido group is labile and can be slowly displaced by hydroxide from solvent, presumably releasing sulfide, yielding a desulfo form of the enzyme. This hydroxide displacement is much faster with reduced rather than oxidized enzyme, when the sulfido ligand becomes protonated to a Mo/W(IV)-SH group, and the formate dehydrogenases as a group typically lose activity rapidly upon reduction. The sulfido group can be removed by reaction of enzyme with cyanide, which releases the sulfur as thiocyanate (Massey & Edmondson, 1970). For enzymes such as xanthine oxidase, which also possesses a catalytically essential Mo=S group, it is possible to reconstitute the activity of naturally occurring or cyanide-treated desulfo enzyme by incubation of the reduced enzyme under anaerobic conditions with sulfide (Wahl & Rajagopalan, 1982). Unfortunately, this procedure is often less effective with the formate dehydrogenases. In some cases maintaining activity can be facilitated by inhibitors. The two most successful examples are azide (typically at a concentration of 3 m*M*), used to protect the *E. coli* FdhF (Axley et al., 1990), and nitrate (at a concentration of 10 m*M*), which has been used with the FdsABG from *C. necator* (Friedebold & Bowien, 1993).

Loss of activity is of course problematic. The case of the molybdenum-containing DMSO reductases from *Rhodobacter sphaeroides* or *R. capsulatus*, both of which possess a serine ligand to the metal rather than cysteine or selenocysteine and a Mo=O rather than Mo=S, presents a cautionary tale. After much confusion as to the structure of the active site metal center, it was eventually demonstrated that one of the two pyranopterin cofactors tended to be displaced from the molybdenum in the course of handling, being replaced in the molybdenum coordination sphere by a Mo=O group (Li, Temple, Rajagopalan, & Schindelin, 2000). The resulting enzyme was not catalytically active, although the displaced pyranopterin could be reinserted into the molybdenum coordination sphere by so-called redox cycling, i.e., reduction with sodium dithionite followed by reoxidation with substrate DMSO (Bray, Adams, Smith, Bennett, & Bailey, 2000). As it turns out, the two forms of the enzyme behaved very differently on Q Sepharose and related materials, and the overall yield of the enzyme was seriously compromised unless the crude cell extract was first redox cycled (Mtei et al., 2011).

There is, to date, no explicit evidence to suggest that pyranopterin dissociation is a problem with any specific formate dehydrogenase, but it must always be borne in mind that this is another possible reason for loss of enzyme functionality. Still, routine redox cycling is not recommended for the formate dehydrogenases due to their instability in the reduced state.

A number of methods can be used to assess cofactor saturation, including inductively coupled plasma mass spectrometry (ICP-MS) that can quantify the amount of molybdenum (and other metals) in a protein sample. When performed carefully, ICP-MS has become a valuable tool in the determination of the metal content in recombinant as well as native proteins. The amount of pyranopterin cofactor present in an enzyme sample is somewhat more difficult to assess, owing to the extreme lability of the cofactor once extruded from a molybdenum enzyme. Still, an HPLC protocol has been developed for quantification of a stable, fluorescent oxidation product of the cofactor, termed Form A, that is, widely used in the field (Johnson, Hainline, Rajagopalan, & Arison, 1984). This method, in conjunction with ICP-MS, can provide an accurate determination of the level of cofactor/metal saturation. A modification of this procedure (G. Schwarz, Institute of Biochemistry, University of Cologne, Personal Communication), for the MPT form of the cofactor is presented below as an ordered protocol for quantitation of Form A ($\varepsilon_{380}=13{,}200\ M^{-1}\ \text{cm}^{-1}$ in 0.1 *M* NAOH):

Oxidation

1. Dissolve 140 μL of enzyme-containing 100 pmol expected Moco content in 100 m*M* Tris/HCl, pH 7.2.
2. Add 17.5 μL 1.5% I_2/2% KI solution plus 1.7 μL of concentrated HCl.
3. Incubate the solution for 30 min at 95°C, then cool to RT, and centrifuge for 2 min at 10,000 rpm to remove denatured protein.
4. Transfer the yellowish supernatant to a new tube.

Dephosphorylation

5. Add 20 μL of freshly prepared 1% ascorbic acid to reduce residual iodine.
6. Add 70 μL of 1 *M* unbuffered Tris plus 5 μL 1 *M* $MgCl_2$ (to raise pH to approximately 8.3 and provide cations required by the alkaline phosphatase).
7. Add 10 units of alkaline phosphatase and incubate ≥30 min at RT in the dark (this cleaves the phosphate to obtain Form A-dephospho; only the dephospho form is retarded by the C18 resin).

HPLC analysis

8. Dilute the sample 1:10 (or any other suitable dilution factor) with ddH_2O and centrifuge for 5 min at 10,000 rpm.

9. Analyze the dephospho-Form A by HPLC, employing a C18 reverse phase column in isocratic mode with 10 m*M* KPi, pH 3.0, containing 10% methanol, detecting the product via fluorescence (excitation at approximately 380 nm, emission at 450 nm), integrating the peaks.
10. Compare the results to a standard (typically bovine xanthine oxidase) treated in the same fashion.

A similar procedure has been used for the preparation and isolation of the dephospho-Form A-GMP from the FdsABG from *R. capsulatus* using xanthine oxidase as a standard (Hartmann & Leimkühler, 2013). It is to be noted, however, that while xanthine oxidase is perfectly suited as a standard for other MPT-containing enzymes, it is less appropriate for the quantification of the bis-MGD form of the cofactor, as the absorbance and fluorescence properties of the dinucleotide form of the cofactor-derived Form A are likely different than the mononucleotide form. It is recommended that DMSO reductase, which also utilizes MGD, be utilized instead (Cobb, Conrads, & Hille, 2005).

Finally, it should be emphasized that the extent of functionality in the molybdenum and tungsten centers of the formate dehydrogenases can be estimated spectrophotometrically, a tool that could be exploited to a greater extent in the literature than it has been to date. A typical protocol for FdsABG involves placing a 50 μL solution of concentrated enzyme in one sidearm of an anaerobic cuvette such as that described earlier, with 50 μL of 400 m*M* sodium formate in the other sidearm. Once 900 μL of buffer is placed in the main body, the cuvette is capped with a rubber septum. A rough anaerobiosis is first obtained with five cycles of alternately evacuating and flushing with Ar on the anaerobic train as described earlier. The septum is then fitted with a small exit needle and the cuvette carefully tilted to maximize surface area for gas exchange, being careful not to introduce the buffer into either sidearm. After 1 h, the contents of the cuvette will be sufficiently anaerobic to continue with the experiment. The buffer is used to zero reference the UV/vis spectrophotometer, if necessary, and the enzyme is then added by tipping the buffer into the sidearm containing the enzyme solution, being careful not to introduce any liquid into the sidearm containing the sodium formate. A spectrum is recorded for the oxidized enzyme (Fig. 2A, blue trace), after which the enzyme solution is tipped into the sidearm containing the formate solution and the formate-reduced spectrum recorded (Fig. 2A, red trace). Finally, a concentrated solution of buffered sodium dithionite is introduced via a gas-tight syringe and the cuvette is incubated until reduction is complete, as judged by the loss of absorbance

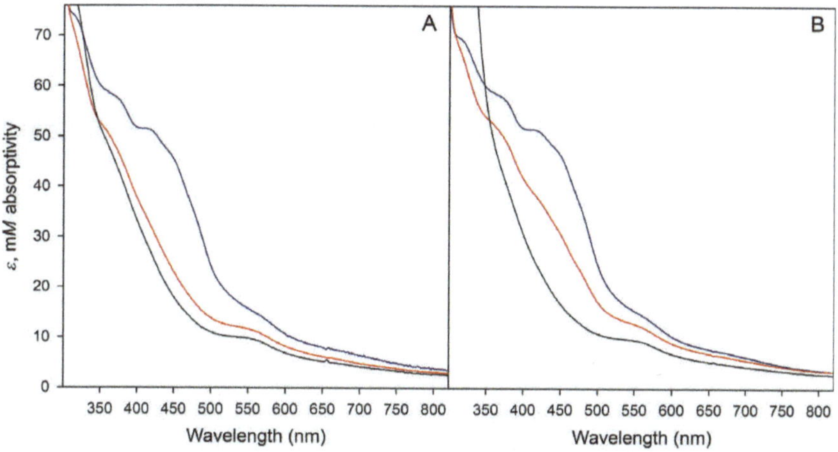

Fig. 2 Assessment of the functionality of the FdsABG formate dehydrogenase. (A) Oxidized (*blue*), formate-reduced (*red*), and sodium dithionite-reduced (*black*) spectra for the wild-type FdsABG. (B) Oxidized (*blue*), formate-reduced (*red*) and sodium dithionite-reduced (*black*) spectra for the recombinant FdsABG. Reduction was performed with 20 m*M* sodium formate and 50 m*M* sodium dithionite in 75 m*M* K-PO_4, pH 7.7, under anaerobic conditions at RT utilizing an anaerobic cuvette of the type depicted in Fig. 1 and described in the text.

at 410 nm—for FdsABG this takes approximately 2–3 h (Fig. 2A, black trace). Formate will reduce only that portion of the enzyme which possesses a functional molybdenum (or tungsten) center, while dithionite will reduce all the enzyme; the extent of the absorbance change seen with formate, as compared with dithionite, is thus directly proportional to the fraction of the enzyme that is functional. The protocol can then repeated with recombinant protein (Fig. 2B). From Fig. 2, the native enzyme is nominally 80% functional, and the recombinant protein is 50% functional, in good agreement with the assessment of functionality by steady-state assay.

The assumption made in the above analysis is that formate fully reduces the functional enzyme, and that dithionite fully reduces both functional and nonfunctional enzyme. Furthermore, it must be borne in mind that both the simpler FdhF enzyme from *E. coli* (which has a single [4Fe–4S] cluster in addition to its molybdenum center, requiring three equivalents for full reduction) and the FdsABG enzyme from *C. necator* (which has five [4Fe–4S] clusters, two [2Fe-2S] clusters, and one equivalent of FMN in addition to the molybdenum center, requiring 11 reducing equivalents for full reduction) require an odd number of reducing equivalents, while formate necessarily donates reducing equivalents in pairs. This being the case, formate inevitably leaves the enzyme incompletely reduced, at least in the short-term (on longer timescales, intermolecular electron transfer between enzyme molecules will eventually lead to full reduction, but this can result in reduction of nonfunctional enzyme, which is to be avoided). To the extent that formate reduces functional enzyme less completely than does dithionite, this method may thus underestimate the functionality of the preparation.

5. Recombinant enzyme preparations

Genomics methodologies have proven extremely successful at identifying genes encoding molybdenum-containing enzymes as well as proteins necessary for their maturation in a great variety of organisms, and a number of different recombinant systems have been exploited for the efficient expression of both wild-type enzymes and variant forms. Heterologous expression of molybdenum- and tungsten-containing proteins, however, presents a number of challenges in addition to those described above for the isolation of native enzymes. The complex nature of the biosynthetic pathway for the pyranopterin cofactor (Mendel, 2013) and the involvement of additional proteins for its subsequent sulfuration and insertion into apoprotein (Iobbi-Nivol & Leimkühler, 2013; Leimkühler & Iobbi-Nivol, 2013) mean that protein expression must be carefully controlled, as protein synthesis can outpace cofactor biosynthesis, with the result that the recombinant apoprotein may only be incompletely reconstituted in the molybdenum or tungsten center. This problem can be serious since the expressed recombinant protein, while lacking the full complement of molybdenum or tungsten, is often replete in the other redox-active cofactors that may be present in the enzyme, seriously complicating the interpretation of experimental results, and making even the determination of specific activity difficult. A number of strategies have been developed for optimizing incorporation of the molybdenum or tungsten center. Slowing the rate of growth of the recombinant cells so as to permit sufficient time for cofactor synthesis—primarily through a combination of low-speed shaking (160–180 rpm) and lower incubation temperatures (16–30°C)—has been found to be beneficial. Addition of molybdate (or tungstate) to the growth medium prior to induction of protein expression is also advisable so as to ensure adequate availability of the metal. The amount of inducer (e.g., IPTG) can also be varied to balance the level of protein expression with endogenous cofactor biosynthesis, and lower concentrations of inducer often lead to higher levels of cofactor saturation (Warelow et al., 2013).

Alternatively, use of a weak promoter can be employed to lower rates of protein synthesis so as to avoid outstripping cofactor biosynthesis (Temple, Graf, & Rajagopalan, 2000).

A major advantage of recombinant methods is the ability to appropriately label the recombinant protein and apply various affinity purification protocols, frequently enabling one-step purification procedures. These have become widely used, and Arnou et al. have published an exhaustive review of various affinity methods for the purification of recombinant proteins (Arnou, Lauritzen, Petersen, & Pedersen, 2006). His-tag methodologies relying on immobilized metal affinity chromatography (IMAC), particularly Ni-NTA and related materials, have been used most frequently in the case of molybdenum- and tungsten-containing enzymes, including various formate dehydrogenases, although the high concentrations of imidazole required to elute His-tagged proteins from these materials can be problematic for metal-containing proteins in general, and molybdenum- or tungsten-containing ones specifically. Placement of the His-tag at the C- or N-terminus (and in the case of multisubunit proteins, which subunit to label) is an important consideration, as is the choice of subunit to be labeled, and must be empirically determined. In the case of the FdsABG formate dehydrogenase, for example, an N-terminal tag on the small FdsG subunit is optimal as it leaves the His-tag well removed spatially from the Mo-containing FdsA subunit (Hartmann & Leimkühler, 2013). Other factors that may dictate the choice of subunit and position labeled in the polypeptide include, for example, where the protein is localized in the cell. For periplasmic proteins that are translocated to the periplasm via the TAT system (as are all periplasmically localized molybdenum- and tungsten-containing enzymes), the His-tag (or other label) must be C-terminal so as to not interfere with recognition of the N-terminal twin-arginine signal sequence that marks these proteins for export to the periplasm. An N-terminal His-tag would in any case be cleaved upon maturation of the enzyme in the periplasm.

The choice of metal used in the IMAC column material can also influence the extent to which purification results in loss of enzymatic activity. The two most widely used metal ions, Ni^{2+} and Co^{2+}, are sufficiently different that while one may result in loss of activity the other may be perfectly benign. In rare cases, it has been found that high concentrations of imidazole itself, used to elute bound enzyme from the IMAC column material, contribute to protein instability and/or depletion of molybdenum from the recombinant protein. This problem can also occur with other metal centers that might be present in the recombinant enzyme. The authors have, for example, observed loss of up to 95% of the heme cofactor in a *b*-type heme protein (part of the mARC-reductive pathway) employing a standard Ni-NTA purification protocol (D. Niks & R. Hille, Unpublished). In some cases, it has been found that the His-tag itself is responsible for the loss of activity, it having been shown, for example, that a C-terminal His-tagged YedY protein is substantially less functional than its unlabeled analogue (Sabaty et al., 2013). Purification of complex His-tagged metalloproteins can be complicated by interference of the metal chelate (as well as the His-tag itself or the imidazole used to elute the recombinant protein from the column) with the active sites of metalloenzymes, and again careful quality control on the purified recombinant protein as to metal and cofactor becomes essential. The extent to which a given resin results in loss of activity can be assessed by standard enzymatic activity assays and protein characterization, as described earlier. In cases where this is found to be a problem, use of other forms of affinity chromatography that do not involve metal chelation (Flag tag, SUMO tag, GST, etc.) should be considered. Finally, it should be borne in mind that in those cases where metal chelating resins themselves are found to lead to considerable loss of activity, native purification methods may be considered as an alternative to other affinity methods, particularly in the cases where protocols have been worked out for the native form of the protein.

Like the biosynthetic pathway for the pyranopterin cofactor, the endogenous machinery for sulfur incorporation into the maturing molybdenum center prior to insertion into apoprotein can be overwhelmed by high expression levels. Coexpression of the gene encoding the sulfurase with that for the structural gene(s) for recombinant formate dehydrogenase (or other molybdenum- or tungsten-containing enzymes requiring the sulfido ligand) has been successfully employed in some cases to circumvent this problem (Schumann et al., 2009). Beyond determining the extent to which a recombinant protein is replete with cofactor is the assessment of whether the cofactor is structurally and functionally intact. In addition to the difference absorption protocol described above for determination of enzyme activity, electron paramagnetic resonance (EPR) spectroscopy can be extremely useful in ascertaining the integrity of a given protein sample. EPR is particularly sensitive to the protein environment of the metal center as well as its structural and functional integrity (particularly when the substrate—formate in the case of the formate dehydrogenases considered here—is used to reduce the enzyme). The oxidized, Mo(VI)- or W(VI)-containing, as-isolated enzyme is typically converted to the EPR-active Mo(V) or W(V) form by partial reduction with, e.g., formate or sodium dithionite, with the amount of the (V) state accumulating ultimately governed by the relative reduction potentials of the M(VI)/M(V) and M(V)/M(IV) couples. In most cases with molybdenum-containing enzymes 20%–50% of the Mo(V) species can be generated; the amount of W(V) is generally lower, approximately 10%. The EPR spectra of functional and a variety of nonfunctional forms of many molybdenum- and tungsten-containing enzymes have been well characterized, including the formate dehydrogenases, and constitute a rigorous test of the integrity of the metal center (Hille, 2010). Diagnosing potentially compromised molybdenum or tungsten centers can be important not only in the characterization of full-length recombinant proteins but also for independently expressed domains or subunits.

6. Summary

Studies of the molybdenum- and tungsten-containing formate dehydrogenases, like any other enzymes, require highly purified proteins of high functionality, and for many of these enzymes it is essential to purify and handle the protein under strictly anaerobic conditions. Use of stabilizing agents such as nitrate or azide for long-term storage is also important. Particularly in the case of structural studies, it is imperative to demonstrate the functionality and integrity of enzyme samples so as to be assured of the catalytic relevance of the structural conclusions. This entails not only pyranopterin and metal quantitation, but spectroscopic analysis as well, including characterization of the absorption spectrum of oxidized enzyme and the spectral changes associated with substrate reduction of the enzyme, as well as EPR spectroscopy of the M(V) state. Use of standardized conditions of anaerobiosis and high [CO_2] concentrations is necessary to accurately assess whether a given molybdenum- or tungsten-containing formate dehydrogenase is able to catalyze the reverse reaction, the reduction of CO_2 to formate, and we reiterate that it seems likely that most if not all of them will eventually be demonstrated to do so.

References

Arnou, J., Lauritzen, C., Petersen, G. E., & Pedersen, J. (2006). Current strategies for the use of affinity tags and tag removal for the purification of recombinant proteins. *Protein Expression and Purification*, *48*(1), 1–13. https://doi.org/10.1016/j.pep.2005.12.002.

Axley, M. J., Grahame, D. A., & Stadtman, T. C. (1990). Escherichia coli formate-hydrogen lyase. Purification and properties of the selenium-dependent formate dehydrogenase component. *Journal of Biological Chemistry*, *265*(30), 18213–18218.

Beinert, H., Orme-Johnson, W. H., & Palmer, G. (1978). Special techniques for the preparation of samples for low-temperature EPR spectroscopy. *Methods in Enzymology, 54*, 111–132. https://doi.org/10.1016/S0076-6879(78)54013-5.

Bray, R. C., Adams, B., Smith, A. T., Bennett, B., & Bailey, S. (2000). Reversible dissociation of thiolate ligands from molybdenum in an enzyme of the dimethyl sulfoxide reductase family. *Biochemistry, 39*(37), 11258–11269. https://doi.org/10.1021/bi0000521.

Cobb, N., Conrads, T., & Hille, R. (2005). Mechanistic studies of *Rhodobacter sphaeroides* Me_2SO reductase. *Journal of Biological Chemistry, 280*(12), 11007–11017. https://doi.org/10.1074/jbc.M412050200.

Foust, G. P., Burleigh, B. D., Mayhew, S. G., Williams, C. H., & Massey, V. (1969). An anaerobic titration assembly for spectrophotometric use. *Analytical Biochemistry, 27*(3), 530–535. https://doi.org/10.1016/0003-2697(69)90066-9.

Friedebold, J., & Bowien, B. (1993). Physiological and biochemical characterization of the soluble formate dehydrogenase, a molybdoenzyme from *Alcaligenes eutrophus. Journal of Bacteriology, 175*(15), 4719–4728. https://doi.org/10.1128/jb.175.15.4719-4728.1993.

Hartmann, T., & Leimkühler, S. (2013). The oxygen-tolerant and NAD+-dependent formate dehydrogenase from *Rhodobacter capsulatus* is able to catalyze the reduction of CO_2 to formate. *FEBS Journal, 280*(23), 6083–6096. https://doi.org/10.1111/febs.12528.

Hartmann, T., Schwanhold, N., & Leimkühler, S. (2015). Assembly and catalysis of molybdenum or tungsten-containing formate dehydrogenases from bacteria. *Biochimica et Biophysica Acta, 1854*(9), 1090–1100. https://doi.org/10.1016/j.bbapap.2014.12.006.

Hille, R. (2010). EPR studies of xanthine oxidoreductase and other molybdenum-containing hydroxylases. In G. Hanson & L. Berliner, Jr., (Eds.), *Biological magnetic resonance*: *Vol. 29. Metals in biology: Applications of high-resolution EPR to metalloenzymes* (pp. 91–121). Berlin: Springer. https://doi.org/10.1007/978-1-4419-1139-1_5.

Iobbi-Nivol, C., & Leimkühler, S. (2013). Bacterial molybdeoenzymes: Old enzymes for new purposes. *FEMS Microbiology Review, 40*(1), 1–18. https://doi.org/10.1093/femsre/fuv043.

Johnson, J. L., Hainline, B. E., Rajagopalan, K. V., & Arison, B. H. (1984). The pterin component of the molybdenum cofactor. Structural characterization of two fluorescent derivatives. *Journal of Biological Chemistry, 259*(9), 5414–5422.

Jormakka, M., Törnroth, S., Byrne, B., & Iwata, S. (2002). Molecular basis of proton motive force generation: Structure of formate dehydrogenase-N. *Science, 295*(5561), 1863–1868. https://doi.org/10.1126/science.1068186.

Leimkühler, S., & Iobbi-Nivol, C. (2013). Molybdenum enzymes, their maturation and molybdenum cofactor biosynthesis in *Escherichia coli. Biochimica et Biophysica Acta, 1827*(8–9), 1086–1101. https://doi.org/10.1016/j.bbabio.2012.11.007.

Li, H.-K., Temple, C., Rajagopalan, K. V., & Schindelin, H. (2000). The 1.3 Å crystal structure of Rhodobacter sphaeroides dimethyl sulfoxide reductase reveals two distinct molybdenum coordination environments. *Journal of the American Chemical Society, 122*(32), 7673–7680. https://doi.org/10.1021/ja000643e.

Maia, L. B., Fonseca, L., Moura, I., & Moura, J. J. G. (2016). Reduction of carbon dioxide by a molybdenum-containing formate dehydrogenase: A kinetic and mechanistic study. *Journal of the American Chemical Society, 138*(28), 8834–8846. https://doi.org/10.1021/jacs.6b03941.

Maia, L. B., Moura, I., & Moura, J. J. G. (2017). Molybdenum and tungsten-containing formate dehydrogenases: Aiming to inspire a catalyst for carbon utilization. *Inorganica Chimica Acta, 455*(SI), 350–363. https://doi.org/10.1016/j.ica.2016.07.010.

Massey, V., & Edmondson, D. E. (1970). Mechanism of inactivation of xanthine oxidase by cyanide. *Journal of Biological Chemistry, 245*(24), 6595–6598.

Mendel, R. R. (2013). The molybdenum cofactor. *Journal of Biological Chemistry*, *288*(19), 13165–13172. https://doi.org/10.1074/jbc.R113.455311.

Mtei, R. P., Lyashenko, G., Stein, B., Rubie, N., Hille, R., & Kirk, M. L. (2011). Spectroscopic and electronic structure studies of a dimethyl sulfoxide reductase catalytic intermediate: Implications for electron- and atom-transfer reactivity. *Journal of the American Chemical Society*, *133*(25), 9762–9774. https://doi.org/10.1021/ja109178q.

Niks, D., Duvvuru, J., Escalona, M., & Hille, R. (2016). Spectroscopic and kinetic properties of the molybdenum-containing, NAD+-dependent formate dehydrogenase from *Ralstonia eutropha*. *Journal of Biological Chemistry*, *291*(3), 1162–1174. https://doi.org/10.1074/jbc.M115.688457.

Oh, J. I., & Bowien, B. (1998). Structural analysis of the *fds* operon encoding the NAD^+-linked formate dehydrogenase from *Ralstonia eutropha*. *Journal of Biological Chemistry*, *273*(41), 26349–26360. https://doi.org/10.1074/jbc. 273.41.26349.

Raaijmakers, H., Macieira, S., Dias, J. M., Texeira, S., Bursakov, S., Huber, R., et al. (2002). Gene sequence and the 1.8Å crystal structure of the tungsten-containing formate dehydrogenase from *Desulfovibrio gigas*. *Structure*, *10*(9), 1261–1272.

Sabaty, M., Grosse, S., Adryanczyk, G., Boiry, S., Biaso, F., Arnoux, P., et al. (2013). Detrimental effect of the 6 His C-terminal tag on YedY enzymatic activity and influence of the TAT signal sequence on YedY synthesis. *BMC Biochemistry*, *14*(28), 14–28. https://doi.org/10.1186/1471-2091-14-28.

Schrapers, P., Hartmann, T., Kositzki, R., Dau, H., Reshke, S., Schulzke, C., et al. (2015). Sulfido and cysteine ligation changes at the molybdenum cofactor during substrate conversion by formate dehydrogenase (FDH) from *Rhodobacter capsulatus*. *Inorganic Chemistry*, *54*(7), 3260–3271. https://doi.org/10.1021/ic502880y.

Schuchmann, K., & Müller, V. (2013). Direct and reversible hydrogenation of CO_2 to formate by a bacterial carbon dioxide reductase. *Science*, *342*(6164), 1382–1385. https://doi.org/10.1126/science.1244758.

Schumann, S., Terao, M., Garattini, E., Saggu, M., Lendzian, F., Hildebrandt, P., et al. (2009). Site directed mutagenesis of amino acid residues at the active site of mouse aldehyde oxidase AOX1. *PLoS One*, *4*(4), 1–9. https://doi.org/10.1371/journal.pone.0005348.

Temple, C. A., Graf, T. N., & Rajagopalan, K. V. (2000). Optimization of expression of human sulfite oxidase and its molybdenum domain. *Archives of Biochemistry and Biophysics*, *383*(2), 281–287. https://doi.org/10.1006/abbi.2000.2089.

Tsai, C. L., & Tainer, J. A. (2018). Robust production, crystallization, structure determination, and analysis of [Fe–S] proteins: Uncovering control of electron shuttling and gating in the respiratory metabolism of molybdopterin guanine dinucleotide enzymes. *Methods in Enzymology*, *599*, 57–196. https://doi.org/10.1016/bs.mie.2017.11.006.

Wahl, R. C., & Rajagopalan, K. V. (1982). Evidence for the inorganic nature of the cyanolyzable sulfur of molybdenum hydroxylases. *Journal of Biological Chemistry*, *257*(3), 1354–1359.

Warelow, T. P., Oke, M., Schoepp-Cothenet, B., Dahl, J. U., Bruselat, N., Sivalingam, G. N., et al. (2013). The respiratory arsenite oxidase: Structure and the role of residues surrounding the rieske cluster. *PLoS One*, *8*(8), 1–10. https://doi.org/10.1371/journal.pone.0072535.

Weiss, M. C., Sousa, F. L., Mrnjavac, N., Neukirchen, S., Roettger, M., Nelson-Sathi, S., et al. (2016). The physiology and habitat of the last universal common ancestor. *Nature Microbiology*, *1*, 16116. https://doi.org/10.1038/NMICROBIL.2016.116.

Yu, X., Niks, D., Mulchandani, A., & Hille, R. (2017). Efficient reduction of CO_2 by the molybdenum-containing formate dehydrogenase from *Cupriavidus necator* (*Ralstonia eutropha*). *Journal of Biological Chemistry*, *292*(41), 16872–16879. https://doi.org/10.1074/jbc.M117.785576.

CHAPTER TWELVE

Production and properties of enzymes that activate and produce carbon monoxide

Rodney Burton, Mehmet Can, Daniel Esckilsen, Seth Wiley, Stephen W. Ragsdale*
Department of Biological Chemistry, University of Michigan Medical School, Ann Arbor, MI, United States
*Corresponding author: e-mail address: sragsdal@umich.edu

Contents

Abstract

The chapter focuses on the methods involved in producing and characterizing two key nickel–iron–sulfur enzymes in the Wood–Ljungdahl pathway (WLP) of anaerobic conversion of carbon dioxide fixation into acetyl-CoA: carbon monoxide dehydrogenase (CODH) and acetyl-CoA synthase (ACS). The WLP is used for biosynthesis of cell material and energy conservation by anaerobic bacteria and archaea, and it is central to several industrial biotechnology processes aimed at using syngas and waste gases for the production of fuels and chemicals. The pathway can run in reverse to allow organisms, e. g., methanogens and sulfate reducers, to grow on acetate. The CODH and ACS intertwine to form a tenacious CODH/ACS complex that converts CO_2, a methyl group, and coenzyme A into acetyl-CoA. CODH also behaves as a modular unit that can

Methods in Enzymology, Volume 613
ISSN 0076-6879
https://doi.org/10.1016/bs.mie.2018.10.005

function as an independent homodimer. Besides coupling to ACS, CODH can interact with hydrogenases to couple CO oxidation to H_2 formation. These enzymes have been purified and characterized from several microbes.

1. Introduction

The chapter is focused on the methods involved in producing and characterizing the catalytic properties of the two key nickel iron–sulfur metalloenzymes in the Wood–Ljungdahl pathway (WLP) of anaerobic conversion of CO_2 to acetyl-coenzyme A (acetyl-CoA). This pathway involves the eight-electron reduction of CO_2 to form acetyl-CoA. In the WLP, these enzymes, carbon monoxide dehydrogenase (CODH), and acetyl-CoA synthase (ACS) intertwine to form a tenacious complex with a 150 Å gas channel that sequesters the CO formed from CO_2 (Eq. 1) at the CODH active site and tunnels it to the A-cluster, the active site of ACS, for catalyzing acetyl-CoA synthesis (Eq. 2). The A-cluster binds and then condenses the three components of acetyl-CoA: CO, a methyl group donated by the methyl-Co(III) state of the corrinoid iron–sulfur protein (CFeSP), and coenzyme A. The WLP can operate in reverse to allow organisms to grow on acetate. Furthermore, CODH is a modular unit that can interact with other enzymes, besides ACS, to fulfill various metabolic roles. For example, it forms a CODH/hydrogenase unit to couple CO oxidation to proton reduction. Much progress has been made on understanding the mechanisms of CODH and ACS based on spectroscopic, crystallographic, kinetic, and electrochemical studies.

$$\text{CODH: } CO_2 + 2H^+ + 2 \text{ electrons} \rightarrow CO + H_2O \quad (1)$$

$$\text{ACS: } CO + CH_3^+ + \text{CoA} \rightarrow CO + CH_3\text{-CO-CoA} \quad (2)$$

These enzymes have been purified and characterized from several microbes and are used as important inspirational models and benchmarks in energy technologies ranging from applications such as fuel cells to fundamental examples of "nearly perfect" catalytic activity.

2. Culturing the microbes

2.1 General considerations about technology for growth of anaerobic cultures

Cultivation of anaerobic cell cultures is as ancient as recorded human civilization in the form of fermentation, and it was on facultative anaerobes that the foundational work of Leeuwenhoek and Pasteur was performed, ushering in the age of modern biochemistry. Anaerobic conditions are essential for culturing and harvesting the oxygen-sensitive microbes, including *Moorella thermoacetica, Clostridium autoethanogenum, Carboxydothermus hydrogenoformans*, and *Methanothermobacter marburgensis*, employed in the Ragsdale Laboratory. We direct the reader to some valuable general anoxic principles and methods (Hungate, 1969; Wolfe, 2011). These principles have been important in developing the anaerobic techniques employed by the Ragsdale Laboratory as described here.

It is critical to maintain low ambient redox potentials in culture medium. As described by Hungate (1969), this requires adding a reducing agent to the culture medium (Hungate, 1969). Commonly used reductants are dithiothreitol (DTT), sodium dithionite, and Ti(III) citrate. To obtain a redox potential of −0.33 V in the absence of a reductant, less than one molecule of O_2 per liter of solution must be present. That is obviously an impossible task; however, reducing agents will act as redox buffers and poise the potential at the desired value. Removal of oxygen is also required because many microbes and metalloproteins react quickly with O_2; furthermore, reducing agents are rapidly oxidized in aerobic medium. Thus, having access to an anaerobic gas chamber, described later, is highly advisable. We also use Schlenk-line technology to thoroughly degas solutions. Trace O_2 levels in compressed gases are removed using O_2 filters (Altech Oxy-trap or a SDG Purity Plus O_2-trap).

Ti(III) citrate is prepared by first mixing 1 mL of 15% $TiCl_3$/HCl solution with 1.6 mL of 1 *M* sodium citrate and then adding 2.5 mL of 2 *M* Tris–HCl buffer at pH 8.0 (Seefeldt & Ensign, 1994). Further neutralization may be needed and can be done using sodium carbonate until no bubbles are observed.

Commonly used anaerobic containment for small liquid microbial precultures includes Hungate-type vials or Balch-type tubes with butyl rubber stoppers and screw cap or alumicap apertures (Hungate, Smith, & Clarke, 1966), stoppered with crimpable aluminum caps (Hungate, 1969), then a bioreactor with fermenter is sufficient for large-scale growth.

2.2 Use of anaerobic chambers

Three types of chambers are used in our laboratory: a Vacuum Atmospheres Company anaerobic gas chamber (Hawthorne, CA), a refrigerated anaerobic chamber (Inert Technologies, Amesbury, MA), and a Coy chamber (Coy Laboratory Products, Grass Lake, MI). The Inert Technologies and Vacuum Atmospheres "hard" chambers are constructed of stainless steel with hard plastic windows and thick gloves and are maintained at or below 2 ppm O_2. There are a total of six chambers in the laboratory varying from one to three persons in size. This allows us to equip different chambers with FPLCs and chromatography columns, spectrophotometers, a 5°C Peltier-cooled storage refrigerator, a sonicator with external cooling waterbath, a weighing scale, an isothermal calorimeter, a benchtop centrifuge, a stopped-flow absorption-fluorescence spectrophotometer, and spectroelectrochemical setup.

The Coy chamber is a soft plastic chamber with an atmosphere of 95% N_2 and 5% H_2 (note that under these conditions, H_2 levels never exceed 4.0%, its flammability limit). The atmosphere is scrubbed by pellets of a palladium catalyst surrounded by absorbent material, which convert the O_2 to H_2O. The Inert Technologies refrigerated hard chamber is equipped with an air conditioning cooling unit to allow an ambient temperature of ~15°C, which is ideal for purifying and storing enzymes from mesophilic organisms such as *C. autoethanogenum*. The Vacuum Atmospheres hard chambers are kept in anoxic N_2, which is continually circulated through BASF catalysts consisting of ~85% aluminum oxide, ~12% copper oxide, with trace nickel monoxide and cobalt oxide. The N_2 is supplied by liquid nitrogen tanks, with ~2–3 ppm oxygen, which is further scrubbed by chamber catalysts to working conditions.

2.3 Cultivation of the microbes

2.3.1 *Growth of* M. thermoacetica

M. thermoacetica (ATCC 39073) has been the model organism for study of the WLP. We routinely grow *M. thermoacetica* under anoxic cultivation conditions in batch cultures on glucose as described (Ljungdahl & Andreesen, 1978), with the following modifications. A typical preparation is described as follows: cells were grown at 55°C in a 10-L cultivation vessel (Eppendorf) using a New Brunswick Scientific BioFlo 110 bioreactor equipped with a MGW Lauda RM6 Brinkman water circulator. The 10 L culture was continuously supplied with CO_2 (99.9%) or a CO_2/N_2 (20%/80%) mixture and agitated at 120 rpm. Growth was monitored through the optical density at 600 nm. We also measured the CO oxidation activity of whole cells using a CO-saturated assay mixture containing 20 m*M* benzyl viologen in 50 m*M* HEPES buffer (pH 8.0). As *M. thermoacetica* is gram (+) (Drake & Daniel, 2004) the purity of the culture during growth was established by gram staining. Cells were harvested by centrifugation under 100% N_2 and stored at −80°C for later use.

M. thermoacetica was originally isolated using glucose as carbon source (earlier). Although this has been the model organism for characterization of the autotrophic WLP, it was not until 1983 (Kerby & Zeikus, 1983) that it was shown to be capable of autotrophic growth on CO_2/H_2 or on CO, demonstrated in a follow-up study a few years later (Daniel, Hsu, Dean, & Drake, 1990). It can grow on other sugars as well as methanol (Hsu, Daniel, Lux, & Drake, 1990), methoxylated aromatics (vanillate and syringate) (Daniel, Wu, & Drake, 1988; Kasmi, Rajasekharan, & Ragsdale, 1994), and carboxylated aromatics (proteocatechuate and *p*-hydroxybenzoate) (Hsu et al., 1990). The methyl and carboxyl groups are stoichiometrically converted to acetyl-CoA (Daniel et al., 1988; Hsu et al., 1990; Kasmi et al., 1994).

2.3.2 Growth of other microbes expressing CODH and ACS

C. hydrogenoformans: The monofunctional CODH and CODH/ACS have been isolated from the gram (+) thermophilic strict anaerobe *C. hydrogenoformans*, which was grown at 65°C on CO as the carbon source, as described (Svetlitchnyi, Peschel, Acker, & Meyer, 2001).

C. autoethanogenum: We are studying the WLP and isolating CODH and CODH/ACS from *C. autoethanogenum*. This organism will grow at 37°C on H_2/CO_2 or CO, as well as various mixes of these gases and syngas (Mock et al., 2015), producing predominantly acetate; it also produces ethanol and has been engineered to produce various industrially relevant chemicals (Kopke et al., 2011; Liew et al., 2017; Norman, Millat, Winzer, Minton, & Hodgman, 2018). *C. autoethanogenum* also grows heterotrophically on sugars (e.g., 4.5 g/L fructose) (Liew et al., 2016; Mock et al., 2015).

Rhodospirillum rubrum: The phototropic bacterium, *R. rubrum*, grows on a variety of carbon and energy sources under aerobic, anaerobic, and fermentative conditions (Ghosh, Hardmeyer, Thoenen, & Bachofen, 1994), although it lacks the WLP. For purification of CODH, *R. rubrum* cells were grown on ammonium malate as N and C sources and then 100% CO was sparged into the culture for 24 h prior to harvesting (Bonam & Ludden, 1987). Carbon monoxide is required to induce the monofunctional CODH (as well as the CO-linked hydrogenase) (Kerby et al., 1992).

Methanogens: The WLP is fundamental to the autotrophic metabolism of methanogens as well as in its growth on acetate (Ferry, 2010). To metabolize acetate, from which 2/3 of the methane in the biosphere is derived, methanogens operate the WLP in reverse. CODH and other proteins of the WLP have been isolated from *Methanosarcina barkeri* (Grahame & Stadtman, 1987), *Methanococcus vannielii* (Demoll, Grahame, Harnly, Tsai, & Stadtman, 1987), and *Methanosarcina thermophila* (Terlesky, Nelson, & Ferry, 1986), grown as described.

2.3.3 Growth of organisms and their heterologous expression of CODH

Jeoung and Dobbek reported expression in *Escherichia coli* and purification of highly active recombinant CODH II from *C. hydrogenoformans* (Jeoung & Dobbek, 2007). As described, two plasmids, one based on the pET28a vector for apo-CODH II expression (pKS2) and the other for iron–sulfur cluster maturation (pKRISC), were transformed into the Rosetta (DE3) cell line. Growth was started aerobically in TB media with slightly high iron and nickel concentrations; then, the culture was switched to anaerobic conditions as CODH and the Isc system were induced by adding (to final concentration) 0.2 m*M* IPTG, 0.5 m*M* $NiCl_2$, 1 m*M* $FeSO_4$, 1 m*M* Na_2S, and 50 m*M* KNO_3. A more detailed discussion on the maturation of metal clusters of CODH has recently been published (Merrouch et al., 2018).

Expression and purification of recombinant CODH I in *E. coli* have also been reported (Inoue, Takao, Fukuyama, Yoshida, & Sako, 2014). While coexpression of CODH I along with a Ni-insertion accessory protein (*CooC3*) yielded an enzyme with an activity of 8060 U/mg, expression without *CooC3* achieved a fourfold lower activity (2270 U/mg).

Heterologous expression in *Clostridium acetobutylicum* of CODHs from *Clostridium carboxidivorans* and *C. ljungdahlii* was reported recently (Carlson & Papoutsakis, 2017; Fast & Papoutsakis, 2018). Characterization of these enzymes has not yet been reported.

3. Purification of CODH, ACS, and CODH/ACS

3.1 General anaerobic technology

General anaerobic protocols for culturing the microbes were explained earlier, but they are also relevant here. Here we will describe procedures for purifying and handling extremely oxygen-sensitive enzymes. All protein purification steps take place under anoxic conditions in anaerobic chambers (earlier) and reducing agents, usually DTT, Ti(III) citrate, and/or dithionite, are used in all solutions to poise and maintain the low redox potentials required for optimal enzyme activity (Wang, Can, Pierce, Ragsdale, & Armstrong, 2013).

3.2 Purification of monofunctional CODH (CooS)

3.2.1 C. hydrogenoformans *CODHI and CODHII*

Here we describe the purification protocol for CODH I and II described by Svetlitchnyi et al. (2001) with the revisions noted as mentioned below.

While the main buffer used in the published procedure contains 50 m*M* Tris–HCl (pH 8.0) and 2 m*M* sodium dithionite, we include 2 m*M* DTT (Buffer A_{Ch}). We have attempted to use methyl viologen (MV) in our columns as an oxygen scavenger; however, this leads to extreme gas formation, which disrupts the chromatography. Throughout the isolation, CO oxidation activity is determined in the various fractions by measuring the absorbance change at 578 nm occurring as a few microliters of the fraction solutions are introduced to a CO-saturated 50 m*M* HEPES buffer (pH 8.0) containing 20 m*M* MV and 2 m*M* DTT, essentially as described earlier (Ragsdale, Clark, Ljungdahl, Lundie, & Drake, 1983). We also examine the fractions by SDS-PAGE because CODH II runs slightly slower than CODH I.

The cell lysis buffer contains buffer A_{Ch} including 0.1 mg/mL of lysozyme, 0.05 mg/mL of DNase I, and 0.2 m*M* phenyl methyl sulfonyl fluoride. While Svetlitchnyi et al. describe an osmotic lysis step with 0.6 *M* sucrose (Svetlitchnyi et al., 2001), we achieve good results by sonication. The first step involves anion exchange chromatography (AEC) with a salt gradient of 0–1 *M* NaCl. The second step uses a hydrophobic interaction chromatography (HIC) column, on which CODH I elutes cleanly before CODH II. In contrast, Svetlitchnyi et al. employ additional HIC, AEC, and size-exclusion chromatography columns (Svetlitchnyi et al., 2001) and found that only one additional HIC is required to obtain protein of ~99% purity. The activities of the purified proteins at 70°C are 15,756 and 13,828 U/mg for CODH I and II, respectively (Svetlitchnyi et al., 2001). These enzymes are highly stable and can be stored in buffer A_{Ch} under anoxic conditions at room temperature without losing any activity for more than a year; alternatively, they can be stored as frozen stocks at −80°C.

3.2.2 Rhodospirillum rubrum *CooS*

Purification of the fully active native CODH from *R. rubrum* can be achieved when cells are grown under CO atmosphere with elevated nickel levels in the growth medium (Ensign, Hyman, & Ludden, 1989). During the purification processes, slightly basic buffers with reductants are used; for instance, 100 m*M* MOPS (pH 7.5) with 5 m*M* sodium dithionite (Bonam & Ludden, 1987). Initial purification steps include sedimentation and resuspension of the chromatophore membranes after disrupting the cells using a homogenizer or a French pressure cell (Ensign, 1995). Heat-shock treatment of the resuspended chromatophores solubilizes the CODH and precipitates large amounts of impurities. After centrifugation, the supernatant is applied to DEAE-cellulose and hydroxylapatite chromatography columns. As a polishing step, preparative native gel electrophoresis or an additional high-resolution Q-sepharose column can be used. The specific activity of the purified protein for CO oxidation is 5200 U/mg at 25°C and pH 7.5 (Ensign et al., 1989).

3.2.3 *Recombinant CODH from* E. coli

The vectors and protocols for recombinant expression of His-tagged *C. hydrogenoformans* CODH I and II were mentioned earlier. These proteins are purified in a single step on an IMAC (Ni-Sepharose) medium containing 1 m*M* sodium dithionite (Jeoung & Dobbek, 2007). The specific activities at 70°C of the purified CODH I and II were 12,229 and 13,828 U/mg, very close to the values reported for the native enzymes.

3.3 Purification of monofunctional Ni-ACS from *E. coli*

Although CODH/ACS has been studied for decades, levels of active ACS did not surpass 40% based on NiFeC spin concentration and various enzymatic assays described earlier. This shortcoming does not affect techniques like electron paramagnetic resonance (EPR) or electron nuclear double resonance (ENDOR) in which Ni_d is EPR silent, and the Ni site in the C-cluster has a different nonoverlapping signal; however, it does compromise the results of many spectroscopic studies, such as Mössbauer or X-ray absorption spectroscopy (XAS) which detect the average metal environment. For example, if the A-cluster is 30% active, the active state of Ni_p represents only 10% (0.3/3) of the total Ni. It is nearly impossible to discern the properties of this small Ni_p population, which is averaged by the remaining 90% of the Ni in the protein. Thus, it was a significant advance when the ACS of the CODH/ACS complex (the α-subunit encoded by *acsB*) was heterologously expressed in *E. coli* and reconstituted to 80%–100% activity A cluster (Can, Giles, Ragsdale, & Sarangi, 2017; Seravalli & Ragsdale, 2008). In a sample of ACS that is 80% active, Ni_p represents at least 40% of the total Ni (0.8/2), with Ni_d making up ~50% of the Ni signal. This development allowed us to obtain a high-resolution analysis of the geometry and electronic structure of the key Ni_p-CO intermediate (Can et al., 2017), providing the first direct validation of the hypothesis that the WLP operates through an organometallic mechanism involving a series of Ni–C bonds (Ragsdale & Wood, 1985).

3.3.1 *Cloning and expression of ACS*

The *acsB* gene encoding ACS from *M. thermoacetica* was cloned into the pET29 (Kan^R and His6-tag; *lac*-inducible) vector for heterologous expression in BL21(DE3) *E. coli* cells. Anaerobic coexpression of the monofunctional ACS with pbd1282 (*A. vinelandii isc*S, *isc*U, *isc*A, *hscB*, *hscA*, *fdx*; Amp^R; *ara* inducible) is used for obtaining proper construction and insertion of the ACS NiFeS cluster (Seravalli & Ragsdale, 2008). To remove the His6-tag, a C-terminal TEV-cleavage site was incorporated into the *acsB*-containing pET29 vector by inserting a GAAAACCTGTATTTCCAGTCG sequence between BamHI and XhoI restriction sites ahead of the His6 sequence, $[CATCAC]_3$.

A 100 mL lysogeny broth (LB) culture of BL21(DE3) *E. coli* cells containing the ACS plasmid and the pdb1282 plasmid is grown overnight in the presence of 100 μg/mL ampicillin (Amp), and 50 μg/mL kanamycin (Kan) is then used to inoculate 1 L of an Amp/Kan-containing LB media on the day of growth. Once the 1 L culture has reached an OD_{600} of 0.7–1.0, the cells are inoculated into 10 L of terrific broth (TB) media containing Kan and Amp in a BIOFLO 110 bioreactor agitated at 250 rpm and 37°C. At an OD_{600} of 0.3–0.4, sterile-filtered arabinose (0.5 g/L) is added to induce expression. Then at an OD_{600} of 0.6, the aerobic solution is vigorously bubbled with high purity N_2 (UHP grade, Cryogenic Gases, Detroit, MI) for at least 45 min, an anaerobically prepared reducing solution (100 mL) consisting of 3.6 g L-cysteine·HCl and 3.6 g Na_2S is added to the bioreactor and, 15 min later, anaerobically prepared 0.5 *M* $Fe^{II}(NH_4)_2(SO_4)_2$ is added to give a final Fe concentration of 100 μ*M*. After 15 further min, anaerobic solutions of 0.5 *M* isopropyl ß-D-1-thiogalactopyranoside (IPTG) and 1.0 *M* $NiCl_2$ are added to give final concentrations of 200 μ*M* IPTG and 200 μ*M* nickel. Cells are grown overnight at 37°C under minimal N_2 flow and are harvested 12–16 h later into anaerobic, air-tight centrifuge bottles, and pelleted at 5000 rpm at 4°C for 10 min. Once cells are pelleted, they are transferred to anaerobic 50-mL Falcon™ tubes, flash frozen with LN_2, and stored in a −80°C freezer.

3.3.2 Lysis, purification, and characterization of ACS

Cell lysis and ACS purification are performed under strictly anaerobic conditions (≤3 ppm O_2). A general purification buffer, buffer A, is composed of 50 m*M* potassium phosphate (KP_i), 300 m*M* NaCl, 10 m*M* imidazole, 5 m*M* β-mercaptoethanol (βME), at pH 8.0. The lysis buffer contains 0.25 mg/mL lysozyme, DNase I, and 0.5 m*M* phenylmethane sulfonyl fluoride (PMSF) in buffer A. For each gram of cell pellet, 4 mL lysis buffer is used to resuspend the cells, which are sonicated (10 s "on"; 20 s "off"; 80% amplitude) for 15 min in a beaker kept at 4°C. The resultant cell lysate is ultracentrifuged for 1 h at 30,000 rpm after which the supernatant is decanted onto a Ni-NTA column equilibrated with buffer A. After washing the Ni-NTA column with three column volumes (CVs) of buffer A, the His6-tagged ACS is eluted with a step gradient of 5% increments from 5% to 30% of buffer B (50 m*M* KP_i, 300 m*M* NaCl, 250 m*M* imidazole, 5 m*M* βME, pH 8) in buffer A. Protein concentration and purity are measured by Rose Bengal dye assay (Elliott & Brewer, 1978) calibrated to a lysozyme standard, and by SDS-PAGE, respectively. Fractions containing pure ACS are pooled and the sample is buffer exchanged into 50 m*M* KP_i, 300 m*M* NaCl, 5 m*M* βME, pH 7.6 to remove imidazole. Sodium citrate is added to the protein solution at a final concentration of 1 m*M*, and then approximately 1 mg TEV protease is added to begin an overnight cleavage reaction. The solution is then reloaded onto the reactivated Ni-NTA column equilibrated with buffer A, and the His6-tag-cleaved ACS, which elutes rapidly in the flow-through fraction, is buffer exchanged into 50 m*M* KP_i pH 7.6, and stored anaerobically at 20°C.

3.3.3 Metal reconstitution of ACS

In order to facilitate proper metal loading in ACS, additions of 5:1 or 6:1 $NiCl_2$:ACS and 1:1 $Fe(NH_4)_2(SO_4)_2$:ACS are added to the pooled ACS from anaerobic 100 m*M* $NiCl_2$ and $Fe(NH_4)_2(SO_4)_2$ stock solutions, βME (2 m*M*, final) is added to the protein solution and the solution is incubated at 45°C for 72 h. Extraneous nickel is then removed from the reconstituted ACS by buffer exchange into a 50 m*M* potassium phosphate buffer containing 100 μ*M* DTT, pH 7.5. The exchange is accomplished by passing the protein (<1 mL) twice through an anaerobic 8.3 mL PD-10 desalting column or by the use of Amicon® pressure filters (10^6-fold) with a 30 kDa cutoff membrane. Amicon ultrafiltration is used to concentrate the protein to 200–500 μ*M*.

Finally, the reconstituted ACS is assayed by EPR spectroscopy to determine the morphology and spin content of the NiFeC EPR signal, metal analysis by ICP-OES/-MS, and measure catalytic activity by the CO/[1-^{14}C]-acetyl-CoA exchange reaction (earlier).

3.4 Purification of *M. thermoacetica* CODH/ACS

The CODH and ACS enzymes form a tight complex and coelute throughout the purification. Thus, CODH/ACS fractions are identified using the CO oxidation activity assay and SDS-PAGE gels. In an early example of this preparation, heat treatment at 67°C for 30 min upon cell disruption was followed with stepwise ammonium sulfate precipitation, and the CODH fractions were solubilized in a pH 7.6 Tris–HCl buffer with 2 m*M* sodium dithionite (Ragsdale, Clark, et al., 1983). In this protocol, chromatographic purification steps included DE32-cellulose resin, hydroxylapatite resin, size exclusion, and DEAE-Sephacel, respectively.

Nowadays in our laboratory, CODH/ACS is purified by FPLC. To each gram of dry cells is added 4 mL of lysis buffer consisting of buffer A_{Mt} (50 mM Tris–HCl pH 7.6 with 2 mM sodium dithionite and 2 mM DTT), 0.5 mM PMSF, 0.25 mg/mL lysozyme, and 10 μg/mL DNase I (Roche: grade II, from bovine pancrease). For 200 mL of lysate, sonication is accomplished at 4°C, running 10 s "on" and 20 s "off" for a total of 15 min. The lysate is centrifuged at 30,000 rpm for 1 h at 4°C. The supernatant is then applied to a 60 mL Q-Sepharose Fast Flow resin (GE Healthcare) equilibrated in buffer A_{Mt}, which is developed with a salt gradient from 0 to 0.6 M NaCl, with CODH/ACS eluting in fractions containing 0.3–0.4 M NaCl.

The next step is hydrophobic chromatography. The CODH/ACS fractions are pooled and ammonium sulfate is added to a final concentration of 0.6 M. The solution is then applied to a column containing 30 mL Phenyl Sepharose 6 Fast Flow (Sigma-Aldrich) resin equilibrated with 0.6 M ammonium sulfate in buffer A_{Mt}, and a decreasing gradient in ammonium sulfate, 0.6–0.0 M, is applied. The CODH/ACS complex elutes in fractions containing 0.2–0.0 M ammonium sulfate. The protein purity judged by SDS-PAGE is higher than 95% and CO oxidation activity varies between 200 and 300 U/mg at 25°C.

4. Characterization of CODH, ACS, and CODH/ACS

4.1 Characterization of CODH

Anaerobic conditions must be used during all experiments to preserve the active states of the clusters and to poise the solutions at the optimal redox condition for the reaction. All bacterial Ni-CODHs, whether they are monofunctional or bound to ACS, form a homodimeric complex containing five iron–sulfur clusters. Each monomer contains a $NiFe_3S_4$ cluster (C-cluster) and a Fe_4S_4 cluster (B-cluster); in addition, a single Fe_4S_4 cluster (D-cluster) links the two monomers covalently (see earlier). Steady-state kinetic analyses of the different classes of CODH have revealed the Michaelis parameters (k_{cat}, K_m, k_{cat}/K_m) and transient kinetic studies have revealed the intermediates involved in the reaction (Fig. 1; a recent review is available, containing mechanistic detail, Can, Armstrong, & Ragsdale, 2014). The B- (Fe_4S_4) and C- ($NiFe_3S_4$) clusters of CODH have distinct spectral signatures (Can et al., 2014; Jeoung & Dobbek, 2007) allowing

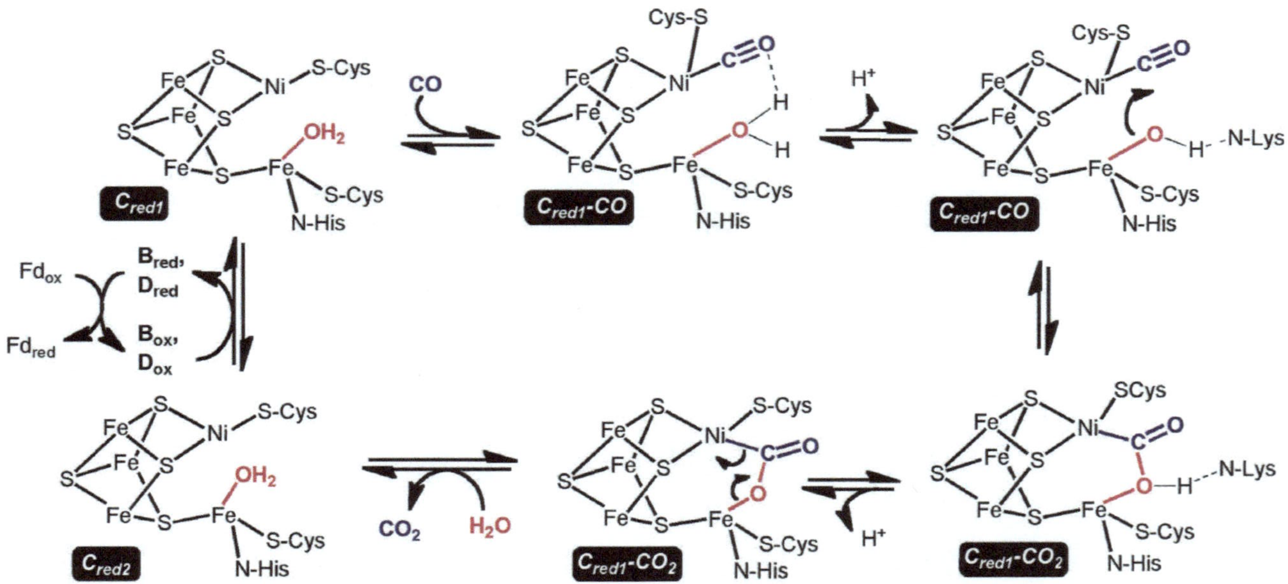

Fig. 1 The CODH mechanism, focusing on reactions at the C-cluster. The two electrons are transferred to the B- and D-clusters en route to external electron acceptors, e.g., ferredoxin.

the use of multiple rigorous methods to dissect the redox and chemical changes that accompany this two-electron reaction. Rapid-freeze-quench (RFQ) EPR studies revealed that CO reacts with the C-cluster in the C_{red1} redox state, converting it rapidly and directly to the C_{red2} state during the two-electron oxidation of CO to CO_2 (Kumar, Lu, Liu, & Ragsdale, 1993). These studies also monitored reduction of the B-clusters and demonstrated their role in electron transfer to external acceptors, although D-cluster reduction has not been experimentally observed under catalytic conditions. Electrochemical studies revealed cooperativity, suggesting that the C_{red1} to C_{red2} conversion is concerted (Wang et al., 2013).

Methods used to dissect the details of the CODH mechanism include spectroscopy (UV–visible, EPR, FTIR, ENDOR, Mössbauer, X-ray absorption, static, and freeze-quenched EPR), kinetics (steady state and transient), electrochemistry (spectroelectrochemistry and direct enzyme immobilized), and computations. Similar methods have been used to characterize ACS. These methods exploit the variations in reduction potential, reduction kinetics, and spectral variations among the clusters due to their unique environments and roles within CODH (see Table 1) and ACS. We focus on the methods that are routinely involved in studying CODH and ACS. We refer the reader to the literature for details of methods we have used to isolate and characterize intermediates in the CODH and ACS mechanisms. We recently published a general review covering these studies (Can et al., 2014).

4.1.1 Steady-state kinetic measurements of CODH

The steady-state reaction has been monitored spectrophotometrically by observing the products and turnover of mediators in the enzyme reaction, by electrochemistry, and by NMR methods, monitoring the CO–CO_2 exchange (see a recent review, Can et al., 2014). Such studies have established values of k_{cat}, K_m (and hence k_{cat}/K_m) for the various substrates in CO oxidation and CO_2 reduction, as well as identified inhibitors and their mode of inhibition. By comparing values of k_{cat}/K_m, it has been established which physiological (e.g., ferredoxin, flavodoxin, rubredoxin, pyruvate ferredoxin oxidoreductase) and nonphysiological (e.g., methyl/benzyl viologen) redox partners couple most effectively to the enzyme.

MV-based assays are conducted by monitoring the time-dependent increase in absorption at 568 nm indicative of one-electron reduced MV; thus two cycles of MV reduction are required for each turnover of CO or CO_2. The assay (Ragsdale, Clark, et al., 1983; Ragsdale, Ljungdahl, &

Table 1 EPR spectra and electrochemical parameters of CODH/ACS

Metal centers	Cluster type	EPR parameters *g* values	Electron transfer kinetics	Reduction potential	Literature
B_{red}	[4Fe4S]	g=2.04, 1.93, 1.89	$90\,s^{-1}$, 5°C $3000\,s^{-1}$, 55°C	B_{ox}/B_{red} −440 mV	Lindahl, Munck, and Ragsdale (1990)
C_{ox}	[Ni-3Fe4S]	EPR silent	Reductive activation; C_{ox} is inactive	C_{ox}/C_{red1} −200 mW	Lindahl, Munck, et al. (1990)
C_{red1}	[Ni-3Fe4S]	g=2.01, 1.81, 1.65	C_{red1}/C_{red2} $400\,s^{-1}$, 5°C $17{,}200\,s^{-1}$, 55°C $2\times10^{8}\,M^{-1}s^{-1}$	C_{red1}/C_{int} −530 mV	Kumar et al. (1993) and Lindahl, Munck, et al. (1990)
C_{red2}	[Ni-3Fe4S]	g=1.97, 1.87, 1.75	Above	Above	Lindahl, Munck, et al. (1990)
D-cluster	[4Fe4S] bridging subunits	Not observed spectroscopically		D_{ox}/D_{red} ?	Doukov, Blasiak, Seravalli, Ragsdale, and Drennan (2008)
A-cluster	[2Ni-4Fe4S]	2.074, 2.028	$A_{ox}/A_{red\text{-}CO}$ $1.1\,s^{-1}$, 55°C	<−600 mV	Kumar et al. (1993)

DerVartanian, 1983) is performed in a 1 mL reaction mixture containing 50 m*M* HEPES pH 8.0, 2 m*M* DTT, 0.02 m*M* sodium dithionite, and 10 m*M* MV. The solution is sparged with CO in an anaerobic cuvette sealed with a rubber septum stopper, then the reaction is initiated by injection of CODH.

4.1.2 EPR and ENDOR measurements

EPR is a critical technique for studying electron transfer between the B- and C-clusters, which have discrete signals, and for measuring the metal cluster spin intensity, which reflects the percentage activity of the protein (Table 1). Our group uses a Bruker (Billerica, MA) Model EMX X-band EPR instrument with a Cold Edge closed cycle helium cryostat capable of achieving temperatures as low as 8 K. The instrument is equipped with a Hewlett-Packard model 5340 automatic frequency counter and a Bruker gaussmeter.

Samples are either frozen in-chamber with liquid nitrogen, or frozen outside the chamber after capping with a septum prior to removal. The time course of reactions on timescales of 30 s and above can be studied by manual freeze quench—running the reaction in an EPR tube and inserting the tube into liquid nitrogen at specified time points.

Pulsed EPR (electron nuclear double resonance, ENDOR) measurements (DeRose, Telser, Anderson, Lindahl, & Hoffman, 1998) using isotopically labeled enzyme and substrates (^{1}H, ^{2}H, ^{14}N, ^{13}C, and ^{57}Fe) indicated that a nucleophilic iron-bound hydroxide ion at the C-cluster reacts with a Ni–CO intermediate to form a Ni–Fe-bridged carboxylate product (Fig. 1). This bound carboxylate further evolves to give CO_2, releasing two electrons and two protons. This description is consistent with other spectroscopic (Can et al., 2014) and crystallographic studies (Jeoung & Dobbek, 2007), although the formation and decay of the Ni–CO species at catalytically relevant rates have never been established by spectroscopic studies.

4.1.3 Transient kinetic experiments

Stopped flow: Rapid kinetics methods enable the reactions as they occur on the enzyme to be monitored and allow the identification of intermediates. Stopped-flow spectrophotometry, with the instrument housed in the anaerobic chamber, is a powerful method to study the reduction of the FeS clusters of CODH. Reduction from the 2+ to the 1+ state results in a significant bleaching in absorbance, with a $\Delta\varepsilon_{420}$ (monomer) of $\sim$20 mM^{-1} cm^{-1} (Svetlitchnyi et al., 2001), the changes stemming predominantly from reduction of the B-cluster. As with any transient kinetic experiments on CODH, the fast rates of reduction challenge the dead time of the instrument given that most mixing experiments are subject to a dead time of $\sim$1.0 ms. Therefore, low temperatures or subsaturating substrate concentrations are often used to follow the enzyme-monitored CO oxidation reaction (reviewed in Can et al., 2014). Our stopped-flow instrument is housed within an anaerobic chamber.

Rapid freeze-quench (RFQ) EPR: Rapid freeze quench (RFQ) involves rapidly (5 ms) mixing two solutions and quenching the reaction by low-temperature freezing (Ballou, 1978). Solutions are placed in syringes; then the reagents are pushed by a ram through a mixing chamber and into a single tubing (of varying lengths) connected to a nozzle that sprays the reaction mixture into an EPR tube immersed in an isopentane bath kept at −140°C. RFQ-EPR methods can assay the rate of substrate binding to the CODH or ACS active site and assess the overall catalytic competence of any species detected. Our RFQ experiments are performed in an anaerobic chamber with a Bio-Logic SFM-2000 rapid mixing unit employing an in-chamber cryobath and an external liquid nitrogen pump. The major limitation of this method is that the CODH reaction is very fast. The steady-state k_{cat} value for the *M. thermoacetica* CODH/ACS is as high as $500\,s^{-1}$ at pH 7.6 and 55°C, the organism's growth temperature (Ragsdale, Clark, et al., 1983; Ragsdale, Ljungdahl, et al., 1983); thus, all kinetically relevant reactions have to occur within this timescale. For the CODH/ACS from *M. thermoacetica*, the rate constant for reduction of the C-cluster was $400\,s^{-1}$ at 5°C, equating to a $t_{1/2}$ of 1.8ms, closer to the dead time of the instrument (Kumar et al., 1993). Any earlier steps, such as binding of CO would have occurred within the mixing time and evade detection. Furthermore, the rate constant for this reaction, if extrapolated to the optimum growth temperature of the organism, would approximate to $17{,}200\,s^{-1}$ at 55°C ($t_{1/2}=40\,\mu s$!).

4.1.4 Spectroelectrochemical and direct electrochemical experiments

Various electrochemical experiments have been performed on CODH. EPR (Lindahl, Munck, et al., 1990) and Mössbauer (Lindahl, Ragsdale, & Münck, 1990) spectroelecrochemical studies revealed the reduction potentials for the B- (−440mV) and C- (−530mV) clusters (Lindahl, Munck, et al., 1990; Lindahl, Ragsdale, et al., 1990). We have used two types of specialized EPR spectroelectrochemical cells for titrating oxygen-sensitive proteins with low-potential metal centers like CODH (Harder, Feinberg, & Ragsdale, 1989; Lindahl, Munck, et al., 1990; Lindahl, Ragsdale, et al., 1990). For EPR spectroelectrochemical studies of this highly O_2-sensitive protein, is important to use a cell (Harder et al., 1989) in which the EPR tube is integrated with the electrochemical titration cell, so that the sample does not need to be transferred from the electrochemical to the spectroscopic cell/tube.

While spectroelectrochemical studies are ideal for examining the properties of specific chromophoric redox centers, it is also possible to examine (Léger et al., 2003) catalytic properties of enzymes with the enzyme directly attached to an electrode surface (also reviewed recently in Can et al., 2014). Over the past decade, we have also described various CODH-catalyzed nanoparticle systems that can rapidly and selectively reduce CO_2 to CO using visible light, most recently in 2018 (Zhang, Can, Ragsdale, & Armstrong, 2018). The studies demonstrate that CODH is remarkably reversible, showing catalytic activity close to the equilibrium redox potential for the CO_2/CO couple (−0.52V), i.e., operating selectively without any significant overpotential.

4.2 Characterization of ACS

Steady-state kinetic methods were central to identifying the ACS activity of the enzyme formerly known as CODH as well as characterizing the mechanism shown in Fig. 2. These experiments revealed the existence of

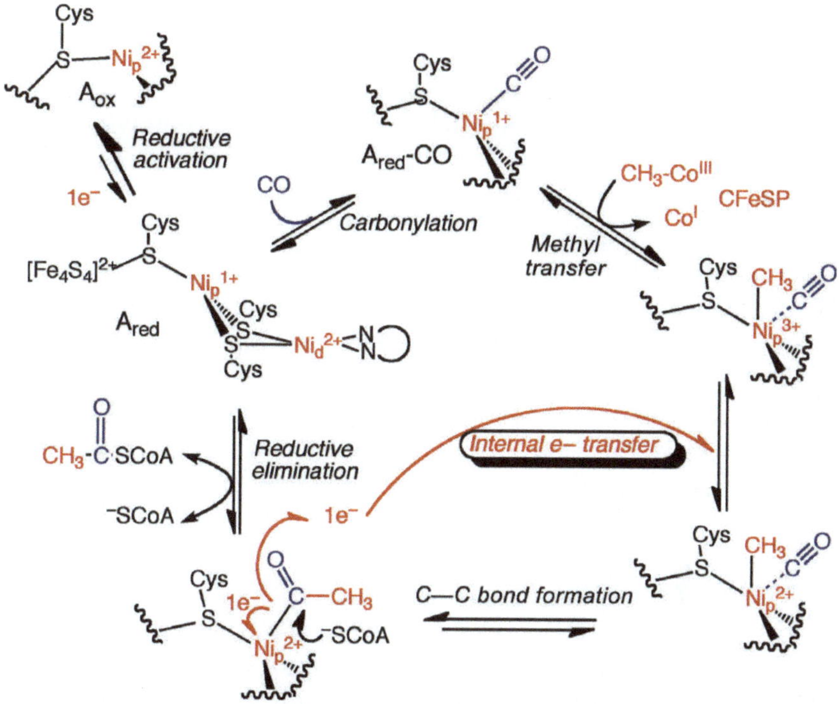

Fig. 2 Proposed organometallic mechanism of anaerobic acetyl-CoA synthesis catalyzed by the A-cluster of ACS. This mechanism, which is supported by recently reviewed kinetic and spectroscopic studies (Can et al., 2014), invokes the formation of the Ni–CO state, which undergoes reaction with a methyl group donated by the methylated CFeSP to form an acetyl-Ni species. Nucleophilic attack by CoA generates acetyl-CoA to regenerate the active state of CODH, ready to undergo another catalytic cycle. This is a simplified mechanism, given that pulse-chase studies demonstrate that CO and the methyl group bind randomly, followed by ordered binding of CoA (Seravalli & Ragsdale, 2008).

enzyme-bound CO, and methyl, acetyl, and CoA-bound intermediates formed during the ACS reaction (Fig. 2). Transient kinetic, spectroscopic, and computational experiments have led to identification of the intermediates, established their kinetic competence, and revealed the changes in Ni_p geometry and electronic structure stemming from these interactions.

4.2.1 Steady-state kinetics

Steady-state kinetic experiments offer some of the best insight into the capability of ACS to produce acetyl-CoA:

$$CO + CH_3\text{-}H_4\text{folate} + CoA \leftrightarrow H_4\text{folate} + \text{acetyl-CoA} \quad (3)$$

$$CO + CH_3\text{-}Co^{III}\text{-CFeSP} + CoA \leftrightarrow \text{CFeSP} + \text{acetyl-CoA} \quad (4)$$

Total synthesis of acetyl-CoA: The total synthesis of acetyl-CoA involves the reaction of CO, CH_3-H_4folate, and CoA to form acetyl-CoA using an assay developed by Hu, Drake, and Wood (1982). This reaction requires CODH/ACS, the CFeSP, and methyltransferase (MeTr); however, when this assay was developed in 1982, enriched fractions containing these enzymes instead of purified proteins were used. The revised method, which uses the purified proteins are described here.

An anaerobic buffer composed of 10 m*M* Tris·HCl, 0.1 *M* NaCl, at pH 7.6 is prepared with 1 m*M* Ti^{III} citrate and 1 m*M* CoA. After adding CO to the buffer to obtain concentrations of 0%–100% saturation, 50 μ*M* ACS (or CODH/ACS), 1 m*M* CH_3-H_4folate, and methyltransferase (MeTr) are added, the reaction is initiated with CFeSP and 100 μL samples are taken and quenched with 100 μL 1 *M* HCl at specified time points. The amount of acetyl-CoA produced is determined by HPLC analysis. Alternatively, the CFeSP can be present in the assay mix prior to adding CODH/ACS (ACS) to initiate the reaction.

CO/acetyl-CoA exchange: The reversibility of the ACS reaction offers a direct way to assay acetyl-CoA synthesis. In 1982, Hu et al. found that [1-^{14}C]-acetyl-CoA can exchange with a headspace of CO (Hu et al., 1982). This assay is now a routine procedure for measuring ACS activity and is convenient because it requires only ACS; thus, it was key to the discovery that ACS (not the CFeSP) is the enzyme that catalyzes acetyl-CoA synthesis (Ragsdale & Wood, 1985). In this reaction (Eq. 5), unlabeled CO exchanges with the ^{14}C-carbonyl group in acetyl-CoA, forming volatile ^{14}CO and leading to a decrease in solution-phase radioactive counts, as quantified by scintillation counting. The [1-^{14}C]-acetyl-CoA method is an excellent way to assess ACS catalytic ability (Raybuck, Bastian, Orme-Johnson, & Walsh, 1988; Raybuck et al., 1991).

$$[1\text{-}^{14}\mathrm{C}]\text{-acetyl-CoA} + {}^{12}\mathrm{CO} \leftrightarrow [1\text{-}^{12}\mathrm{C}]\text{-acetyl-CoA} + {}^{14}\mathrm{CO} \qquad (5)$$

To conduct the CO/acetyl-CoA exchange assay, a 1 mL anaerobic reaction solution is prepared in a 10–15 mL vial containing 300 m*M* MES, pH 6.5, 20 m*M* KCl, 200 μ*M* MV, 2 m*M* Ti^{III} citrate, and 200 μ*M* unlabeled acetyl-CoA. The solution headspace is purged with pure CO for 5–10 min, 5 μL of a 1 mg/mL ACS sample (5 μg ACS) is added, and the reaction solution is incubated at 55°C for 5 min. The assay is initiated by adding 0.05 μCi of [1-^{14}C]-acetyl-CoA solution; 40 μL samples are quenched into scintillation vials containing 100 μL 1 *M* HCl at various time points, 4 mL scintillation liquid is added to the samples, and the radioactivity remaining in the [1-^{14}C]-acetyl-CoA is determined by scintillation counting.

The CO exchange rate is calculated as a function of exchange velocity for a given concentration of unlabeled acetyl-CoA, or AcCoA, as shown in Eq. (6) (Grahame, 2011; Raybuck et al., 1988), where F is the isotopic fraction exchanged at a given time, t. Mathematically, $F=(([A_0^*]-[A^*])/([A_0^*]-[A_\infty^*]))$, and A^* is the remaining 1-[^{14}C]-acetyl-CoA in solution at time t; A_0^* the initial cpm at time = 0 min; and A_∞^* the cpm at time = ∞, or the count at isotopic equilibrium (Grahame, 2011; Raybuck et al., 1988). Assuming the reaction proceeds to 100% completion, the isotopic equilibrium for [^{14}CO] is expected to be far greater than the isotopic equilibrium for [1-^{14}C]-acetyl-CoA, simplifying Eq. (6) to Eq. (7). Rearranging Eq. (7), the CO exchange rate can be assessed (Eq. 8) by plotting the natural log of the fraction as a function of time.

$$v^* = -([CO][AcCoA]/([CO]+[AcCoA]))^*(1/t)^*[\ln(1-F)] \quad (6)$$

$$v^* = -[AcCoA]^*(1/t)^*[\ln(1-F)] \quad (7)$$

$$-((t)^*(v^*)/[AcCoA]) = [\ln(1-F)] \quad (8)$$

Methylation of ACS using CH_3-CFeSP: The only known direct methyl donors to ACS are methyl-CFeSP and methylated corrinoids. Both methyl iodide and methyl-H_4folate can generate active methyl-Co(III); however, they do not generate the active methylated form of ACS. Thus, when methyl iodide or methyl-H_4folate is used as the methyl donor for acetyl-CoA synthesis or ACS, the reaction involves: (1) transfer of a methyl group from methyl iodide or methyl-H_4folate to the Co(I)-state of the protein bound or free corrinoid (usually achieved using reduction by Ti(III) citrate), followed by (2) transfer of the methyl group from CH_3-Co(III) to ACS (Seravalli, Brown, & Ragsdale, 2001). Thus, use of methyl-H_4folate or methyl iodide as methyl donor, as described (Menon & Ragsdale, 1998), requires only catalytic amounts of the CFeSP, which undergoes turnover at each cycle of the reaction. For use in these reactions, the CFeSP is purified essentially as described (Ragsdale, Lindahl, & Münck, 1987), except that FPLC chromatography is now used. To determine the fractions containing CFeSP, samples from the chromatographic column are mixed with cyanide and proteins are precipitated by heating at 95°C to release the cyanide-bound cofactor. After centrifugation, the pink cyanocobalamin-containing fractions are revealed by absorption bands at 367, 540, and 580 nm.

When the CH_3-CFeSP is used directly as the methyl donor, the reaction is performed with near- or above-stoichiometric levels of CH_3-CFeSP relative to those of ACS (Menon & Ragsdale, 1999). This reaction, which is fast and requires transient kinetic methods, is described later.

Methylation of ACS using methyl-cobinamide: We have also used free methylcobalamin, methyl-cobinamide, and methyl-(Me_3-benzimidazolyl)cobamide to methylate ACS (Seravalli et al., 2001). To synthesize cobinamide, dried cyanocobalamin is mixed with anhydrous CF_3SO_3H (triflic acid). After overnight incubation, this extremely acidic solution is neutralized by mixing with 1 *M* dibasic sodium phosphate solution containing a trace amount of cyanide (Zou, Evans, & Brown, 1995). The final mixture is applied to a reverse phase chromatography column (Porapak Rxn) and the cobinamide is purified using an acetonitrile gradient. Cobinamide is lyophilized and then solubilized in water to a final 20 m*M* concentration. Methyl-cobinamide is generated by reducting cobinamide with a fivefold excess of Ti(III) citrate and reacting with excess methyl iodide (40% in methanol) for 1 h; after which a pure product is obtained by gel filtration on a BioGel P3 (BioRad) column. Similar to methylation by CH_3-CFeSP, methylation of ACS can be achieved by reducing ACS with a threefold excess of Ti(III) citrate and mixing with methylcobinamide. The base-off methyl-Co species are significantly more reactive than methyl-cobalamin (Seravalli et al., 2001).

Reaction of ACS with CoASH: ACS can catalyze an exchange reaction between CoA and acetyl-CoA. This reaction was first followed using [3′-^{32}P]-coenzyme A (Ramer, Raybuck, Orme-Johnson, & Walsh, 1989) and involves cleavage of the C–S bond of acetyl-CoA to form the Ni-acyl intermediate, which rebinds to the radiolabeled CoA. Samples from the assay solution at various time points are separated by isocratic solvent HPLC, and radioactive counts measured for the corresponding CoA peaks. Similar to the CO/[1-^{14}C]-AcCoA to ^{14}CO exchange assay, the acetyl-CoA/[3′-^{32}P]-CoA exchange assay continues until the reaction reaches isotopic equilibrium. In place of ^{32}P, the CoA/acetyl-CoA exchange reaction can also be followed using 3′-dephospho-CoA (Bhaskar, Demoll, & Grahame, 1998), instead of the radiolabeled substrate. The K_m for dephospho-CoA is similar to that of CoA.

4.2.2 Reaction of ACS with CO

EPR spectroscopy is one of the most powerful methods for studying ACS. While the A-cluster of as-isolated ACS is EPR silent, exposure of the protein to CO in the presence of a reductant, e.g., dithionite, Ti(III)-citrate, DTT, reveals a unique EPR spectrum with *g* values of 2.074 and 2.028

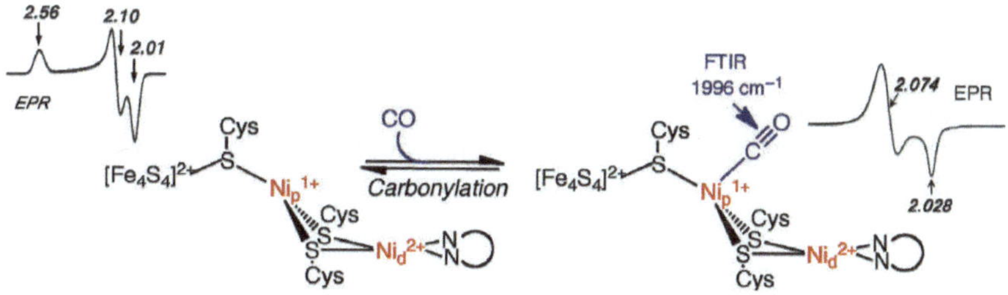

Fig. 3 Binding of CO to the Ni(I) form of ACS to form the catalytically active Ni(I)-CO state. Included are EPR and FTIR spectral details for the Ni–CO state of the A-cluster and the EPR spectrum of the Ni(I) species which is also generated upon photolysis.

(Fig. 3). Many experiments have shown this to be a key intermediate in the ACS mechanism (Can et al., 2014). By measuring the splitting of the lines in the EPR spectrum observed upon substitutions with ^{57}Fe, ^{61}Ni, and ^{13}CO, this signal was proposed to originate from a Ni–CO bond formed by the complex of CO with this spin-coupled cluster and became known as the "NiFeC signal" (Ragsdale & Kumar, 1996; Ragsdale, Ljungdahl, et al., 1983; Ragsdale, Wood, & Antholine, 1985). Pulsed EPR (ENDOR) (Fan, Gorst, Ragsdale, & Hoffman, 1991) and Mössbauer (Lindahl, Munck, et al., 1990; Lindahl, Ragsdale, et al., 1990) methods have also been used to describe the electronic state of ACS and its interactions with substrates. The catalytic competency of ACS samples can be measured quantitatively by determining the double integrated signal intensity (the spin concentration) of the NiFeC signal in comparison to that of a Cu^{II} perchlorate standard measured under the same conditions (Fee, 1978). The amount of the NiFeC EPR signal is a direct measure of enzyme activity. The form of ACS that binds CO was proposed to be Ni_p^I, based on EPR and FTIR spectroscopy following low-temperature photolysis of the NiFeC species (Bender et al., 2010). This Ni(I) species is unstable and immediately converts back to the NiFeC species with an activation energy of only 1 kJ/mol, indicating the existence of an alcove that sequesters CO near Ni_p, a hypothesis consistent with crystallographic studies of CODH/ACS (Doukov et al., 2008).

Although FTIR is a commonly used technique to study small molecules, it is less commonly used in the study of proteins because of the strong and broad absorbance bands of water in the $1645\,cm^{-1}$ region, which overlap with the amide I band of proteins and some side chain bands. However, metal–CO bands have been studied in various proteins like hydrogenase, myoglobin, CODH, and ACS because the CO stretch is discretely located in the $2000\,cm^{-1}$ region, significantly away from other contaminating bands.

As shown in Fig. 3, the CO stretching frequency at 1995 cm^{-1} allowed us to demonstrate that CO was bound to the A-cluster, validate photolysis of the CO, and determine the energy barrier of CO rebinding to the ACS A-cluster, as described earlier (Bender et al., 2010; George, Seravalli, & Ragsdale, 2005; Kumar & Ragsdale, 1992). Preparation of 300 μL FTIR samples is accomplished by buffer exchanging ACS sample into an anaerobic D_2O buffer of 50 m*M* KP_i, pH 7.7 by Amicon pressure filtration. The ACS sample is then concentrated to 1 m*M*, 300 μL of 2 m*M* sodium dithionite is added to reduce the protein, and the sample is purged with CO gas for 10 min. The ACS sample is then anaerobically transferred to an FTIR CaF_2 window setup, which is immediately subjected to FTIR data acquisition.

XAS has been used to probe the Ni and Fe sites in CODH and ACS. The first direct demonstration of the Ni-CO bond in ACS was obtained through XAS studies (Can et al., 2017), with DFT calculations providing a powerful complement to the experimental studies.

4.2.3 Single turnover and transient kinetics to study ACS

General aspects of stopped-flow and RFQ-EPR methods are described earlier in the CODH section. Using RFQ-EPR, rates of CO binding to ACS to form Ni_p-CO and its condensation with the methylated CFeSP to form Ni_p acetyl have been measured by growth and decay of the paramagnetic NiFeC EPR signal over time (Seravalli, Kumar, & Ragsdale, 2002). Exponential curves are plotted based on the intensity at NiFeC as a function of time, and the kinetic rate constants derived from the exponential curves.

RFQ has been used in conjunction with other spectroscopic methods, such as Mössbauer or ENDOR. The same unit can also be used for rapid chemical quench. In this case, the reaction is quenched by spraying the solution into a chemical denaturant (generally base or acid).

Reaction of ACS with CO: The binding of CO to ACS is studied by RFQ-EPR by measuring the rate of formation of the NiFeC signal. In a typical experiment (Seravalli et al., 2002), a solution containing 100 μ*M* ACS is rapidly mixed with a buffered solution of CO and the reaction is quenched by spraying the mixture into liquid isopentane.

Stopped-flow FTIR has also been used to measure the rate of CO binding to ACS, providing further evidence that the NiFeC species is a catalytically competent intermediate in the ACS mechanism. The reaction is performed as described (Bender et al., 2010; George et al., 2005).

Reaction of ACS with CH_3-CFeSP: This reaction is fast and must be monitored by rapid kinetic methods. The most convenient option is stopped-flow UV–visible spectrophotometry by monitoring the spectral changes at

390 nm (formation of Co^{1+}) and 450 nm (depletion of methyl-Co^{3+}). First, the CFeSP needs to be methylated, and a convenient methyl donor is methyl iodide used as described (Lu, Harder, & Ragsdale, 1990). Approximately 50 μ*M* CFeSP in 100 m*M* Tris–HCl (pH 7.6) buffer is reduced by addition of 0.625 m*M* Ti(III) citrate. Methyl iodide is added (250 μ*M*, final) at 55°C in the dark, then after about 15 min, CH_3-CFeSP is buffer exchanged into 100 m*M* Tris–HCl (pH 7.6) buffer. Reduced Co(I)- and CH_3-Co(III) states of the CFeSP can be distinguished by UV–visible spectroscopy since reduced CFeSP shows a sharp band at 390 nm ($\varepsilon = 40\,cm^{-1}\,mM^{-1}$) and CH_3-CFeSP shows a broad band at 450 nm ($\Delta\varepsilon = 16.2\,cm^{-1}\,mM^{-1}$). Alternatively, if $^{14}CH_3I$ is used to prepare CH_3-CFeSP, the methylation can be followed by measuring the radioactivity after separating the ACS and CFeSP by phenyl-sepharose chromatography and measuring the $^{14}CH_3$ group associated with each protein (Lu et al., 1990). Methylation of the CFeSP with methyl-H_4folate is performed as described (Ragsdale et al., 1987).

ACS is methylated by adding Ti(III) citrate and rapidly mixing with CH_3-CFeSP in the stopped-flow instrument. The reaction is monitored by following the changes in the absorption spectrum associated with demethylation of CH_3-CFeSP (Seravalli et al., 2002). In a typical reaction, 10–100 μ*M* CODH/ACS (or ACS) is mixed with 10–100 μ*M* CH_3-CFeSP and the reaction is monitored for a few seconds. Additionally, the impact of other ACS substrates can be studied by mixing 100 μ*M* ACS including CO and/or CoA in different concentrations with 100 μ*M* CH_3-CFeSP.

Methylation of ACS-CO: Because binding of CO and the methyl group follow a random mechanism, enzyme-monitored acetyl-CoA synthesis can be followed by measuring the depletion of the NiFeC signal. In a typical freeze-quench experiment (Seravalli et al., 2002), a solution containing 200 μ*M* ACS incubated with CO is rapidly mixed with 100 μ*M* CH_3-CFeSP (or other cobalamin-based methyl donor). CoA (1 m*M*) can be included in this reaction. The sample is sprayed at different reaction times over a 10 s time frame into a freezing bath of isopentane as described earlier. Most of the NiFeC decay occurs within the first 0.5 s. When CoA is included, the amount of synthesized acetyl-CoA can be measured by HPLC.

5. Closing and future directions

This is an opportune time to review the methods used to characterize the two key enzymes in the WLP. We now have heterologous and homologous genetic systems for expressing CODH and ACS. Strains of acetogens

(*Clostridium ljungdahlii, Clostridium acetogenum, Clostridium ragsdalii*, etc.) have been developed for biotechnological applications. Combining mutagenesis and directed evolution with mechanistic studies offer a powerful toolbox for gaining a better understanding of the mechanisms of these two enzymes as well as for developing more powerful biotechnological platforms to activate CO_2 and to convert syngas and various waste gas mixtures into desired fuels and chemicals.

In terms of mechanism, we can look forward to the use of mutagenesis coupled to the methods described in this chapter to understand the early steps in the mechanisms of CO oxidation and CO_2 reduction that have yet not been detected. We have recently obtained validation of the organometallic ACS mechanism by XAS analysis of the Ni(I)–CO complex (Can et al., 2017); the metal-CH_3 and metal-Ni intermediates still require characterization. Photoactive Ru and Ni bimetallic models have been developed that selectively reduce CO_2, i.e., do not reduce H^+ to evolve H_2 (Schneider, Manesis, Stevenson, & Shafaat, 2018). Similarly, effective biomimetic ACS models have been developed, allowing comparison of the properties of the model with those of the enzyme (Manesis, O'Connor, Schneider, & Shafaat, 2017; Manesis & Shafaat, 2015). Recent studies coupling mutagenesis with spectroscopic and kinetic studies are also uncovering the importance of the intermolecular channel in enzymatic activity. We hope that the methods used in characterizing CODH and ACS described here will spur researchers to discover more of the unique important properties of these proteins.

Acknowledgments

Work in the Ragsdale lab on the Wood–Ljungdahl Pathway has been funded by NIH (R37-GM39451). S.W.R. is personally grateful to all members of the Ragsdale lab and all our lineages for the progress on understanding the mechanisms of CODH and ACS and the WLP in general.

References

Ballou, D. P. (1978). Freeze-quench and chemical-quench techniques. *Methods in Enzymology*, *54*, 85–93.

Bender, G., Stitch, T. A., Yan, L., Britt, R. D., Cramer, S. P., & Ragsdale, S. W. (2010). Infrared and EPR spectroscopic characterization of a Ni(I) species formed by photolysis of a catalytically competent Ni(I)-CO intermediate in the acetyl-CoA synthase reaction. *Biochemistry*, *49*, 7.

Bhaskar, B., DeMoll, E., & Grahame, D. A. (1998). Redox-dependent acetyl transfer partial reaction of the acetyl-CoA decarbonylase/synthase complex: Kinetics and mechanism. *Biochemistry*, *37*, 14491–14499.

Bonam, D., & Ludden, P. W. (1987). Purification and characterization of carbon monoxide dehydrogenase, a nickel, zinc, iron–sulfur protein, from *Rhodospirillum rubrum*. *Journal of Biological Chemistry*, *262*, 2980–2987.

Can, M., Armstrong, F. A., & Ragsdale, S. W. (2014). Structure, function, and mechanism of the nickel metalloenzymes, CO dehydrogenase, and acetyl-CoA synthase. *Chemical Reviews*, *114*, 4149–4174.

Can, M., Giles, L. J., Ragsdale, S. W., & Sarangi, R. (2017). X-ray absorption Spectroscopy reveals an organometallic Ni–C bond in the CO-treated form of acetyl-CoA synthase. *Biochemistry*, *56*, 1248–1260.

Carlson, E. D., & Papoutsakis, E. T. (2017). Heterologous expression of the *Clostridium carboxidivorans* CO dehydrogenase alone or together with the acetyl coenzyme A synthase enables both reduction of CO_2 and oxidation of CO by *Clostridium acetobutylicum*. *Applied and Environmental Microbiology*, *83*, e00829-17.

Daniel, S. L., Hsu, T., Dean, S. I., & Drake, H. L. (1990). Characterization of the H_2- and CO-dependent chemolithotrophic potentials of the acetogens *Clostridium thermoaceticum* and *Acetogenium kivui*. *Journal of Bacteriology*, *172*, 4464–4471.

Daniel, S. L., Wu, Z., & Drake, H. L. (1988). Growth of thermophilic acetogenic bacteria on methoxylated aromatic acids. *FEMS Microbiology Letters*, *52*, 25–28.

DeMoll, E., Grahame, D. A., Harnly, J. M., Tsai, L., & Stadtman, T. C. (1987). Purification and properties of carbon monoxide dehydrogenase from *Methanococcus vannielii*. *Journal of Bacteriology*, *169*, 3916–3920.

DeRose, V. J., Telser, J., Anderson, M. E., Lindahl, P. A., & Hoffman, B. M. (1998). A multinuclear ENDOR study of the C-cluster in CO dehydrogenase from *Clostridium thermoaceticum*: Evidence for HxO and histidine coordination to the [Fe4S4] center. *Journal of the American Chemical Society*, *120*, 8767–8776.

Doukov, T. I., Blasiak, L. C., Seravalli, J., Ragsdale, S. W., & Drennan, C. L. (2008). Xenon in and at the end of the tunnel of bifunctional carbon monoxide dehydrogenase/acetyl-CoA synthase. *Biochemistry*, *47*, 3474–3483.

Drake, H. L., & Daniel, S. L. (2004). Physiology of the thermophilic acetogen *Moorella thermoacetica*. *Research in Microbiology*, *155*, 869–883.

Elliott, J. I., & Brewer, J. M. (1978). The inactivation of yeast enolase by 2,3-butanedione. *Archives of Biochemistry and Biophysics*, *190*, 351–357.

Ensign, S. A. (1995). Reactivity of carbon monoxide dehydrogenase from *Rhodospirillum rubrum* with carbon dioxide, carbonyl sulfide, and carbon disulfide. *Biochemistry*, *34*, 5372–5381.

Ensign, S. A., Hyman, M. R., & Ludden, P. W. (1989). Nickel-specific, slow-binding inhibition of carbon monoxide dehydrogenase from *Rhodospirillum rubrum* by cyanide. *Biochemistry*, *28*, 4973–4979.

Fan, C. L., Gorst, C. M., Ragsdale, S. W., & Hoffman, B. M. (1991). Characterization of the Ni–Fe–C complex formed by reaction of carbon monoxide with the carbon monoxide dehydrogenase from *Clostridium thermoaceticum* by Q-band ENDOR. *Biochemistry*, *30*, 431–435.

Fast, A. G., & Papoutsakis, E. T. (2018). Functional expression of the *Clostridium ljungdahlii* acetyl-CoA synthase in *Clostridium acetobutylicum* as demonstrated by a novel in vivo CO exchange activity, on the way to heterologous installation of a functional Wood–Ljungdahl pathway. *Applied and Environmental Microbiology*, *84*, e02307-17.

Fee, J. A. (1978). Transition metal electron paramagnetic resonance related to proteins. *Methods in Enzymology*, *49*, 512–528Academic Press.

Ferry, J. G. (2010). How to make a living by exhaling methane. *Annual Review of Microbiology*, *64*, 453–473.

George, S. J., Seravalli, J., & Ragsdale, S. W. (2005). EPR and infrared spectroscopic evidence that a kinetically competent paramagnetic intermediate is formed when acetyl-coenzyme A synthase reacts with CO. *Journal of the American Chemical Society*, *127*, 1.

Ghosh, R., Hardmeyer, A., Thoenen, I., & Bachofen, R. (1994). Optimization of the sistrom culture medium for large-scale batch cultivation of *Rhodospirillum rubrum* under semiaerobic conditions with maximal yield of photosynthetic membranes. *Applied and Environmental Microbiology*, *60*, 1698–1700.

Grahame, D. A. (2011). Methods for analysis of acetyl-CoA synthase: Applications to bacterial and archaeal systems. *Methods in Enzymology*, *494*, 189–217. Elsevier.

Grahame, D. A., & Stadtman, T. C. (1987). Carbon monoxide dehydrogenase from *Methanosarcina barkeri*. Disaggregation, purification, and physicochemical properties of the enzyme. *Journal of Biological Chemistry*, *262*, 3706–3712.

Harder, S. R., Feinberg, B. A., & Ragsdale, S. W. (1989). A spectroelectrochemical cell designed for low temperature electron paramagnetic resonance titration of oxygen-sensitive proteins. *Analytical Biochemistry*, *181*, 283–287.

Hsu, T., Daniel, S. L., Lux, M. F., & Drake, H. L. (1990). Biotransformations of carboxylated aromatic compounds by the acetogen *Clostridium thermoaceticum*: Generation of growth-supportive CO_2 equivalents under CO_2-limited conditions. *Journal of Bacteriology*, *172*, 212–217.

Hu, S.-I., Drake, H. L., & Wood, H. G. (1982). Synthesis of acetyl coenzyme A from carbon monoxide, methyltetrahydrofolate, and coenzyme A by enzymes from *Clostridium thermoaceticum*. *Journal of Bacteriology*, *149*, 440–448.

Hungate, R. E. (1969). A roll tube method for cultivation of strict anaerobes. *Methods in Microbiology*, *3B*, 117–132.

Hungate, R. E., Smith, W., & Clarke, R. T. (1966). Suitability of butyl rubber stoppers for closing anaerobic roll culture tubes. *Journal of Bacteriology*, *91*, 908–909.

Inoue, T., Takao, K., Fukuyama, Y., Yoshida, T., & Sako, Y. (2014). Over-expression of carbon monoxide dehydrogenase-I with an accessory protein co-expression: A key enzyme for carbon dioxide reduction. *Bioscience, Biotechnology, and Biochemistry*, *78*, 582–587.

Jeoung, J. H., & Dobbek, H. (2007). Carbon dioxide activation at the Ni,Fe-cluster of anaerobic carbon monoxide dehydrogenase. *Science*, *318*, 1461–1464.

Kasmi, A. E., Rajasekharan, S., & Ragsdale, S. W. (1994). Anaerobic pathway for the conversion of the methyl group of aromatic methyl ethers to acetic acid by *clostridium thermoaceticum*. *Biochemistry*, *33*, 11217–11224.

Kerby, R. L., Hong, S. S., Ensign, S. A., Coppoc, L. J., Ludden, P. W., & Roberts, G. P. (1992). Genetic and physiological characterization of the *Rhodospirillum rubrum* carbon monoxide dehydrogenase system. *Journal of Bacteriology*, *174*, 5284–5294.

Kerby, R., & Zeikus, J. G. (1983). Growth of *Clostridium thermoaceticum* on H_2/CO_2 or CO as energy source. *Current Microbiology*, *8*, 27–30.

Kopke, M., Mihalcea, C., Liew, F., Tizard, J. H., Ali, M. S., Conolly, J. J., et al. (2011). 2,3-Butanediol production by acetogenic bacteria, an alternative route to chemical synthesis, using industrial waste gas. *Applied and Environmental Microbiology*, *77*, 5467–5475.

Kumar, M., Lu, W.-P., Liu, L., & Ragsdale, S. W. (1993). Kinetic evidence that CO dehydrogenase catalyzes the oxidation of CO and the synthesis of acetyl-CoA at separate metal centers. *Journal of the American Chemical Society*, *115*, 11646–11647.

Kumar, M., & Ragsdale, S. W. (1992). Characterization of the carbon monoxide binding site of carbon monoxide dehydrogenase from *Clostridium thermoaceticum* by infrared spectroscopy. *Journal of the American Chemical Society*, *114*, 8713–8715.

Léger, C., Elliott, S. J., Hoke, K. R., Jeuken, L. J., Jones, A. K., & Armstrong, F. A. (2003). Enzyme electrokinetics: Using protein film voltammetry to investigate redox enzymes and their mechanisms. *Biochemistry*, *42*, 8653–8662.

Liew, F., Henstra, A. M., Kpke, M., Winzer, K., Simpson, S. D., & Minton, N. P. (2017). Metabolic engineering of *Clostridium autoethanogenum* for selective alcohol production. *Metabolic Engineering*, *40*, 104–114.

Liew, F., Martin, M. E., Tappel, R. C., Heijstra, B. D., Mihalcea, C., & Kopke, M. (2016). Gas fermentation-A flexible platform for commercial scale production of low-carbon-fuels and chemicals from waste and renewable feedstocks. *Frontiers in Microbiology*, 7, 694.

Lindahl, P. A., Munck, E., & Ragsdale, S. W. (1990). CO dehydrogenase from Clostridium thermoaceticum. EPR and electrochemical studies in CO_2 and argon atmospheres. *Journal of Biological Chemistry*, *265*, 3873–3879.

Lindahl, P. A., Ragsdale, S. W., & Münck, E. (1990). Mössbauer studies of CO dehydrogenase from *Clostridium thermoaceticum*. *Journal of Biological Chemistry*, *265*, 3880–3888.

Ljungdahl, L. G., & Andreesen, J. R. (1978). Formate dehydrogenase, a selenium-tungsten enzyme from *Clostridium thermoaceticum*. *Methods in Enzymology*, *53*, 360–372.

Lu, W. P., Harder, S. R., & Ragsdale, S. W. (1990). Controlled potential enzymology of methyl transfer reactions involved in acetyl-CoA synthesis by CO dehydrogenase and the corrinoid/iron-sulfur protein from *Clostridium thermoaceticum*. *Journal of Biological Chemistry*, *265*, 3124–3133.

Manesis, A. C., O'Connor, M. J., Schneider, C. R., & Shafaat, H. S. (2017). Multielectron chemistry within a model nickel metalloprotein: Mechanistic implications for acetyl-CoA synthase. *Journal of the American Chemical Society*, *139*, 10328–10338.

Manesis, A. C., & Shafaat, H. S. (2015). Electrochemical, spectroscopic, and density functional theory characterization of redox activity in nickel-substituted azurin: A model for acetyl-CoA synthase. *Inorganic Chemistry*, *54*, 7959–7967.

Menon, S., & Ragsdale, S. W. (1998). Role of the [4Fe-4S] cluster in reductive activation of the cobalt center of the corrinoid iron-sulfur protein from *Clostridium thermoaceticum* during acetate biosynthesis. *Biochemistry*, *37*, 5689–5698.

Menon, S., & Ragsdale, S. W. (1999). The role of an iron–sulfur cluster in an enzymatic methylation reaction. Methylation of CO dehydrogenase/acetyl-CoA synthase by the methylated corrinoid iron–sulfur protein. *Journal of Biological Chemistry*, *274*, 11513–11518.

Merrouch, M., Benvenuti, M., Lorenzi, M., Léger, C., Fourmond, V., & Dementin, S. (2018). Maturation of the [Ni-4Fe-4S] active site of carbon monoxide dehydrogenases. *Journal of Biological Inorganic Chemistry*, *23*, 613–620.

Mock, J., Zheng, Y., Mueller, A. P., Ly, S., Tran, L., Segovia, S., et al. (2015). Energy conservation associated with ethanol formation from H_2 and CO_2 in *Clostridium autoethanogenum* involving electron bifurcation. *Journal of Bacteriology*, *197*, 2965–2980.

Norman, R. O., Millat, T., Winzer, K., Minton, N. P., & Hodgman, C. (2018). Progress towards platform chemical production using *Clostridium autoethanogenum*. *Biochemical Society Transactions*, *46*, 523–535.

Ragsdale, S. W., Clark, J. E., Ljungdahl, L. G., Lundie, L. L., & Drake, H. L. (1983). Properties of purified carbon monoxide dehydrogenase from *Clostridium thermoaceticum*, a nickel, iron–sulfur protein. *Journal of Biological Chemistry*, *258*, 2364–2369.

Ragsdale, S. W., & Kumar, M. (1996). Nickel-containing carbon monoxide dehydrogenase/acetyl-CoA synthase. *Chemical Reviews*, *96*, 25.

Ragsdale, S. W., Lindahl, P. A., & Münck, E. (1987). Mössbauer, EPR, and optical studies of the corrinoid/Fe-S protein involved in the synthesis of acetyl-CoA by *Clostridium thermoaceticum*. *Journal of Biological Chemistry*, *262*, 14289–14297.

Ragsdale, S. W., Ljungdahl, L. G., & DerVartanian, D. V. (1983). ^{13}C and ^{61}Ni isotope substitutions confirm the presence of a nickel(III)–carbon species in acetogenic CO dehydrogenases. *Biochemical and Biophysical Research Communications*, *115*, 7.

Ragsdale, S. W., & Wood, H. G. (1985). Acetate biosynthesis by acetogenic bacteria: Evidence that carbon monoxide dehydrogenase is the condensing enzyme that catalyzes the final steps of the synthesis. *Journal of Biological Chemistry*, *260*, 3970–3977.

Ragsdale, S. W., Wood, H. G., & Antholine, W. E. (1985). Evidence that an iron–nickel–carbon complex is formed by reaction of CO with the CO dehydrogenase from *Clostridium thermoaceticum*. *Proceedings of the National Academy of Sciences of the United States of America*, *82*, 6811–6814.

Ramer, S. E., Raybuck, S. A., Orme-Johnson, W. H., & Walsh, C. T. (1989). Kinetic characterization of the [3′-^{32}P]-coenzyme A/acetyl coenzyme A exchange catalyzed by a three-subunit form of the carbon monoxide dehydrogenase/acetyl-CoA synthase from *Clostridium thermoaceticum*. *Biochemistry*, *28*, 4675–4680.

Raybuck, S. A., Bastian, N. R., Orme-Johnson, W. H., & Walsh, C. T. (1988). Kinetic characterization of the carbon monoxide–acetyl-CoA (carbonyl group) exchange activity of the acetyl-CoA synthesizing CO dehydrogenase from *Clostridium thermoaceticum*. *Biochemistry*, *27*, 7698–7702.

Raybuck, S. A., Ramer, S. E., Abbanat, D. R., Peters, J. W., Orme-Johnson, W. H., Ferry, J. G., et al. (1991). Demonstration of carbon–carbon bond cleavage of acetyl coenzyme A by using isotopic exchange catalyzed by the CO dehydrogenase complex from acetate-grown *Methanosarcina thermophila*. *Journal of Bacteriology*, *173*, 3.

Schneider, C. R., Manesis, A. C., Stevenson, M. J., & Shafaat, H. S. (2018). A photoactive semisynthetic metalloenzyme exhibits complete selectivity for CO_2 reduction in water. *Chemical Communications (Cambridge)*, *54*, 4681–4684.

Seefeldt, L. C., & Ensign, S. A. (1994). A continuous, spectrophotometric activity assay for nitrogenase using the reductant titanium(III) citrate. *Analytical Biochemistry*, *221*, 379–386.

Seravalli, J., Brown, K. L., & Ragsdale, S. W. (2001). Acetyl coenzyme A synthesis from unnatural methylated corrinoids: Requirement for "Base-Off" coordination at cobalt. *Journal of the American Chemical Society*, *123*, 1786–1787.

Seravalli, J., Kumar, M., & Ragsdale, S. W. (2002). Rapid kinetic studies of acetyl-CoA synthesis: Evidence supporting the catalytic intermediacy of a paramagnetic NiFeC species in the autotrophic Wood–Ljungdahl pathway. *Biochemistry*, *41*, 1807–1819.

Seravalli, J., & Ragsdale, S. W. (2008). Pulse-chase studies of the synthesis of acetyl-CoA by carbon monoxide dehydrogenase/acetyl-CoA synthase: Evidence for a random mechanism of methyl and carbonyl addition. *Journal of Biological Chemistry*, *283*, 8384–8394.

Svetlitchnyi, V., Peschel, C., Acker, G., & Meyer, O. (2001). Two membrane-associated NiFeS-carbon monoxide dehydrogenases from the anaerobic carbon-monoxide-utilizing eubacterium *Carboxydothermus hydrogenoformans*. *Journal of Bacteriology*, *183*, 5134–5144.

Terlesky, K. C., Nelson, M. J., & Ferry, J. G. (1986). Isolation of an enzyme complex with carbon monoxide dehydrogenase activity containing corrinoid and nickel from acetate-grown *Methanosarcina thermophila*. *Journal of Bacteriology*, *168*, 1053–1058.

Wang, V. C., Can, M., Pierce, E., Ragsdale, S. W., & Armstrong, F. A. (2013). A unified electrocatalytic description of the action of inhibitors of nickel carbon monoxide dehydrogenase. *Journal of the American Chemical Society*, *135*, 2198–2206.

Wolfe, R. S. (2011). Techniques for cultivating methanogens. In S. W. Ragsdale & A. C. Rosenzweig (Eds.), *Methods in enzymology: Vol. 494* (pp. 1–22): Elsevier. 2011/03/16 ed.

Zhang, L., Can, M., Ragsdale, S., & Armstrong, F. (2018). Fast and selective photoreduction of CO_2 to CO catalyzed by a complex of carbon monoxide dehydrogenase, TiO_2 and Ag nanoclusters. *ACS Catalysis*, *8*, 2789–2795.

Zou, X. A., Evans, D. R., & Brown, K. L. (1995). Efficient and convenient method for axial nucleotide removal from vitamin-B-12 and its derivatives. *Inorganic Chemistry*, *34*, 1634–1635.

CHAPTER THIRTEEN

Genetic techniques for studies of methyl-coenzyme M reductase from *Methanosarcina acetivorans* C2A

Dipti D. Nayak[a,b], William W. Metcalf[a,b,*]
[a]Carl R. Woese Institute for Genomic Biology, University of Illinois, Urbana, IL, United States
[b]Department of Microbiology, University of Illinois, Urbana, IL, United States
*Corresponding author: e-mail address: metcalf@life.illinois.edu

Contents

Abstract

Methanogenic archaea generate methane as a by-product of anaerobic respiration using CO_2, C_1 compounds (like methanol or methylated amines), or acetate as terminal electron acceptors. Methanogens are an untapped resource for biotechnological advances related to methane production as well as methane consumption. However, key biological features of these organisms remain poorly understood. One such feature

Methods in Enzymology, Volume 613
ISSN 0076-6879
https://doi.org/10.1016/bs.mie.2018.10.012

is the enzyme methyl-coenzyme M reductase (referred to as MCR), which catalyzes the last step in the methanogenic pathway and results in methane formation. Gene essentiality has limited genetic analyses of MCR thus far. Therefore, studies of this important enzyme have been limited to biochemical and biophysical techniques that are especially laborious and often reliant on sophisticated instrumentation that is not commonly available. In this chapter, we outline our recently developed CRISPR–Cas9-based genome editing tools and describe how these tools have been used for the introduction of a tandem affinity purification tag at the chromosomal *mcr* locus in the model methanogen, *Methanosarcina acetivorans* C2A. We also report a protocol for rapid affinity purification of MCR from *M. acetivorans* C2A that will enable high-throughput studies of this enzyme in the future.

1. Introduction

A group of microorganisms within the Archaea, collectively referred to as methanogens, are the predominant source of methane on Earth (Thauer, Kaster, Seedorf, & Buckel, 2008). Methanogens are prevalent in anoxic environments and generate ca. 1 gigaton of methane annually, which accounts for 70%–80% of the annual emissions of this potent greenhouse gas (Schaefer et al., 2016; Thauer et al., 2008). As such, the contribution of methanogens to climate change and the global carbon cycle are apparent. In addition, because methane is a clean burning renewable fuel with high calorific value, methanogens have immense potential to serve as biocatalysts for clean energy-related technological advances (Wood, 2017). However, the absence of tools for high-throughput studies of methanogens has severely crippled efforts toward their use in biotechnological applications. In this chapter, we describe genetic techniques that were recently developed to aid and accelerate studies of MCR, an enzyme of importance for biotechnology and renewable energy, from the methanogenic archaeon *Methanosarcina acetivorans* C2A.

Even though methanogenesis as a metabolic trait is limited to members of the archaeal branch of the tree of life, these organisms are considerably diverse (Liu & Whitman, 2008; Spang & Ettema, 2017; Thauer et al., 2008). Notably, sources of methanogens range from Antarctic lakes where the temperatures are below the freezing point of water (Franzmann, Springer, Ludwig, Conway De Macario, & Rohde, 1992) to "black smokers" at the seabed where temperatures are well above the boiling point of water (Bult et al., 1996). Depending on the strain, the substrate breadth can vary as well; while some methanogens can only grow on $H_2 + CO_2$ or formate, others use methylated compounds (such as methanol, methylated amines, methylated sulfides) or even acetate as methanogenic substrates (Liu & Whitman, 2008; Thauer et al., 2008). This versatility is an especially desirable trait from a biotechnological standpoint. Despite this apparent diversity, the core methanogenic process is highly conserved, especially the last step that leads to methane formation, which is catalyzed by an enzyme called methyl-coenzyme M reductase (referred to as MCR henceforth).

Since its initial discovery (Gunsalus & Wolfe, 1980), MCR has become the subject of biochemical and biophysical research spanning decades, laboratories, and continents (Ermler, Grabarse, Shima, Goubeaud, & Thauer, 1997; Prakash, Wu, Suh, & Duin, 2014; Thauer, 1998; Wongnate et al., 2016). Structural studies have revealed that MCR is a hexamer comprised of three subunits (α, β, γ) in a $\alpha_2\beta_2\gamma_2$ configuration (Ermler et al., 1997; Grabarse, Mahlert, Shima, Thauer, & Ermler, 2000) (Fig. 1). Each molecule of MCR contains two active sites that lie in a buried pocket within the α subunits (Fig. 1). It has been postulated that the two active sites, which are

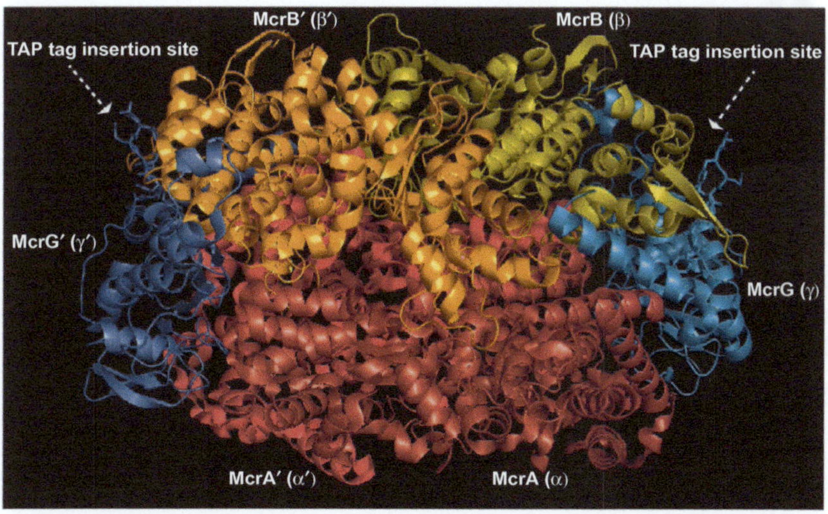

Fig. 1 Structure of methyl coenzyme-M reductase from *Methanosarcina barkeri* (PDB accession number: 1e6y). The α and α' subunits are colored in shades of pink as indicated; the β and β' subunits are colored in yellow and orange, respectively; the γ and γ' subunits are colored in shades of blue as indicated. The N-terminus of the γ and γ' subunits is highlighted to show the loop where the tandem affinity purification (TAP) tag is inserted. Note: the amino-acid identity of the α, β, and γ subunits between *M. barkeri* and *M. acetivorans* is $\geq$90%.

ca. 50 Å apart, might be coupled in a manner similar to a two-stroke engine (Goenrich, Duin, Mahlert, & Thauer, 2005). Within each active site, methyl-coenzyme M (CH_3-S-CoM; methyl-2-sulfanylethanesulfonate) is reduced by coenzyme B (CoB; 7-thioheptanoylthreonine phosphate) to form methane and the mixed heterodisulfide of CoB and CoM (CoM-S-S-CoB) as follows:

$$CH_3 - S - CoM + HS - CoB \rightleftarrows CH_4 + CoM - S - S - CoB$$

The active site of MCR is known as Factor 430 (F_{430}): it contains a porphyrinoid cofactor coordinated to a central Ni atom and the reduced Ni(I) form of F_{430} is essential for catalysis (Moore et al., 2017; Zheng, Ngo, Owens, Yang, & Mansoorabadi, 2016). The low redox potential of the Ni(I)/Ni(II) couple (ca. −650 mV) renders F_{430} especially sensitive to oxidative inactivation, which makes MCR recalcitrant to mechanistic analyses (Goubeaud, Schreiner, & Thauer, 1997). However, recent studies led by Steve Ragsdale's group at the University of Michigan have provided evidence in support of a reaction mechanism that leads to the formation of a methyl radical and a Ni (II)-thiolate intermediate (Wongnate et al., 2016).

Genetic studies of MCR have been scant, only barely keeping pace with other advancements (Bokranz, Baumner, Allmansberger, Ankel-Fuchs, & Klein, 1988; Weil, Cram, Sherf, & Reeve, 1988). We posit two critical hurdles that may have impeded genetic studies of MCR. First, heterologous expression of MCR in a nonnative host such as *Escherichia coli* is likely to be extremely challenging for many reasons: (a) currently unknown electron donors required to maintain F_{430} in the reduced Ni (I) state might be absent, (b) the host might not encode genes for the synthesis of CoM, CoB, or F_{430}, and (c) F_{430} is especially oxygen-labile. Second, all known methanogens are obligate for this metabolic trait (i.e., biomass production and energy conversation are strictly and singularly coupled to methane production) (Spang & Ettema, 2017; Thauer et al., 2008). Thus, not only is MCR universally conserved, but also it is essential for the growth and survival of these organisms. Therefore, gene essentiality is likely to have impeded genetic analyses of MCR even in genetically tractable methanogens. Overall, the dearth of methods for genetic analyses of MCR is a critical hurdle that must be overcome to harness the biotechnological potential of methanogens. To this end, in this chapter we describe how Cas9-based genome editing tools can be used to genetically manipulate the chromosomal *mcr* operon in *M. acetivorans* C2A to introduce a tandem affinity purification (TAP) tag for rapid purification of this important enzyme.

2. CRISPR–Cas9 genome editing in *Methanosarcina* spp.

CRISPR–Cas9-based genome editing has been widely used for genome manipulation in a broad range of organisms, but mostly within the eukaryotic domain (Hsu, Lander, & Zhang, 2014; Li et al., 2015; Peters et al., 2015) (Tables 1 and 2). This technique relies on two components: (a) Cas9 and (b) single guide (sg) RNA (Barrangou & van Pijkeren, 2016). Cas9, originally derived from *Streptococcus pyogenes*, is an RNA-guided DNA endonuclease (Doudna & Charpentier, 2014). The sgRNA has two components: (a) a 20-bp region of homology to a chromosomal locus of interest (also referred to as the target sequence) and (b) an 80-bp scaffold that adopts a secondary structure to enable the sgRNA to bind to Cas9 (Jinek et al., 2012). Upon binding the DNA strand complementary to the target sequence, with a NGG protospacer adjacent motif (PAM) at the 3′ end, the sgRNA triggers Cas9 to generate a double-stranded break (DSB) at the chromosomal locus of interest (Fig. 2). This lethal DSB can be repaired by the native homology-dependent repair mechanism when a homology repair (HR) template to

Table 1 List of plasmids for genome editing in *Methanosarcina* spp.

Plasmid	Features	Source
pAMG40	Vector for fosmid retrofitting that contains pC2A and λattP	Guss et al. (2008)
pJK027A	Vector with P*mcrB*(*tetO1*) promoter fusion to *uidA* that contains ΦC31attB and λattB	Guss et al. (2008)
pDN201	pJK027A-derived plasmid with p*mcrB*(tetO1) promoter fusion to *S. pyogenes cas9*	Nayak and Metcalf (2017)
pDN206	Cointegrate of pDN201 and pAMG40	Nayak and Metcalf (2017)
pDN303	pDN201-derived plasmid with a synthetic fragment containing P*mtaCB1* promoter fusion to a sgRNA targeting *mcrG*	Nayak et al. (2017)
pDN305	pDN303-derived plasmid containing a repair template to introduce a tandem affinity purification tag (containing a 3 × FLAG tag and a Twin-Strep tag) at the N-terminus of *mcrG*	Nayak et al. (2017)
pDN307	Cointegrate of pDN303 and pAMG40	Nayak et al. (2017)
pDN309	Cointegrate of pDN305 and pAMG40	Nayak et al. (2017)

Table 2 List of *Methanosarcina acetivorans* strains discussed in this chapter

Strain	Genotype	Construction details	Source
WWM60	Δ*hpt*::*PmcrB-tetR*	—	Guss et al. (2008)
WWM1054	Δ*hpt*::*PmcrB-tetR*, N-terminal TAP tag (3 × FLAG and Twin-Strep tag) upstream of *mcrG*	WWM60 was transformed to Pur^R with pDN247; plasmid-cured strain was isolated by colony purifying Pur^R transformants on solid medium with 8ADP	Nayak et al. (2017)

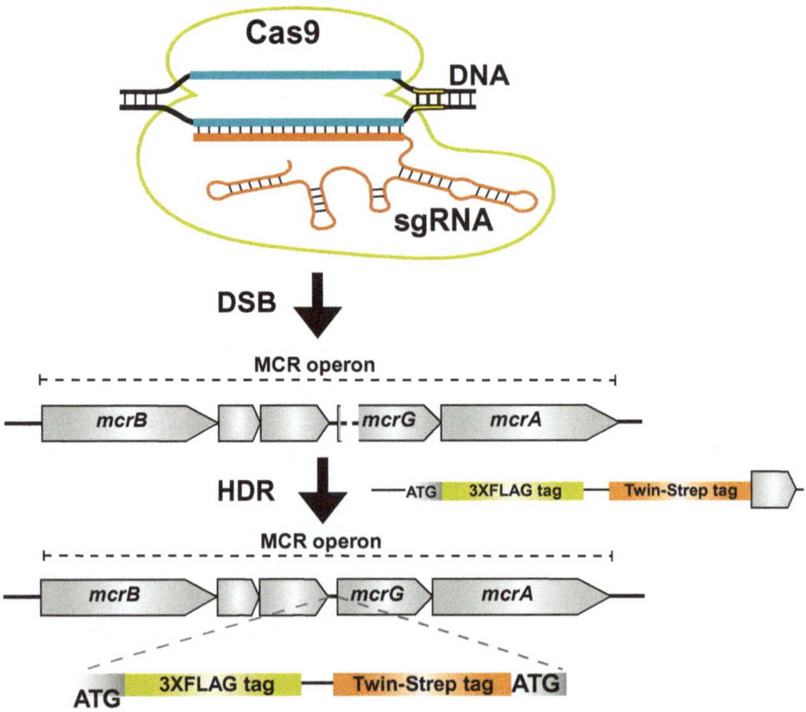

Fig. 2 A schematic overview of Cas9-based genome editing to introduce a tandem affinity purification (TAP) tag at the N-terminus of the *mcrG* locus (locus tag: MA4547) in *M. acetivorans* C2A. Upon binding the target sequence (in *blue*) flanked by an NGG protospacer adjacent motif (PAM; in *yellow*), Cas9 generates a double-strand break at the *mcrG* locus. Homology-dependent repair (HDR) with a repair template containing the TAP tag flanked by regions of the homology surrounding the DSB leads to its introduction at the N-terminus of *mcrG*.

introduce the desired mutation is provided (Fig. 2). Importantly, the introduced mutation must alter or remove the target sequence to prevent additional rounds of cleavage by Cas9.

In a recent study (Nayak & Metcalf, 2017), we described the development of a Cas9-based platform for genetic manipulation of the model archaeon *M. acetivorans* C2A. Since then, we have implemented this platform in another closely related methanogen, *Methanosarcina barkeri* Fusaro as well (unpublished data). The Cas9-based genetic toolbox can be used in conjunction with preexisting genetic tools and provides several advantages over prior techniques:

2.1 Advantages of CRISPR–Cas9 genome editing

- Generation of a single mutant in *M. acetivorans* C2A using this technique takes only ca. 3–4 weeks compared to ca. 8–12 weeks using the previously established genetic technique that requires chromosomal integration/double-crossover of a mutagenic plasmid (Nayak & Metcalf, 2017). Note: The mutant generation time can vary depending on the strain, growth substrate, and the phenotype. The estimate provided above is for the deletion of the *ssuC* locus in *M. acetivorans* C2A using trimethylamine (TMA) as the growth substrate.
- Mutant generation is more efficient. Using the Cas9-based genome editing technique, nearly 100% of transformants contain the desired mutation on the chromosome (Nayak & Metcalf, 2017). In contrast, only ca. 50% of the transformants generated after the double-crossover step using the previous technique typically contain the desired mutation on the chromosome.
- Multiple mutations can be introduced simultaneously. Successful introduction of up to three different mutations without compromising the efficiency and speed of mutant generation has been achieved (Nayak & Metcalf, 2017).
- Gene essentiality can be reliably established. By running appropriate transformation controls for the Cas9-based plasmids, a researcher can establish whether a gene of interest is essential.
- Manipulation of essential genes is feasible. Manipulating essential genes to introduce single nucleotide polymorphisms or affinity tags for purification by the chromosomal integration/double-crossover technique is cumbersome, especially for genes arranged in an operon, due to polar effects or dosage issues induced upon homologous recombination and integration of the mutagenic plasmid. The Cas9-based technique eliminates these hurdles for manipulation of essential genes.

2.2 Features of the Cas9-containing vector (pDN201)

All mutagenic plasmids for Cas9-based genome editing of *Methanosarcina* spp. are derived from a Cas9-containing vector called pDN201 (Nayak & Metcalf, 2017) (Fig. 3). Relevant features of this vector for growth and manipulation in *E. coli* are as follows:

- The vector contains the F origin of replication and the *cat* (chloramphenicol acetyltransferase) cassette that confers chloramphenicol resistance in *E. coli*.
- The vector contains the λ attB attachment site for Gateway cloning.

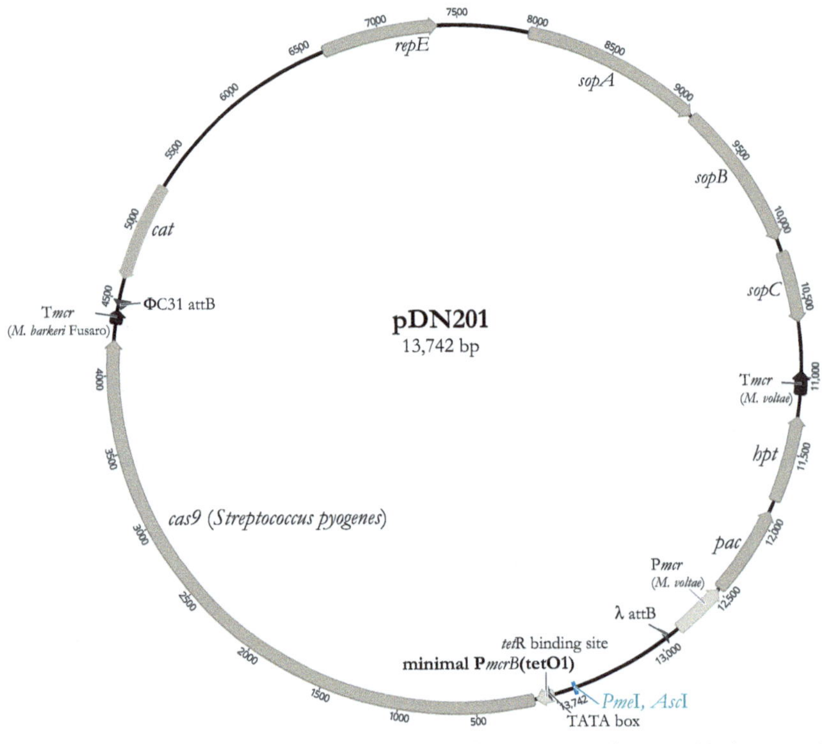

Fig. 3 Plasmid map of the base vector (pDN201) for genome editing in *Methanosarcina* spp. The plasmid contains the *cas9* ORF from *Streptococcus pyogenes* under the control of a tetracycline-inducible promoter P*mcrB*(tetO1). The sgRNA and the homology repair template can be cloned into the *Asc*I and *Pme*I sites, respectively. *cat*, chloramphenicol acetyltransferase; *repE*, replication initiation protein from the *E. coli* F plasmid; *sopA*, *sopB*, *sopC*, plasmid partitioning proteins from the *E. coli* F plasmid; T*mcr*, the terminator of the *mcr* operon; *hpt*, hypoxanthine phosphoribosyltransferase; *pac*, puromycin acetyltransferase; *tetR*, tetracycline responsive repressor from Tn10.

The appropriate *E. coli* host is a derivate of DH10B, WM4489, engineered to provide copy-number control through regulation of the *trf33* gene by a rhamnose-inducible promoter (Kim et al., 2012). For plasmid purification, rhamnose should be added to a final concentration of 10–20 m*M* in the outgrowth medium.

Relevant features of this vector for growth and maintenance in *Methanosarcina* spp. (Fig. 3) are as follows:

- The vector contains an operon encoding the *hpt-pac* cassette. The *hpt* (hypoxanthine phosphoribosyltransferase) locus confers sensitivity to 8-aza-2,6-diaminopurine (8-ADP) and can be used to cure the mutagenic plasmid from cells, and the *pac* (puromycin transacetylase) locus confers puromycin-resistance in *Methanosarcina* spp.
- The vector contains the ΦC31 attB attachment site for chromosomal integration in an appropriate host.
- The vector contains the native *cas9* ORF (without any codon optimization) from *S. pyogenes* under the control of a tetracycline-inducible strong promoter P*mcrB*(tetO1) (Guss, Rother, Zhang, Kulkarni, & Metcalf, 2008). Note: As reported in our previous study, addition of tetracycline did not change the efficiency of genome editing (Nayak & Metcalf, 2017), thus the *Methanosarcina* host strain does not need to encode the *tetR* repressor gene.

The appropriate *Methanosarcina* host strain for Cas9-based genome editing should have the following features:

- The native *hpt* locus should be inactivated by a chromosomal deletion. Note: The *hpt* locus can serve as a neutral locus for addition of genes.
- For chromosomal integration of the mutagenic plasmid, the host strain should encode the ΦC31 *int* gene and also contain the ΦC31attB attachment site for site-specific recombination on the chromosome. Note: we do not recommend chromosomal integration of the mutagenic plasmid to avoid inactivation of Cas9 in the host strain.
- For autonomous replication of the mutagenic plasmid, the host strain should *not* encode the ΦC31 *int* gene or contain any ΦC31 attachment site. Note: The base vector pDN201 does not contain genetic elements for autonomous replication in *Methanosarcina* spp. For autonomous replication, the pDN201-derived vector needs to be retrofitted with pAMG40 (containing the pC2A backbone) (Guss et al., 2008) as described in Section 3.5. The vector pDN206, a cointegrate of pDN201 and pAMG40, is routinely used as a positive control in genome editing experiments.

3. Construction of mutagenic plasmids to introduce a TAP tag at the *mcr* locus in *M. acetivorans* C2A

3.1 *In silico* design of sgRNA construct

The sgRNA construct (Fig. 4) has the following components:

- Promoter. We use the promoter of the *mtaCB1* operon, encoding a methanol-specific methyltransferase from *M. acetivorans* C2A to drive the expression of the sgRNA. The promoter elements of this operon were mapped previously in the Metcalf group (Bose & Metcalf, 2008). The transcription start site was mapped to position 537, 295 (G) on the strand of the chromosome and the putative TATA box (TATAT) was mapped to the region between 537, 320–537, 325 on the strand of the chromosome. The *mtaCB1* operon is induced ca. 100-fold in the presence of methanol (Bose & Metcalf, 2008) in the growth medium, and to preserve binding sites for regulatory elements we use the region from 537, 660 to 537, 284 on the strand of the *M. acetivorans* C2A chromosome as the promoter for the sgRNA. Note: As reported in our previous study, addition of methanol does not change the efficiency of genome editing (Nayak & Metcalf, 2017); thus, genome editing can be performed on any growth substrate.
- Target sequence. In general, the following rubric is used for identifying/designing the 20 bp target sequence (or protospacer) based on guidelines provided in previous studies (Cobb, Wang, & Zhao, 2015):
  - 3′ Protospacer adjacent motif should be NGG and should *not* be included in the target sequence. Note: In our experience, including the PAM in the target sequence is the most common error in the design of sgRNAs
  - Last 12 nt (3′ end) of the target sequence (commonly referred to as the seed sequence) + PAM [test all combinations of NGG and NAG, as Cas9 can also use NAG as the PAM (Ran et al., 2013)] should be unique in the genome to prevent off-target matches
  - Preferably, the target sequence should be on the noncoding strand if within the coding sequence of a gene
  - The target sequence(s) should be <500 bp from the HR template

  Note: There are many tools currently available to automate target sequence design for Cas9-based genome editing of a gene/region of interest. We use the "Finding CRISPR sites" tool within the Geneious bioinformatics platform (Kearse et al., 2012) to design the target sequence and identify off-target-binding sites within the genome of the appropriate *Methanosarcina* spp.

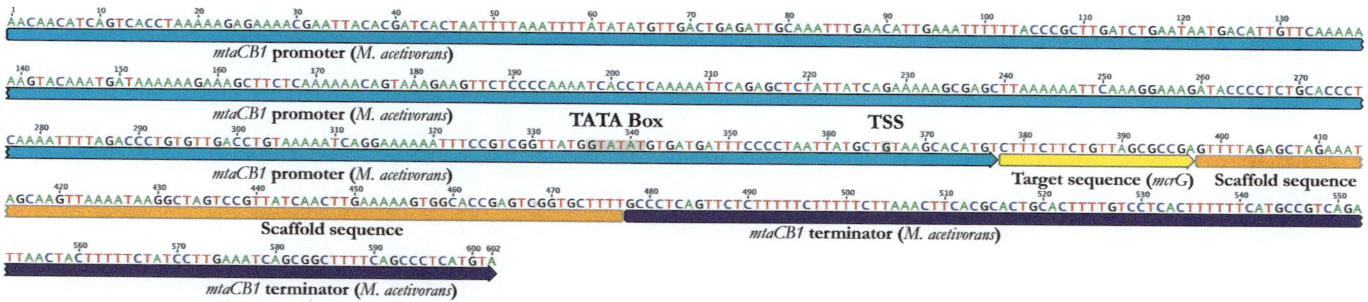

Fig. 4 Design of the synthetic construct for the expression of the single guide (sg) RNA to introduce a tandem affinity purification (TAP) tag at the N-terminus of the *mcrG* locus (locus tag: MA4547) in *M. acetivorans* C2A. The *mtaCB1* promoter (*blue*) and terminator (*purple*) sequences from *M. acetivorans* C2A drive and terminate the expression of the sgRNA, respectively. The sgRNA contains a 20 bp target sequence within the *mcrG* ORF (in *yellow*) and an 80 bp scaffold sequence (in *orange*) that binds to Cas9. *mtaCB1*, genes encoding methanol-specific methyltransferase 1; *TSS*, transcription start site.

- Target sequence to introduce a TAP tag at the *mcr* locus in *M. acetivorans* C2A. We chose to insert the TAP tag at the N-terminus of the *mcrG* locus within the *mcr* operon and used the "Finding CRISPR sites" tool within the Geneious platform to identify candidate target sequence(s) in the 250 bp region after the start codon of *mcrG* (5, 598, 895-5, 599, 144 on the strand of the *M. acetivorans* C2A chromosome). The target sequence was selected as described earlier. Note: An additional constraint for the design of this specific target sequence was to ensure that the seed sequence contains an amino acid with degenerate codons for the design of the HR template.
- Scaffold sequence. The target sequence is fused with 80 bp scaffold region (Fig. 4) derived from Ran et al. (2013) as follows: GTTTT AGAGC TAGAA ATAGC AAGTT AAAAT AAGGC TAGTC CGTTA TCAAC TTGAA AAAGT GGCAC CGAGT CGGTG CTTTT.
- Terminator. We use the putative terminator of the *mtaCB1* operon, encoding a methanol-specific methyltransferase, from *M. acetivorans* C2A to terminate the expression of the sgRNA (Fig. 4). Analysis of reads from RNA-sequencing of methanol-growth *M. acetivorans* C2A indicates that *mta*CB1 transcript terminates at position 534, 768 on the stand of the chromosome (unpublished data). We use the region from 534, 719 to 537, 843 on the strand of the *M. acetivorans* C2A chromosome as the terminator for the sgRNA.

  For each genome editing experiment, the entire sgRNA construct, from the promoter to the terminator, is synthesized as a double-stranded DNA fragment. Appropriate 30-bp overlaps at the 5′ and 3′ end of the synthetic construct are added to aid cloning using the Gibson assembly technique (see Section 3.2). We use the gblocks gene fragments service from Integrated DNA Technologies (IDT, Coralville, IA, USA) to order the synthetic constructs.

3.2 Construction of a pDN201-derived vector with sgRNA

A standardized protocol for construction of a pDN201-derived vector with the sgRNA of interest is outlined below. The sgRNA-containing plasmid is a useful intermediate, as it can serve as a negative control in genome editing experiments (see Section 3.6 for further details)

1. The freezer stock/agar stab of WM7959 (WM4489/pDN201) is streaked out on 1.5% LB agar plates supplemented with chloramphenicol (10 μg/mL) and incubated overnight at 37°C for single colonies.

2. A single colony of WM7989 is inoculated in a test tube containing 5 mL of LB (or SOB) supplemented with chloramphenicol (10 μg/mL) and 10–20 m*M* Rhamnose. Incubation is carried out on a roller/shaker at 37°C overnight.
3. pDN201 plasmid DNA is extracted using any plasmid purification kit. The DNA concentration and quality are checked using a Nanodrop (or another device/technique as available).
4. 1 μg of pDN201 plasmid DNA is linearized with *Asc*I (New England Biolabs, Ipswich, MA, USA) per manufacturer's instructions and gel-purified using any gel extraction kit.
5. The synthetic DNA fragment containing the sgRNA is reconstituted in Tris–EDTA buffer at pH 8.0, per manufacturer's instructions.
6. A Gibson assembly is performed with the linearized pDN201 backbone and synthetic DNA fragment containing the sgRNA using a HiFi DNA assembly master mix (New England Biolabs, Ipswich, MA, USA) per manufacturer's instructions. Note: We recommend performing an *in silico* Gibson assembly reaction to determine the sequence of the overlaps for Gibson assembly. The overlaps for Gibson assembly can be added at the 5′/3′ end of the construct to be synthesized as a double-stranded DNA fragment.
7. ca. 60 μL electrocompetent WM4489 cells is transformed with ca. 1 μL HiFi assembly. A protocol for making electrocompetent cells is available through many resources online for, e.g., the Protocols section of the New England Biolabs website.
8. Appropriate dilutions of the transformation reaction are plated on prewarmed 1.5% LB agar plates supplemented with chloramphenicol (10 μg/mL), and incubated overnight at 37°C.
9. Finally, we screen resulting colonies for insert using primers outside the *Asc*I cloning site, and sequence plasmids from colonies that test positive for insert. A freezer stock of the appropriate *E. coli* strains is generated for future use.

3.3 Design of HR Template

General guidelines for the design of a HR template are as follows:

- Genome editing efficiency depends on the size of the homology region provided. The editing efficiency doubled when the size of the upstream and downstream flanks for gene deletion increased from ca. 500 to 1000 bp (Nayak & Metcalf, 2017). Note: we have *not* tested the genome editing efficiency of fragments <500 bp.

- Genome editing efficiency depends on the distance of the HR template from the target sequence. We observed that templates more than 500 bp from the target sequence drastically reduce the efficiency of genome editing (Nayak & Metcalf, 2017). Note: We recommend designing multiple sgRNAs for the generation of deletions >1000 bp in size.

The TAP tag sequence was designed to contain a 3× FLAG tag for quantifying MCR protein levels using western blot with commercially available anti-FLAG antibodies, followed by the Enterokinase cleavage site (DDDDK), and a 2× Strep tag for affinity-purification of MCR (Nayak, Mahanta, Mitchell, & Metcalf, 2017) using the Strep-tactin Superflow Plus resin (QIAGEN, Germantown, MD, USA). In the HR template, the TAP tag sequence is translationally fused to the N-terminus of *mcrG* and the codon for Serine at positions 5, 599, 106 of the *mcrG* coding sequence was changed from TCC to TCA to prevent targeting after the insertion of the TAP tag. A ca. 100 bp region upstream and downstream of the TAP tag is also included in the HR template. The chromosomal regions were amplified by PCR using the genomic DNA from *M. acetivorans* C2A as the template and the TAP tag was synthesized as a double-stranded DNA fragment.

3.4 Cloning the HR template to generate the genome editing plasmid

The protocol for introducing the HR template in the pDN201 derivative(s) containing the desired sgRNA(s) is similar to the protocol described earlier in Section 3.2. Notable differences are as follows:

1. 1 μg of plasmid DNA is linearized with *Pme*I (New England Biolabs, Ipswich, MA, USA) per manufacturer's instructions and the linearized plasmid DNA is purified using any gel extraction kit.
2. A Gibson assembly is performed with the linearized vector backbone and HR template(s) using a HiFi DNA assembly master mix (New England Biolabs, Ipswich, MA, USA) per manufacturer's instructions. Note: We recommend performing an *in silico* Gibson assembly reaction to determine the sequence of the overlaps for Gibson assembly. Overlaps for Gibson assembly are added at the 5′ end of primers if the HR template is to be amplified by PCR, or at the 5′/3′ end of the fragment for synthesis.
3. Resulting colonies are screened for inserts using primers outside the *Pme*I cloning site. Plasmids from colonies that test positive for insert(s) are sequenced. A freezer stock of the appropriate *E. coli* strains is generated for future use.

3.5 Retrofitting plasmids with pAMG40

We recommend retrofitting pDN201-derived plasmids that will be used for transformations with pAMG40. The resulting cointegrate would be able to autonomously replicate in *Methanosarcina* spp. using the pC2A origin of replication and can be readily cured after genome editing. A detailed description of the vector pAMG40 can be found in Guss et al. (2008), but relevant features for this chapter are as follows:

- It contains the entire pC2A plasmid from *M. acetivorans* C2A. A specific region within pC2A that contains the origin of replication remains unknown.
- It contains the λ attP attachment site for Gateway cloning.
- It encodes the *aph* (aminoglycoside phosphotransferase) cassette that confers kanamycin resistance in *E. coli*.

A standardized protocol for retrofitting the pDN201-derived plasmids with pAMG40 using the Gateway cloning technique is as follows:

- Follow steps 1–3 in the protocol outlined in Section 3.2 to purify plasmid DNA. Note: *E. coli* strains with pDN201-derived plasmids should be cultivated in medium supplemented with chloramphenicol (10 μg/mL), whereas WM3357 (WM1788/pAMG40) should be cultivated in medium supplemented with kanamycin (25 μg/mL).
- Perform Gateway cloning with an equimolar ratio of the pDN201-derived plasmid and pAMG40 using the Gateway BP Clonase II Enzyme Mix (Thermo Fisher Scientific, Waltham, MA, USA) per manufacturer's instructions. Note: We typically incubate the BP clonase reaction at room temperature for ca. 1 h and subsequently inactivate BP clonase with 1 μL Proteinase K (Thermo Fisher Scientific, Waltham, MA, USA) at 37°C for 30 min.
- Transform ca. 60 μL electrocompetent WM4489 cells with ca. 1 μL of the BP clonase reaction.
- Plate appropriate dilutions of the transformation reaction on prewarmed 1.5% LB agar plates supplemented with chloramphenicol (10 μg/mL) and kanamycin (25 μg/mL) and incubate overnight at 37°C.
- Verify cointegrates by analyzing the restriction digest pattern with appropriate restriction enzymes. Note: Cointegrates do not need to be verified by Sanger sequencing as the BP clonase reaction does not contain an error-prone DNA amplification step.
- Generate a freezer stock of the appropriate *E. coli* strains for future use.

3.6 Transformation of *Methanosarcina* spp. with genome editing plasmids

We recommend using WWM60 (Δ *hpt*::p*mcrB-tetR*) (Guss et al., 2008) for genome editing experiments with *M. acetivorans* C2A and setting up appropriate controls as follows:

- Positive control: We recommend using pDN206 (the cointegrate of pDN201 and pAMG40) as a positive control for the transformation reaction.
- Negative control: We recommend using a cointegrate of pAMG40 and the pDN201-derived plasmid with the appropriate sgRNA as a negative control for genome editing as well as a no-DNA negative control for the transformation reaction.

A liposome-mediated transformation reaction as described in detail in Metcalf, Zhang, Apolinario, Sowers, and Wolfe (1997) is used to introduce mutagenic plasmids in *Methanosarcina* spp. For *M. acetivorans* C2A, we use a 10 mL culture volume [mid-late exponential phase culture in high-salt (HS) medium with 50 m*M* TMA as the growth substrate] per transformation reaction. For, *M. barkeri* Fusaro, we use a 30 mL culture volume (mid-late exponential phase culture in HS medium with 125 m*M* methanol as the growth substrate) per transformation reaction. See Wolfe (2011) for details of the HS medium and techniques for anaerobic growth of *Methanosarcina* spp. A standardized protocol for mutant generation using the Cas9-based genome editing technique is as follows:

- Perform liposome-mediated transformation and plate cells on 1.5% HS agar plates supplemented with the appropriate growth substrate and 2 μg/mL puromycin. Incubate plates at 37°C as previously described in Metcalf, Zhang, and Wolfe (1998) for ca. 2 weeks.
- Screen puromycin resistant (Pur^R) transformants for the desired mutation by designing appropriate primers (outside the HR template) and amplifying the chromosome locus by PCR. To screen for the TAP tag at the N-terminus of *mcrG* we designed the following forward (TAC CCA TTC AAT GAC TTC TGC) and reverse (GCAC AGA TTG AAA TGC ACA AG) primers. A 1207-bp band is observed if the TAP tag is present, whereas a 1033-bp band is observed for the WT locus. Note: We typically screen 4–5 Pur^R transformants per mutagenic plasmid and commonly observe that all Pur^R transformants test positive for the desired mutation.
- Streak out 3 Pur^R transformants that test positive for the desired mutation on HS agar plates supplemented with the appropriate growth

substrate and 20 μg/mL 8-ADP. Incubate plates at 37°C as previously described in Metcalf et al. (1998) for ca. 1 week. Screen single colonies for the desired mutation by PCR amplification of the chromosomal locus.

- Pick 2–3 independently derived 8-ADP^R isolates that test positive for the desired mutation and inoculate in 10 mL liquid HS medium supplemented with the appropriate growth substrate. Incubate cultures at 37°C.
- When cultures reach mid-late exponential phase, transfer a 100-fold dilution to 10 mL liquid HS medium supplemented with the appropriate growth substrate and 2 μg/mL puromycin to test if the mutagenic plasmid is cured. If cultures are Pur^S (i.e., do not grow in medium supplemented with puromycin), the plasmid has been cured. Note: Measure the optical density at 600 nm (OD_{600}) in liquid HS medium with 2 μg/mL puromycin for up to 1 week. If no change in OD_{600} is observed after a week, score the cultures Pur^S.
- Confirm that the mutagenic plasmid has been cured by PCR amplification of the *repA* gene from pC2A in 8ADP^R, Pur^S isolates using the forward primer (CTG CAA TAC ACT TTC TTG TCC) and reverse primer (TCA TCC ACT TTG AAG GGA GAA G). If pC2A$^+$, a 1045-bp band will be observed, whereas no band will be observed if pC2A is absent. It is imperative to set up a positive control (such as pAMG40) for this PCR reaction.
- Promptly generate −80°C freezer stocks of one or two independently derived 8ADP^R, Pur^S isolates with the desired mutation.

4. Affinity purification of TAP-tagged MCR from *M. acetivorans* C2A

Specific details regarding the cultivation and growth of strictly anaerobic microorganisms are provided in Wolfe (2011). The following protocol describes a technique for aerobic purification of TAP-tagged MCR from *M. acetivorans* C2A. Using this affinity purification technique ca. 1 mg of MCR protein can be reliably purified from 250 mL of culture in 1–2 h (Fig. 5). Aerobically purified MCR is inactive but can be used for several applications such as thermal stability measurements (Nayak et al., 2017) and X-ray crystallography.

A standardized protocol for affinity purification of MCR is described below.

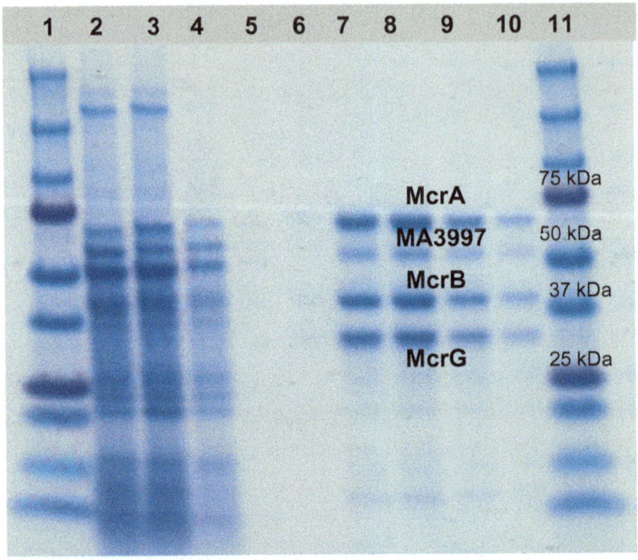

Fig. 5 SDS-PAGE analysis of affinity-purified methyl coenzyme-M reductase (MCR) from *M. acetivorans* C2A. Lanes 1, 11: Protein standard; Lane 2: Crude lysate; Lane 3: Cleared lysate; Lane 4: Flow through; Lane 5: Wash 1; Lane 6: Wash 2; Lane 7: Elution fraction 1; Lane 8: Elution fraction 2; Lane 9: Elution fraction 3; Lane 10: Elution fraction 4. The three subunits of MCR copurified with a fourth band, identified by LC-MS analysis as the hypothetical protein encoded by MA3997.

4.1 Prepare buffers for aerobic purification of MCR

Buffer NP (1 L)—Will be used as the lysis buffer and wash buffer

- Weigh 6.9 g $NaH_2PO_4 \cdot H_2O$ (final concentration = 50 m*M* NaH_2PO_4)
- Weight 17.53 g NaCl (final concentration = 300 m*M* NaCl)
- Dissolve in ca. 500 mL ultrapure H_2O
- Adjust pH to 8.0 using a 5 *M* NaOH stock solution
- Raise volume to 1 L with ultrapure H_2O
- Filter-sterilize and store at 4°C

Buffer NPD (100 mL)—Will be used as the elution buffer

- Weigh 53.5 mg of desthiobiotin (final concentration = 2.5 m*M* desthiobiotin)
- Dissolve in 100 mL Buffer NP
- Filter-sterilize and store at 4°C

4.2 Cell culture and lysis

- Thaw a cryopreserved stock of WWM1054 (Genotype: TAP tag fused to the N-terminus of *mcrG* in WWM60) (Nayak et al., 2017) at room temperature and inoculate 2 mL in a Balch tube (Chemglass Life Sciences, Vineland, NJ, USA) containing 10 mL HS medium supplemented with 50 m*M* TMA as the sole growth substrate.

- Incubate Balch tube (Chemglass Life Sciences, Vineland, NJ, USA) at 37 °C. Acclimate WWM1054 to HS medium supplemented with 50 m*M* TMA for 10–15 generations prior to cultivation for protein purification.
- Transfer a 100-fold dilution of a late-exponential phase culture of WWM1054 to an anaerobic culture bottle (such as an "Och" bottle) (Chemglass Life Sciences, Vineland, NJ, USA) containing 250 mL HS medium supplemented with 50 m*M* TMA as the sole growth substrate. Monitor growth by measuring OD_{600} at regular intervals. Note: It is important to "vent" the cultures during the growth phase to depressurize the anaerobic culture bottles.
- Harvest cultures in mid-exponential phase and collect the cell pellet by centrifugation (3000 × *g*) for 15 min at 4°C.
- Discard supernatant and resuspend cell pellet in 5 mL buffer NP for lysis.
- Sonicate for three bursts of 10 s with 30-s intervals. Keep cells on ice during sonication. Add a small amount of DNase and mix gently to reduce viscosity of the lysate. Store 20 μL of crude lysate at −20°C for SDS-PAGE analysis.
- Clear the crude lysate by centrifugation (17,500 × *g*) for 30 min at 4°C. Store 20 μL of cleared lysate at −20°C for SDS-PAGE analysis and use the remainder for protein purification.
- Equilibrate the cleared lysate to room temperature prior to purification.

4.3 Protein purification

- Equilibrate a gravity column containing 2 mL Strep-tactin Superflow Plus slurry (a 50% suspension) (QIAGEN, Germantown, MD, USA) with 4 mL buffer NP at room temperature. Note: Vortex the Strep-tactin Super Flow Plus slurry for 10 s before transferring to the gravity column.
- Load the cleared lysate on the column and collect 20 μL of the flow-through for SDS-PAGE analysis.
- Wash the column twice with 4 mL of buffer NP two times. Collect 20 μL of each wash for SDS-PAGE analysis. Note: After the second wash, the resin should have a bright yellow color due to the oxidized Ni(II) form of F_{430} in MCR.

- Elute the protein in four fractions with 500 μL buffer NPD per fraction. Use 20 μL of each fraction for SDS-PAGE analysis. Note: The highest concentration of the protein is typically observed in the second or third fraction.
- Measure the protein concentration using a Coomassie Plus (Bradford) assay kit (Pierce Biotechnology, Thermo-Scientific, Rockford, IL, USA) with BSA (bovine serum albumin) as the standard per the manufacturer's instructions.

4.4 Buffers for SDS-PAGE analysis

20% SDS (w/v)

- Weigh out 50 g SDS
- Dissolve in ultrapure H_2O and raise volume to 250 mL
- Store at room temperature

10 × SDS running buffer

- Weigh out 30.25 g Tris Base (final concentration will be 25 m*M*)
- Weigh out 144.1 g glycine (final concentration will be 192 m*M*)
- Dissolve in ultrapure H_2O and raise volume to 1 L
- Store at room temperature

1 × SDS running buffer (prepare fresh on the day of and discard after single use)

- 100 mL 10 × SDS running buffer
- 5 mL 20% SDS
- Fill to 1 L with ultrapure water

Sample buffer (prepare fresh on the day of and discard after single use)

- Add 20 μL β-mercaptoethanol (BME) to 380 μL Laemmli sample buffer (Bio-Rad, Hercules, CA). Note: Handle BME in a fumehood
- Vortex to mix

4.5 SDS-PAGE analysis of affinity-purified MCR

- Add 10 μL 2 × sample buffer to an equal volume of each sample and boil for 10 min
- Load 10 μL of boiled sample in wells of a 4%–20% Mini-Protean TGX denaturing SDS-PAGE gel (Bio-Rad, Hercules, CA)
- Run gel at 70 V until the dye front reaches the bottom of the gel
- Wash gel with ultrapure H_2O for 15 min on a gently shaking platform. Repeat the wash step for a total of three washes.

- Stain with Gel code Blue Stain Reagent (Thermo Fisher Scientific, Waltham, MA) and destain with ultrapure H_2O per manufacturer's instructions.

In ongoing experiments, we are adapting this protocol for affinity purification of MCR under anaerobic conditions to retrieve active protein for kinetic and mechanistic analysis.

Acknowledgments

The authors acknowledge the Division of Chemical Sciences, Geosciences, and Biosciences, Office of Basic Energy Sciences of the U.S. Department of Energy through Grant DE-FG02-02ER15296 (to W.W.M.); the Carl R. Woese Institute for Genomic Biology postdoctoral fellowship (to D.D.N.); and the Simons Foundation Life Sciences Research Foundation postdoctoral fellowship (to D.D.N.). Any opinions, findings, and conclusions or recommendations expressed in this material are those of the authors and do not necessarily reflect the views of the funding sources.

References

Barrangou, R., & van Pijkeren, J. P. (2016). Exploiting CRISPR-Cas immune systems for genome editing in bacteria. *Current Opinion in Biotechnology*, *37*, 61–68. https://doi.org/10.1016/j.copbio.2015.10.003. Elsevier Ltd.

Bokranz, M., Baumner, G., Allmansberger, R., Ankel-Fuchs, D., & Klein, A. (1988). Cloning and characterization of the methyl coenzyme M reductase genes from *Methanobacterium thermoautotrophicum*. *Journal of Bacteriology*, *170*(2), 568–577. https://doi.org/10.1128/jb.170.2.568-577.1988.

Bose, A., & Metcalf, W. W. (2008). Distinct regulators control the expression of methanol methyltransferase isozymes in *Methanosarcina acetivorans* C2A. *Molecular Microbiology*, *67*(3), 649–661. https://doi.org/10.1111/j.1365-2958.2007.06075.x.

Bult, C. J., White, O., Olsen, G. J., Zhou, L., Fleischmann, R. D., Sutton, G. G., et al. (1996). Complete genome sequence of the methanogenic archaeon, *Methanococcus jannaschii*. *Science*, *273*, 1058–1073. https://doi.org/10.1126/science.273.5278.1058.

Cobb, R. E., Wang, Y., & Zhao, H. (2015). High-efficiency multiplex genome editing of *Streptomyces* species using an engineered CRISPR/Cas system. *ACS Synthetic Biology*, *4*(6), 723–728. https://doi.org/10.1021/sb500351f.

Doudna, J. A., & Charpentier, E. (2014). The new frontier of genome engineering with CRISPR-Cas9. *Science*, *346*(6213), 1258096. https://doi.org/10.1126/science.1258096.

Ermler, U., Grabarse, W., Shima, S., Goubeaud, M., & Thauer, R. K. (1997). Crystal structure of methyl-coenzyme M reductase: The key enzyme of biological methane formation. *Science*, *278*(5342), 1457–1463.

Franzmann, P. D., Springer, N., Ludwig, W., Conway De Macario, E., & Rohde, M. (1992). A methanogenic archaeon from Ace lake, Antarctica: *Methanococcoides burtonii* sp. nov. *Systematic and Applied Microbiology*, *15*(4), 573–581. https://doi.org/10.1016/S0723-2020(11)80117-7.

Goenrich, M., Duin, E. C., Mahlert, F., & Thauer, R. K. (2005). Temperature dependence of methyl-coenzyme M reductase activity and of the formation of the methyl-coenzyme M reductase red2 state induced by coenzyme B. *Journal of Biological Inorganic Chemistry*, *10*(4), 333–342.

Goubeaud, M., Schreiner, G., & Thauer, R. K. (1997). Purified methyl-coenzyme M reductase is activated when the enzyme-bound coenzyme F430 is reduced to the nickel(I) oxidation state by titanium(III) citrate. *European Journal of Biochemistry/FEBS*, *243*, 110–114.

Grabarse, W., Mahlert, F., Shima, S., Thauer, R. K., & Ermler, U. (2000). Comparison of three methyl-coenzyme M reductases from phylogenetically distant organisms: Unusual amino acid modification, conservation and adaptation. *Journal of Molecular Biology*, *303*(2), 329–344. https://doi.org/10.1006/jmbi.2000.4136.

Gunsalus, R. P., & Wolfe, R. S. (1980). Methyl coenzyme M reductase from *Methanobacterium thermoautotrophicum*. Resolution and properties of the components. *Journal of Biological Chemistry*, *255*(5), 1891–1895.

Guss, A. M., Rother, M., Zhang, J. K., Kulkarni, G., & Metcalf, W. W. (2008). New methods for tightly regulated gene expression and highly efficient chromosomal integration of cloned genes for *Methanosarcina* species. *Archaea*, *2*(3), 193–203. https://doi.org/10.1155/2008/534081.

Hsu, P. D., Lander, E. S., & Zhang, F. (2014). Development and applications of CRISPR-Cas9 for genome engineering. *Cell*, *157*(6), 1262–1278. https://doi.org/10.1016/j.cell.2014.05.010.

Jinek, M., Chylinski, K., Fonfara, I., Hauer, M., Doudna, J. A., & Charpentier, E. (2012). A programmable dual-RNA-guided DNA endonuclease in adaptive bacterial immunity. *Science*, *337*, 816–822. https://doi.org/10.1126/science.1225829.

Kearse, M., Moir, R., Wilson, A., Stones-Havas, S., Cheung, M., Sturrock, S., et al. (2012). Geneious Basic: An integrated and extendable desktop software platform for the organization and analysis of sequence data. *Bioinformatics*, *28*(12), 1647–1649. https://doi.org/10.1093/bioinformatics/bts199.

Kim, S. Y., Ju, K. S., Metcalf, W. W., Evans, B. S., Kuzuyama, T., & Van Der Donk, W. A. (2012). Different biosynthetic pathways to fosfomycin in *Pseudomonas syringae* and *Streptomyces* species. *Antimicrobial Agents and Chemotherapy*, *56*(8), 4175–4183. https://doi.org/10.1128/AAC.06478-11.

Li, Y., Pan, S., Zhang, Y., Ren, M., Feng, M., Peng, N., et al. (2015). Harnessing Type I and Type III CRISPR-Cas systems for genome editing. *Nucleic Acids Research*, *44*(4), 1–12. https://doi.org/10.1093/nar/gkv1044.

Liu, Y., & Whitman, W. B. (2008). Metabolic, phylogenetic, and ecological diversity of the methanogenic archaea. *Annals of the New York Academy of Sciences*, *1125*, 171–189. https://doi.org/10.1196/annals.1419.019.

Metcalf, W. W., Zhang, J. K., Apolinario, E., Sowers, K. R., & Wolfe, R. S. (1997). A genetic system for Archaea of the genus *Methanosarcina*: Liposome-mediated transformation and construction of shuttle vectors. *Proceedings of the National Academy of Sciences of the United States of America*, *94*(6), 2626–2631.

Metcalf, W. W., Zhang, J. K., & Wolfe, R. S. (1998). An anaerobic, intrachamber incubator for growth of Methanosarcina spp. on methanol-containing solid media. *Applied and Environmental Microbiology*, *64*(2), 768–770.

Moore, S. J., Sowa, S. T., Schuchardt, C., Deery, E., Lawrence, A. D., Ramos, J. V., et al. (2017). Elucidation of the biosynthesis of the methane catalyst coenzyme F 430. *Nature*, *543*(7643), 78–82. https://doi.org/10.1038/nature21427.

Nayak, D. D., Mahanta, N., Mitchell, D. A., & Metcalf, W. W. (2017). Post-translational thioamidation of methyl-coenzyme M reductase, a key enzyme in methanogenic and methanotrophic archaea. *eLife*, *6*(I), e29218. https://doi.org/10.7554/eLife.29218.

Nayak, D. D., & Metcalf, W. W. (2017). Cas9-mediated genome editing in the methanogenic archaeon *Methanosarcina acetivorans*. *Proceedings of the National Academy of Sciences of the United States of America*, *114*(11), 2976–2981. https://doi.org/10.1073/pnas.1618596114.

Peters, J. M., Silvis, M. R., Zhao, D., Hawkins, J. S., Gross, C. A., & Qi, L. S. (2015). Bacterial CRISPR: Accomplishments and prospects. *Current Opinion in Microbiology*, *27*, 121–126. https://doi.org/10.1016/j.mib.2015.08.007.

Prakash, D., Wu, Y., Suh, S., & Duin, E. C. (2014). Elucidating the process of activation of methyl-coenzyme M. *Journal of Bacteriology*, *196*(13), 2491–2498. https://doi.org/10.1128/JB.01658-14.

Ran, F. A., Hsu, P. D., Wright, J., Agarwala, V., Scott, D. A., & Zhang, F. (2013). Genome engineering using the CRISPR-Cas9 system. *Nature Protocols*, *8*(11), 2281–2308.

Schaefer, H., Fletcher, S. E. M., Veidt, C., Lassey, K. R., Brailsford, G. W., Bromley, T. M., et al. (2016). A 21st-century shift from fossil-fuel to biogenic methane emissions indicated by $^{13}CH_4$. *Science*, *354*(6281), 80–84.

Spang, A., & Ettema, T. J. G. (2017). Archaeal evolution: The methanogenic roots of Archaea. *Nature Microbiology*, *2*, 17109. https://doi.org/10.1038/nmicrobiol.2017.109.

Thauer, R. K. (1998). Biochemistry of methanogenesis: A tribute to Marjory Stephenson. *Microbiology*, *144*, 2377–2406.

Thauer, R. K., Kaster, A., Seedorf, H., & Buckel, W. (2008). Methanogenic archaea: Ecologically relevant differences in energy conservation. *Nature Reviews. Microbiology*, *6*, 579–592. https://doi.org/10.1038/nrmicro1931.

Weil, C. F., Cram, D. S., Sherf, B. A., & Reeve, J. N. (1988). Structure and comparative analysis of the genes encoding component C of methyl coenzyme M reductase in the extremely thermophilic archaeabacterium *Methanothermus fervidus*. *Journal of Bacteriology*, *170*(10), 4718–4726.

Wolfe, R. S. (2011). Techniques for cultivating methanogens. *Methods in Enzymology*, *494*, 1–22. Elsevier Inc. https://doi.org/10.1016/B978-0-12-385112-3.00001-9.

Wongnate, T., Sliwa, D., Ginovska, B., Smith, D., Wolf, M. W., Lehnert, N., et al. (2016). The radical mechanism of biological methane synthesis by methyl-coenzyme M reductase. *Science*, *352*(6288), 953–958. https://doi.org/10.1126/science.aaf0616.

Wood, T. K. (2017). Metabolic manipulation of methanogens for methane machinations. *Microbial Biotechnology*, *10*(1), 9–10. https://doi.org/10.1111/1751-7915.12425.

Zheng, K., Ngo, P. D., Owens, V. L., Yang, X., & Mansoorabadi, S. O. (2016). The biosynthetic pathway of coenzyme F430 in methanogenic and methanotrophic archaea. *Science*, *354*(6310), 339–342. https://doi.org/10.1126/science.aag2947.

CHAPTER FOURTEEN

Navigating methane metabolism: Enzymes, compartments, and networks

David A. Collins, Marina G. Kalyuzhnaya*
Department of Biology, San Diego State University, San Diego, CA, United States
*Corresponding author: e-mail address: mkalyuzhnaya@sdsu.edu

Contents

Methods in Enzymology, Volume 613
ISSN 0076-6879
https://doi.org/10.1016/bs.mie.2018.10.010

Abstract

Microbial methane utilization—a key node in the global carbon cycle—controls and often eliminates emission of methane into the atmosphere. The diversity and distribution of microbes capable of methane oxidation is astounding. However, from a biochemical point of view, only a very narrow set of unique enzymes underlies their metabolic capabilities. Despite this restriction, the successful integration of the enzymes into nonmethanotrophs, if judged by the ability of the trait to grow on methane, remains to be achieved. Failures and small victories with heterologous expression have highlighted a set of challenges linked to structure, compartmentalization, and regulation of the methanotrophic metabolic network. A better understanding of how these challenges are handled by cells of native methane-consuming bacteria is required. In this chapter we focus on key experimental aspects of working with native methanotrophic bacteria, including routine cultivation strategies, lab-scale bioreactor setups, genetic alteration, imaging, and basic *-omic*-level approaches.

1. Culturing methanotrophs

1.1 Growth medium

The original nitrate mineral medium (NMS) described by Whittenbury, Phillips, and Wilkinson (1970) is always a good starting point for the growth or isolation of many methanotrophic strains. However, several aspects of methanotrophic physiology must be considered for further optimization. In addition to obvious modifications, such as pH, salinity/osmolarity, and temperature, several factors can impact and improve growth parameters. Media commonly used for methanotrophic bacteria growth are summarized in Table 1. Some key minerals are typically supplemented as trace elements (Table 1). They include copper, which is an essential element for optimal growth and activity of membrane-bound methane monooxygenase (Semrau, DiSpirito, Gu, & Yoon, 2018; Sirajuddin & Rosenzweig, 2015) as well as electron transport systems, such as terminal cytochrome oxidases; iron, which is critical for methane and methanol oxidation and respiration (Gayazov & Mshenskii, 1991; Semrau et al., 2018); molybdenum and tungsten, which control the type of formate oxidation system used (Chistoserdova, Laukel, Portais, Vorholt, & Lidstrom, 2004); lanthanides, which modulate the expression and activity of two forms of methanol dehydrogenase (Chistoserdova & Kalyuzhnaya, 2018; Chu & Lidstrom, 2016; Gu, Haque, DiSpirito, & Semrau, 2016). The range of key microelements is also listed in Table 1.

While numerous growth trials are always a good way to optimize cultivation medium, genome mining of methanotrophic bacteria enables

Table 1 Composition of media routinely used for cultivating methanotrophic bacteria

Component[a]	K_s/K_i^1 (mg) [ion]	Whittenbury[2]	SCP fermentation[3] LD[#]//HD[$]	DSMZ[4] #632 NMS	Alkaliphiles[5–6]	Acidophiles[7]	*Verrucomicrobia*[8]
Nitrogen source							
KNO_3	—	1	—	1	1	1	5 m*M*
NH_4Cl	380/0.7 [NH_4^+]	0.5	As needed^	—	—	—	—
Sulfur source							
$MgSO_4 \times 7H_2O$	—	1	0.2//0.8	1	0.2	0.1	—
Na_2SO_4	—	—	—	—	—	—	0.14
K_2SO_4	—	—	—	—	—	—	0.35
Other							
$CaCl_2$	—	0.15	0.02//0.08	0.1	0.015	0.015	0.02
$LaCl_3 \times 7H_2O$[b]	—	—	—	—	—	**0.06**	
NaCl	—	—	—	—	1–30	—	—
Fe(III) EDTA	−/0.1 [Fe^{2+}]	4 mg		4 mg	—	—	—
$MgCl_2 \times 6H_2O$	—	—		—	—	—	0.2 m*M*
1000 × Trace element solution[c]		0.5 mL L^{-1}	—	0.5 mL L^{-1}	1 mL L^{-1}	1 mL L^{-1}	1 mL L^{-1}
Phosphorous source and pH buffering							
KH_2PO_4	700/0.8	0.1	0.2//0.8	0.272	0.109	0.2	—
Na_2HPO_4	[PO_4^{3-}]	0.1	0.12//0.5	0.717	0.114	—	0.14

Continued

Table 1 Composition of media routinely used for cultivating methanotrophic bacteria—cont'd

Component	K_s/K_i^1 (mg) [ion]	Whittenbury[2]	SCP fermentation[3] LD#//HD$	DSMZ[4] #632 NMS	Alkaliphiles[5–6]	Acidophiles[7]	*Verrucomicrobia*[8]
High pH buffering: carbonate buffer							
$NaHCO_3$	—	—		—	3.02	—	—
Na_2CO_3	—	—		—	0.42	—	—
Low pH buffering							
H_3PO_4	—					As needed	
H_2SO_4 (1 *M*)	—						$1\,mL\,L^{-1}$
Final pH	—	6.8		6.8	8.6–9	4.5–5	2.7
Solid media							
Bacto agar	—	12.5		12.5	14.6		—
1000 × trace element solution[c]							
Na_2EDTA	—	3		0.5	5	5	—
$FeSO_4 \times 7H_2O$	–/0.1 [Fe^{2+}]	1.1	8//40	0.2	2	2	1.39
$ZnSO_4 \times 7H_2O$	–/0.042 [Zn^{2+}]	—	0.5//2	0.01	0.3	0.1	0.3
$ZnCl_2$		0.042	—	—	—	—	—
$MnCl_2 \times 4H_2O$	–/1.72 [Mn^{2+}]	0.04	0.03//1.2	0.03	0.03	0.03	1
$CoCl_2 \times 6H_2O$	–/0.046 [Co^{2+}]	0.19	0.1//0.4	0.02	0.2	0.02	0.24

$CuSO_4 \times 5H_2O$	–/0.054 [Cu^{2+}]	—	2//8	—	1.2	0.1	2.5
$CuCl_2 \times 2H_2O$		0.002	—	—	—	—	—
$Na_2O_4W \times 2H_2O$		—	—	—	**0.3**	—	—
$NiCl_2 \times 6H_2O$		0.024	—	0.002	0.05	0.02	0.23
$Na_2MoO_4 \times 2H_2O$	–/0.12 [MoO_4^{2-}]	0.018	0.3//1.2	0.003	0.05	0.03	0.24
H_3BO_3		0.3	—	0.03	0.03	—	—
$CeCl_3$		—	—	—	—	—	**0.25**
H_2SO (mL)			0.05//0.2				
Nitrilotriacetic acid (NTA)		—	—	—	—	—	9.5

[a]Salt are listed as g per litter, unless otherwise stated.
[b]Addition of La is optional; it could be used alone or with $CaCl_2$.
[c]Prepared as a 1000 × stock solution in 1 L volume of ddH_2O and added to media before sterilization. Chelating agents such as EDTA and NTA should be dissolved into solution first before adding additional salts.
In bold: cultures grown with the compound listed exhibit increased growth. [1]Gyazov (1992). [2]Whittenbury et al. (1970); [3]Mshenskii (1979), media optimized for high-density fermentation [#]for low cell density (LD, 5 g CDW/L) or [$]for high density (HD, 25–30 g CDW/L); ammonium is used as a source of N and pH stabilizing agent (pH 6.6). [4]DSMZ GmbH (2007). NMS medium; [5]Kalyuzhnaya et al. (2001); [6] Akberdin, Thompson, Hamilton, et al. (2018); Akberdin, Thompson, & Kalyuzhnaya (2018); [7]Dedysh, Panikov, and Tiedje (1998); [8]Pol et al. (2007). CDW, cell dry weight; SCP, single cell.

additional strategies for further improvements. For example, many species of alkaliphilic methanotrophic bacteria accumulate formate in their growth medium when grown on standard NMS medium. The concentration of formate in the growth medium of *Methylomicrobium* spp. grown with methane can reach 5–10 m*M*. Genome mining indicated that the strain possesses a single tungsten-dependent formate dehydrogenase (Akberdin, Thompson, Hamilton, et al., 2018). Incorporation of tungsten in the culture medium almost completely eliminated formate excretion and improved overall growth and carbon utilization efficiency (Akberdin, Thompson, Hamilton, et al., 2018).

1.2 Solid medium cultivation

When possible, solid media cultivation can provide a reliable approach for routine cultivation, verification of purity, counting colony-forming units, or isolating mutants. Some species of methanotrophic bacteria are highly sensitive to agar contaminants, but use of Bactoagar (BD Difco) typically eliminates that problem. Since methane is a gas at ambient temperature, the cultivation of methane-consuming bacteria requires special techniques and dedicated instrumentation, which help to control the gas supply in incubation chambers. Anaerobic jars, which are designed for growth of anaerobic and microaerobic organisms, are typically used. Anaerobic jar refill systems, such as Anoxomat (Advanced Instruments), can be adapted for plate incubation in desired environments. The typical atmosphere for incubation of plates comprises 10%–20% CH_4 and air.

1.3 Batch cultivation

Liquid cultures are used for media optimization, validation of mutant phenotypes, or initial trials for sourcing nutrients (e.g., biogas, coal gas, nitrogen fixation). Additionally, batch cultures are used as inocula for bioreactor continuous cultures and for experiments involving growth rates, protein expression, genomic DNA extraction, and initial transcriptomic studies. However, due to the gaseous nature of methane, which requires the use of closed vials, batch culture experiments are not optimal for investigating the kinetics of methane consumption or estimating biomass yield and carbon conversion efficiencies. The limitations of the batch cultures are due to the highly dynamic changes of both methane and oxygen concentrations in closed settings. Thus, in order to obtain steady-state data, continuous cultures should be used.

1.4 Continuous bioreactor cultivation

Numerous studies have been carried out to optimize the cultivation of methanotrophic bacteria in continuous culture (summarized by Mühlemeier, Speight, & Strong, 2018). Continuous cultivation requires additional optimizations (Gilman et al., 2017; Mshensky, 1979; Zhivotchenko, Nikonova, & Jørgensen, 1995), and some examples of growth media optimized for single cell protein production with low (up to 5 g CDW/L) and high (up to 30 g CDW/L) cell densities are provided in Table 1. Furthermore, the cultivation setup itself poses several challenges. Most of the systems have been custom-built and, due to the flammable nature of methane, should be replicated with caution. One optimal solution for routine lab cultivation is the use of safe mixtures of methane and oxygen, which typically range between 0%–5% CH_4 and 0%–3.5% oxygen (see Ma, 2015 for additional recommendations). A continuous culture setup diagram is shown in Fig. 1. In general, any bioreactor system can be converted to a gas-fermentation system with just a few modifications. First, the gas inputs should be controlled by calibrated mass-transfer systems in order to provide a reliable estimation of substrate consumption. The concentrations of methane and oxygen in input and output gases should be monitored by gas chromatography or sensor systems (i.e., BlueSens; https://www.bluesens.com/products/all-gas-sensors/).

1.5 Routine protocol for continuous bioreactor cultivation

The protocol described below is adapted for working with a mini-parallel bioreactor system, such as DASbox (Eppendorf). The system could be connected to a custom-built gas-distribution system, which controls the gas mixture input. Only nonflammable mixtures of methane (<5% CH_4) and oxygen (<3.5% O_2) should be used, and output gases must be connected to exhaust vents. The DASbox system offers single use plastic vessels, which are handy for small-scale analyses of minerals (Cu, Fe, La, etc.), effects on cell growth, and/or methane oxidation. Four parallel experiments could be carried in one run, providing sufficient statistical data for analysis. Examples of a continuous culture parameters setup are shown in Fig. 2.

1.6 Protocol 1. Setting up bioreactor batch and continuous culture experiments

1.6.1 Step 1. Preparing media and the bioreactor vessel

Media for batch culture must be sterilized by autoclaving within the bioreactor vessel for large-scale experiments. All probes [oxygen probe (DO),

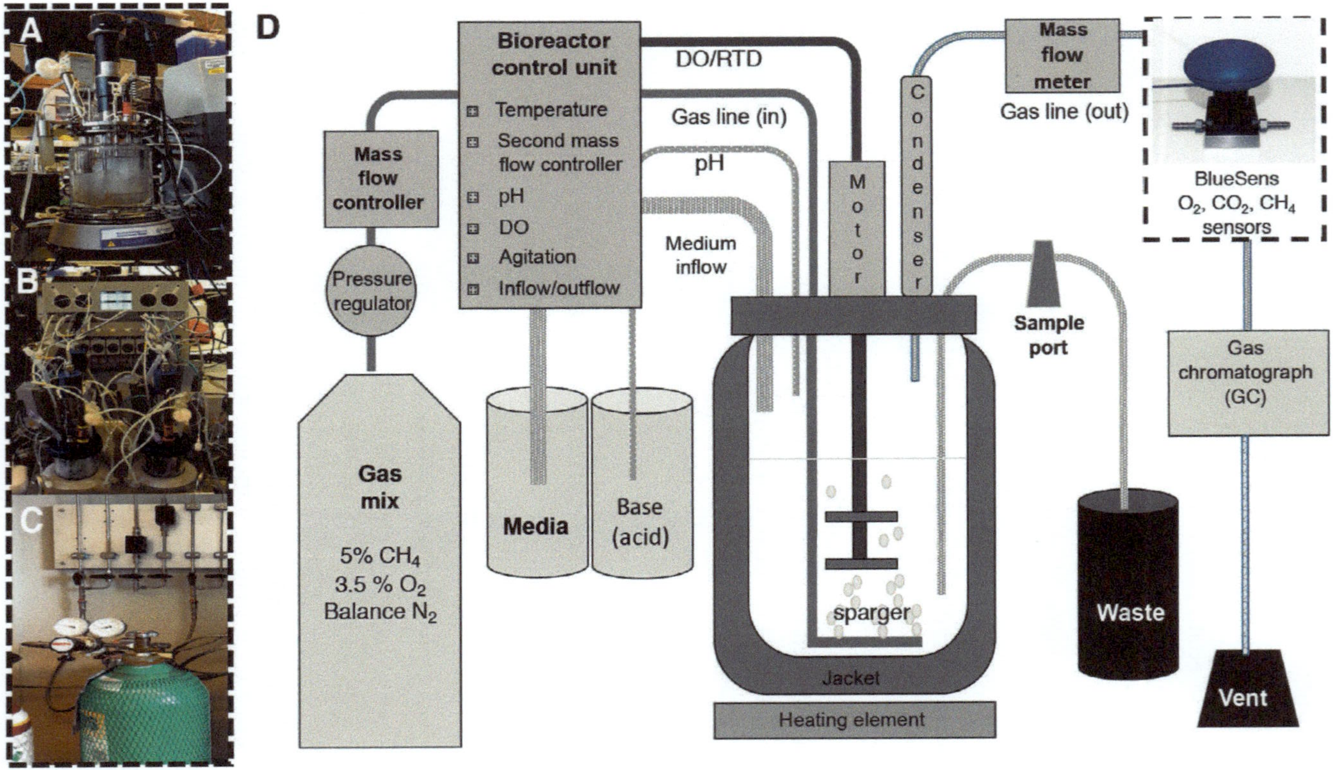

Fig. 1 Typical lab-scale bioreactor setup for cultivating methane-consuming bacteria. (A, B) Benchtop fermenter and bioreactor unit, such as Eppendorf New Brunswick BioFlo/CelliGen 115 3 L (A) or Eppendorf DASbox parallel benchtop bioreactors (B). (C) Dedicated gas mass-transfer system panel with pressure regulators, mass flow controllers, shut-off units, connected to gas mix tanks. (D) Schematic components of a continuous culture bioreactor. (E) Gas composition continuous analysis with BlueSens.

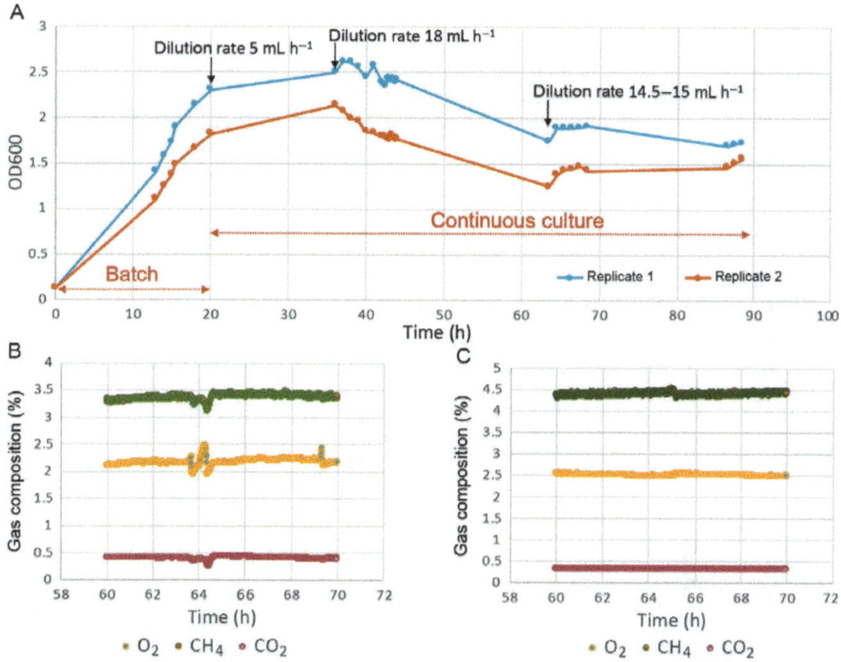

Fig. 2 Continuous cultivation of *M. alcaliphilum* 20Z^R in a DASbox bioreactor system. (A) Optical density of cultures grown in DASbox mini bioreactor (0.5 L working volume; 200 mL culture), as batch (0 – 20 h) and chemostat mode (20–90 h). Steady state was reached at 60–80 h. (B, C) O_2 and CH_4 consumptions and CO_2 production in steady state for bioreactor replicate 1 (B) and 2 (C).

pH, redox] must be sterilized in situ with the media and should be carefully covered with silicon caps as well as aluminum foil to protect the electronic connections. All pH and redox probes and pump flow rates must be calibrated prior to sterilization. The pH probe should be calibrated by attaching the probe to the control panel and following the calibration parameters on the control panel. Briefly, low pH 4.0 and high pH 10.0 buffers are measured by the probe and reading value set at the known pH of each buffer, and the slope between the two points is used to determine intermediate pHs. After calibration the probe is cleaned and installed into the reactor before sterilization. Peristaltic pumps onboard the unit control the supply of additional media, waste outflow, or the addition of acids/bases during an experiment. Each pump must be calibrated using the same type and diameter of tubing that will be used during the experiment, prior to autoclaving them with the media. Pumps are calibrated by pumping at 100% output for each pump for 30 s to 1 min, and measuring the volume pumped. The flow rate in mL/h can then be calculated for each pump. Any ports (sample, addition, harvest, and level controllers) of the vessel should be closed and capped in foil as well. The gas supply line should have a 0.2 μm filter attached and be sealed in foil without blocking the filter to prevent overpressurization during autoclaving. Autoclaving should be performed at 121 °C and 100 kPa (15 psi) above ambient pressure for at least 2 h to sterilize the vessel and media. Any other media (base, acid, buffers, and trace solutions) for continuous culture phases should also be autoclaved with the same settings and duration. Supply lines should be autoclaved with the media as well, so they can be connected to media inflow pumps directly into the fermenter.

Comments: Vessels for the DASbox system are available as disposable, presterilized units which benefit from being virgin materials for trace mineral experiments or comparative conditions.

1.6.2 Step 2. Preparing the run

It is first necessary to connect the motor and set up the agitation rate. Due to the nature of the carbon source, high agitation is essential. The system should run at 500–1000 rpm. Dissolved oxygen (DO) is measured during fermentation to adjust gas flow into the reactor. The DO probe must be calibrated prior to activating the gas flow. The probe is autoclaved with the vessel and calibrated afterward once the gas flow lines are connected to the bioreactor. To calibrate the probe, the reading value of the unattached probe is defined as 0% DO and the probe then is connected to the control panel to begin measuring DO. The agitation is set to maximum speed, and to deliver at least 1 SLPM (standard liter per minute) of room air through the sparger into the media for several hours (or overnight) to saturate the media with oxygen. Once the probe reading has stabilized the value is set to 100% DO to finalize calibration. After the gas is connected, the DO parameter will drop due to O_2 stripping, but it should stabilize at 10%–20% DO for a standard gas mixture. A gas mix for standard experiments consists of 5% methane, 3.5% oxygen, nitrogen balance, and is delivered from the tank at no more than 30 psi through mass flow controllers to the control unit. The control unit additionally is supplied with 30 psi room air and the ratio between the two gases can be controlled on the control panel to adjust the amount of oxygen delivered to the culture. During an experiment, the gas flows from the control unit, to the culture through a 0.2 μm filter and is dispersed into bubbles by the sparger. The gas composition can be analyzed using an automated sampling gas chromatograph (GC) or by real-time monitoring using a dedicated BlueSens sensor for each gas to be measured: O_2, CO_2, CH_4 (https://www.bluesens.com). Initial data on flow rate and gas composition for at least two agitation rates (500, 750, or 1000 rpm) when operating with an uninoculated bioreactor should be collected. The data will be used as blank parameters for gas consumption calculations.

1.6.3 Step 3. Inoculation

Once all probes and conditions on the bioreactor are stable, an inoculum is added aseptically through the addition port of the vessel. For standard experiments a 1:100 inoculum is prepared beforehand (i.e., 150 mL to 1.5 L batch culture) and added to the bioreactor at the start of the experiment.

1.6.4 Step 4. Monitoring cell growth

Parameters such as pH, DO, gas flow, and temperature will be monitored and displayed on the bioreactor control panel. Additionally, samples of cells should be withdrawn to monitor optical density and possible contamination. Check for contamination at least daily during an experiment by removing a sample from the culture, such as during measurement of optical density. Plate an aliquot on a solid media that will permit growth of contaminating microbes but not the reactor organism, such as the base media of the reactor supplemented with 10% rich media such as Difco nutrient broth (NB) media (VWR). Incubate the plate at the temperature used for bioreactor culture for 2–3 days to ensure no contaminant CFUs. Finally, the volume of media remaining in the reactor should be recorded daily along with media inflow/outflow volumes. The gas consumption data must be collected during the steady-state phase of the bioreactor run. Exhaust gases should be analyzed for changes in composition compared with inflow gas composition and normalized to the decrease in flow rate. Similar to inflow gases, an automated sampler connected to a GC system or real-time BlueSens sensor can be used to determine the consumption of O_2 or CH_4 and production of CO_2 during the experiment. At 25°C and 14.696 psi, the densities of methane and oxygen are 0.6569 and 1.3088 g/L, respectively. The typical output of the continuous culture is shown in Fig. 2.

1.6.5 Step 5. Completing a bioreactor run

When an experiment is finished, biomass and spent media are harvested by centrifugation and frozen for downstream analyses. Any remaining culture should be decontaminated and the bioreactor vessel components disassembled and cleaned. Probes should be washed and stored in manufacture's recommended storage solutions. Waste collection vessels should be emptied and decontaminated with bleach before discarding large volumes.

2. Genetic manipulation of methanotrophs

The most recent tools for genetic alterations in methanotrophic bacteria have recently been summarized by Henard and Guarnieri (2018). In general, each genetic modification starts with a set of routine steps including the identification of the chromosomal locus for insertion or deletion and the design of primers for PCR amplification of genomic DNA that are of sufficient length and specific for the desired locus (which can be confirmed by a BLASTn of the primer sequence), the construction of a genetic vector, and the incorporation of the genetic construct via conjugation or electroporation. The procedure used for gene transfer might require further optimization, including selection of the recipient and donor strains. For conjugation-based protocols, the use of *E. coli* S17-1 strain has proven to be successful for many species of methanotrophic bacteria (Ojala, Beck, & Kalyuzhnaya, 2011; Puri et al., 2015; Simon, Priefer, & Pühler, 1984). In some cases the transfer of plasmid could also be facilitated by a triparental conjugation with strains harboring a pRK2013-helper plasmid (Figurski & Helinski, 1979). A counterselection strategy for eliminating *E. coli* cells after conjugation should also be developed. Initial screening of isolates for antibiotic resistance is a good starting point. We found that many methanotrophs are naturally resistant to chloramphenicol and ampicillin (Ojala et al., 2011). However, if resistance is not observed, strains resistant to rifamycin can be selected for by a simple plating of cell biomass (typically 10^9–10^{10} cells) on optimal medium supplemented with the antibiotic. The procedure will typically yield 2–5 colonies which have a mutation in RNA polymerase and thus the resistance will be stable over generations. Many methanotrophic strains show surprisingly high immunity against foreign DNA, which can severally restrict application of genetic approaches to studying cell physiology. Given below is a step-by-step protocol for selecting methanotrophic traits with improved accessibility for genetic alterations. We also describe protocols for constructing a GFP-tagged protein and deleting a gene. Similar approaches could be used for other genetic modifications, such as a promoter swap, changing a ribosome binding site, or the integration of novel genes or gene clusters.

2.1 Protocol 2. Selection of traits with improved accessibility for genetic alterations

A selection strategy is depicted in Fig. 3.

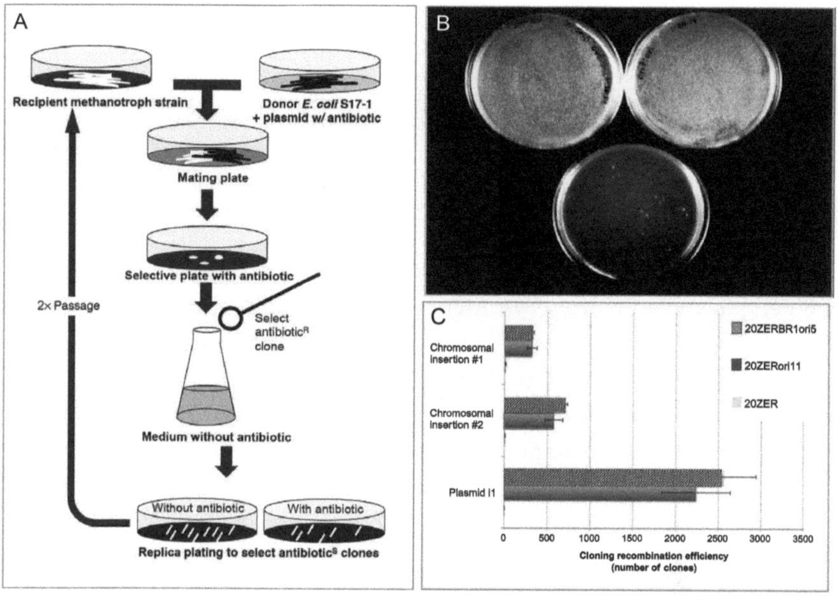

Fig. 3 Generation of strains with improved gene-acquisition capabilities. (A) Schematic overview of the protocol for clone selection. (B) Plates with Kan^R clones generated in *Methylomicrobium alcaliphilum* Strain 20ZERBR1-ori5 (*top left*), Strain 20ZER-ori7 (*top right*), and Strain 20ZER (*bottom*). (C) Cloning efficiency for transfer of pDrive ori 1 plasmid, and generation of recombinant clones using pCM184:EctR or pCM184:mch.

2.1.1 Step 1. Generating initial transconjugants

Typically, this step can be accomplished by conjugation (described below). Conjugation with *E. coli* S17-1 harboring an antibiotic-resistance plasmid, typically kanamycin, such as pDrive-ori, pCM132, pCM66, or pAPW78, can be carried out as described by Ojala et al. (2011). This first round results in a few antibiotic-resistant (e.g., Kan^R) colonies.

2.1.2 Step 2. Plasmid curing

The clones are then transferred two times on selective media (Ojala et al., 2011) to ensure purity. Subsequently, the antibiotic-resistance vector should be cured from the strain by two successive transfers on growth medium without antibiotic. Samples ofliquid culture are subsequently plated on solid medium without antibiotic, and resulting individual clones are tested for antibiotic resistance.

2.1.3 Step 3. Final selection

Clones generated during the first round should be used as a recipient culture in the second round of conjugation. Generally, Kan^S clones obtained after the second round of selection display at least >100-fold higher rate of gene transfer via conjugation and electroporation compared to the original strain (Fig. 3). The growth parameters of the new variants are similar to those of the original strain.

3. Protocol 3. Genetic alterations

It is generally assumed that homologous recombination is the main mechanism for making genetic alterations in methanotrophic bacteria. The genetic tools for the constructions of mutants include suicide vectors, such as pCM184 (Marx & Lidstrom, 2002), pCM433 (Marx, 2008), and ZeoCassette (Yan, Chu, Puri, Fu, & Lidstrom, 2016). While it has never been studied specifically, it has been observed that upstream and downstream flanks for gene/genetic region incorporation or deletion should be in the range of 600–800 bp. Although various strategies can be used for plasmid construction, we found that Gibson assembly (Gibson et al., 2009) is one of the fastest routes to achieve that goal. The resulting constructs can then be incorporated into methanotrophic cells via conjugation or electroporation (Table 2).

3.1 Example 1. Recombination-mediated mutagenesis by suicide vectors in methanotrophs

3.1.1 Step 1. Flanking region preparation

PCR amplify the upstream and downstream homology region (flanks) of the target gene. The optimal range of the flanking regions should be 600–800 bp. The primers should include at least 20 bp homology to the cloning site of the fragment (e.g., vector being used) or a restriction enzyme recognition site.

3.1.2 For insertions

PCR amplify the insertion sequence, such as GFP, with a forward primer that has 20 bp 5′ complementarity to the 3′ sequence of the gene of interest. This primer is the reverse complement to the reverse primer for the upstream region. The reverse primer for the insertion should have 5′ complementarity for the gene of interest's downstream homology region, and is the reverse complement of the downstream region forward primer.

3.1.3 Step 2. Assembly

Gel purify PCR products and assemble all fragments using a gene assembly protocol, based on Gibson assembly, such as NEBuilder® HiFi DNA Assembly Master Mix (NEBiolabs). After the final plasmid is validated by sequencing, proceed with transfer into a methanotroph strain by conjugation or electroporation.

Table 2 Commonly used genetic tools for methanotrophic bacteria

Plasmid/ construct	Applications	Genus	Selectable markers [Reference]
pAWP89, pAWP78	Broad-host range, replicating plasmids with constitutive gene expression	*Methylomicrobium*, *Methylobacter*, *Methylomonas*	Kanamycin [1,2]
pCM433	*oriT*-mobilized allelic exchange	*Methylomicrobium*	Tetracycline [3], or kanamycin [1], or gentamycin [4] Sucrose
ZS	Linear DNA construct for gene knockout with selection (Zeo) and counterselection (*sacB*) cassettes	*Methylomicrobium*	Zeocin and sucrose [2]
pKiKo	*oriT*-mobilized site-specific gene knock-in, gene knockout	*Methylomicrobium*	Kanamycin [O. Demidenko, Unpublished]
pCM184	Broad-host range *cre-lox* gene knockout	*Methylobacterium*, *Methylocella*	Ampicillin, kanamycin, and tetracycline [5–6]
pK18mobsacB	Allelic exchange, gene knockout	*Methylosinus*	Kanamycin and sucrose [7–8]
pCAH01	Replicating pAWP78-derived broad-host range aTc-inducible gene expression	*Methylomicrobium*	Ampicillin and kanamycin [9]

[1] Puri et al. (2015); [2] Yan et al. (2016); [3] Marx (2008); [4] Henard et al. (2018); [5] Marx and Lidstrom (2002); [6] Crombie and Murrell (2011); [7] Semrau et al. (2013); [8] Haque, Gu, DiSpirito, and Semrau (2016); [9] Henard et al. (2016).

3.2 Conjugation

3.2.1 Step 1. Recipient culture preparation

Prepare a lawn of methanotrophic culture on an optimal mineral media and incubate with methane until a sufficient density of biomass has grown (3–5 days).

3.2.2 Step 2. Donor preparation

Prepare chemically competent *E. coli* S17-1 using a high-efficiency method, such as rubidium chloride (Green & Rogers, 2013), and transform them with a plasmid such as pCM433, pCM184, or pAWP89 containing the desired sequences (for insertion or deletion). Resistant clones are then grown as a lawn on rich selective media (LB agar plate with streptomycin and specific selectable marker on the plasmid being used). This lawn should be started as an overnight culture before mating.

3.2.3 Step 3. Mating

Conjugation is performed by mixing the donor *E. coli* and recipient methanotrophic culture on mating media (mineral medium, such as NMS or P, supplemented with 10% LB media or nutrient broth). Scrape a full loopful of *E. coli* S17-1 and smear it in a small area on mating media (approximately 1–2 cm in diameter) and then, with a new loop, scoop up an equal amount of biomass from methanotrophic culture and smear it in the same area as the *E. coli*. The two strains should be evenly mixed together on the mating media to ensure adequate cell-to-cell contact between the two species. Mating plates with the two strains are then incubated at 30°C with methane to support growth of both species for at least 2 days, but no more than 4 days.

3.2.4 Step 4. Selection

Cell biomass should be scraped off the mating plate and transferred to selective media. The entire mating mixture should be transferred to 2–3 plates and spread into a thin layer with a cell spreader. Allow resistant colonies to grow by incubating with methane at 30°C for 5–7 days on mineral medium containing the selective antibiotic. Resulting clones can be purified from residual donor *E. coli* by patch plating colonies onto fresh medium with an appropriate antibiotic for the recipient strain, such as rifamycin. Selected colonies should then be screened by PCR or phenotype depending on the plasmid being used.

3.2.5 Step 5. Counterselection (if applicable)

Resistant colonies postconjugation that are merodiploid for the homology sequence introduced can then be counterselected to generate double crossover mutants. For example, in the case of pCM433 which contains *sacB*, selected resistant clones should be transferred onto optimum growth medium plates containing 2%–2.5% sucrose and incubated with methane, allowing for colonies to grow. Patch these sucrose-resistant colonies onto media with and without a selective antibiotic. Clones which are sucrose-resistant and antibiotic-sensitive should then be screened by PCR for the desired deletion or insertion as well as sequenced to confirm the desired double crossover mutation. For insertions, use primer sets that amplify from the chromosome and the inserted sequence. For deletions, use primers which span the deleted sequence. It is good practice to further validate the resulted constructs by sequencing the amplified fragments.

3.3 Electroporation

Electroporation can also be used to rapidly introduce plasmids and linear PCR constructs into methanotrophs. Several protocols are now described (Yan et al., 2016). The protocol below was further optimized for use with salt-tolerant (halophilic) methanotrophs. The media routinely used to grow halophilic strains contain excessive amounts of salts that will cause arcing in the cuvette during electroporation. Thus, halophilic cultures must be grown on media having the lowest salt concentration possible. Additionally, DNA products to be transferred should be purified and dissolved in distilled water rather than an elution buffer. A video of the protocol could be found at http://sci.sdsu.edu/kalyuzhlab/protocol-9-electroporation-of-20-zr/.

3.3.1 Step 1. Cell preparation

Grow cell cultures (75–100 mL) until the OD is between 0.3 and 0.6 at 600 nm, usually 1 day if subcultured from a liquid culture or 2 days from a plate. Collect cells by centrifugation at 4700 rpm for 15 min at 4°C, remove, and discard the supernatant by decanting without disturbing the pellet. Resuspend the pellets in 10 mL of cold molecular grade water (milliQ water, or for some strains a 10%–15% glycerol solution in water) and collect cells again by centrifugation. Repeat this wash step two more times, then resuspend the pellet in 100–150 μL cold ddH_2O, which yields 2–3 50 μL of transformation reactions, and place on ice.

3.3.2 Step 2. Electroporation

Transfer 50 μL of cell suspension into a microfuge tube. Add DNA (less than 6 μL of DNA containing 300–500 ng DNA) to a transformation tube and mix gently by stirring with pipette tip. Allow DNA to incubate with cells for at least 30 min on ice. Transfer to an ice-cold 0.1 cm gap electroporation cuvette and electroporate with 1.5 kV, 200 Ω, 25 μF (time constant is usually 3.5–5 ms) and immediately add 1 mL of room-temperature optimal growth media.

3.3.3 Step 3. Recovery

Transfer 1 mL electroporation reaction into 10 mL of growth media in 125 mL bottles. Seal vials, add 40–50 cc methane, and incubate the cultures at optimal temperature in a shaking incubator at 200 rpm for overnight (approximately 14–16 h).

3.3.4 Step 4. Selection

Centrifuge cells for 10–15 min at room temperature at 4700 rpm, decant supernatant, and resuspend cells in 0.5–1 mL growth media. Plate an equal volume of each suspension on 3–4 selective plates with the appropriate antibiotic. Allow for resistant colonies to grow on selective plates and proceed with characterization by colony PCR, counterselection, or sequencing, depending on the purpose of the transformation.

The efficiency of plasmid transfer via electroporation is typically lower then via conjugation, with about 10–200 clones observed for suicide vectors and expression plasmids, respectively.

3.4 Example: Generating a GFP translational fusion with a putative S-layer protein in *M. alcaliphilum*

In order to construct a transcriptional fusion between a C-terminus of a putative S-layer protein *Methylomicrobium alcaliphilum*: MALCv4_0971, 1102328-1108873, 6546 bp (2181aa), and sfGFP (Pédelacq, Cabantous, Tran, Terwilliger, & Waldo, 2006), two sets of primers were designed: forward primers Malcv4_0971F′1 and Malcv4_0971F′2 (lowercase) with 5′ complementarity to the suicide vector pCM433 (uppercase) or GFP (uppercase, bold); reverse primers Malcv4_0971R′1 and Malcv4_0971R′2 (lowercase) with 5′ complementarity to GFP coding sequence (uppercase, bold) or to pCM433 (uppercase).

Malcv4_0971F′1: 5′ GTGCCACCTGACGTCTAGATCTcgatgtgtt gagcaacttcactg 3′

Malcv4_0971R′1: 5′ **CAGCTCTTCGCCTTTACGCAT**agcgagg acgaagttatcaac 3′

Malcv4_0971F′2: 5 ′**GTACAAATAATCTAGACTAG**tcccactttt aaagtgag 3′

Malcv4_0971R′2: 5′ GCTGGATCCTCTAGTGAGCTCgccatcctg ttgagctggc 3′

GFP F′: 5′ gttgataacttcgtcctcgct**ATGCGTAAAGGCGAAGAGCTG** 3′

GFP R′: 5′ ctcactttaaaagtggga**CTAGTCTAGATTATTTGTAC** 3′

pCM433 F′1: 5′ gccagctcaacaggatggcGAGCTCACTAGAGGATC CAGC 3′

pCM433 R′1: 5′ cagtgaagttgctcaacacatcgAGATCTAGACGTCAGG TGGCAC 3′

The upstream and downstream MALCv4_0971 flanking regions are amplified from *M. alcaliphilum* 20Z^R DNA using Phusion (NEBiolabs),

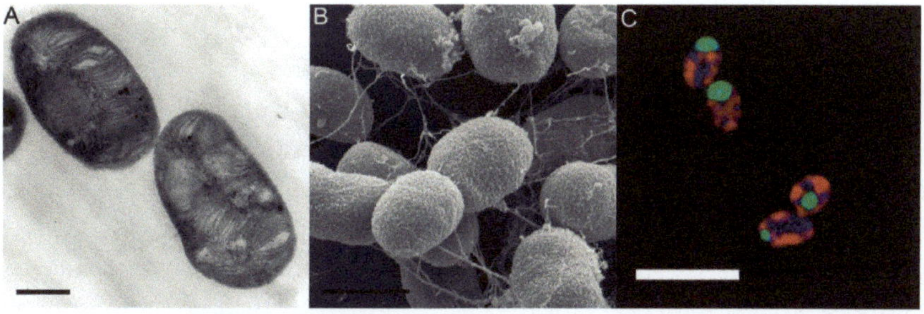

Fig. 4 Microscopy. (A) Transmission electron micrograph of *M. alcaliphilum* 20Z^R showing characteristic intracytoplasmic membranes filling a large portion of cell volume. (B) Scanning electron micrograph of *M. alcaliphilum* 20Z^R which has a rough, pitted cell exterior due to surface layer protein (SLP). (C) Epifluorescent micrograph of *M. alcaliphilum* 20Z^R staining for DNA with DAPI, intracytoplasmic membranes with Fm4-64, and expressing a putative S-layer protein-sfGFP C-terminal translational fusion.

a high-fidelity polymerase. The sfGFP fusion insert and the suicide vector pCM433 backbone were amplified using GFP F′ and GFP R′ and pCM433 F′1 and pCM433 R′1, respectively. The PCR products were separated by gel electrophoresis and purified. The assembly of the fragments was carried out using a NEBuilder® HiFi DNA assembly mix, according to the NEBlab manufacturer's directions. The reaction mixture was transformed into competent *E. coli* S17-1 and kanamycin-resistant clones were selected. The clones were regrown in liquid culture, and then the plasmid was isolated and sequenced to confirm correct assembly. The resultant strain was used for biparental mating with *M. alcaliphilum* 20Z^R. Sucrose-resistant, kanamycin-sensitive clones were confirmed as double crossover mutants by PCR and sequencing. The clones were examined by epifluorescent microscopy to localize the putative S-layer protein (see Section 4.4 for details). The image of *M. alcaliphilum* 20Z^R cells harboring the sfGFP fusion is shown in Fig. 4C.

4. Microscopy methods

The unique cell structure of methanotrophic bacteria is commonly associated with their metabolic capabilities. Over the years numerous techniques have been developed and tested on methanotrophic bacteria to study the connection between cellular protein localization and in vivo function (Fig. 4). Here we summarize the most common strategies to generate fixed or in vivo images of methanotrophic cells.

4.1 Protocol 4. Transmission electron microscopy

Transmission electron microscopy (TEM) is one of the most widely used techniques for the investigation of methanotrophic cell structure. Methanotrophs are known for their ability to produce intracytoplasmic membranes (ICMs), which are envisioned as dedicated compartments for methane oxidation. The structure and amount of ICMs can be linked to the type of methane oxidation system (i.e., membrane-bound methane monooxygenase) and cell physiological state (starvation vs active growth). Additional information obtained from TEM might include the amount of carbon storage compounds (polyhydrobutyrate or glycogen), motility (flagella and pili), accumulation of polyphosphates, and surface structures (S-layers, nanowires, fibrils). Cells from both liquid culture and solid medium can be examined for ultrastructure such as the formation and dynamics of ICMs, surface layer proteins, and storage granules. Cells are fixed and then prepared for embedding in resin to stabilize sections for microscopy using a well-established protocol (Hayat, 1981).

4.1.1 Step 1. Primary fixation

Caution: Aldehydes are toxic and carcinogenic; arsenic buffers are also toxic. Cell cultures of adequate density (10^6–10^8 cell/mL) should be mixed with 4% glutaraldehyde and 0.2 *M* Caco buffer (1:1, v:v) and incubated for a period of 1 h to overnight on ice or at 4°C. After incubation, wash cells in 0.1 *M* Caco buffer.

4.1.2 Step 2. Secondary fixation

Caution: Osmium compounds are volatile, toxic, and carcinogenic; arsenic buffers are also toxic. Mix washed primary fixed cells 1:1 (v:v) with 2% osmium tetroxide in 0.2 *M* Caco buffer and incubate for 1–2 h on ice or 4°C. The cells should turn black. Wash cells 3 × in water by spinning down gently in a tabletop centrifuge, and resuspending in 1 mL filtered deionized distilled (dd)H_2O. Fixing samples with osmium should be carried out in a chemical fume hood.

4.1.3 Step 3. Embedding

Gently spin down fixed cell biomass at low *g*-force and remove supernatant by pipette. Mix cells in 50 μL of 1%–2% agarose in water. Quickly dispense cells in small aliquots as beads or a layer of agarose onto a clean glass slide and allow agar to solidify.

4.2 Alternative protocol: Vapor fixation

Colonies grown on agar plates can be processed for TEM without disturbing them by using vapor fixation. Draw a circle on the lid above the colony and invert the plate. Open the plate and dispense a small droplet (50 μL) of stock 8% glutaraldehyde solution in the marked area of the lid beneath the colony. Allow to fix overnight. Pipette off the fixative after 12 h and dispose of properly. Repeat with the stock 4% osmium solution for an additional 12 h overnight fixation. A margin of agar containing the fixed colony can then be excised and placed into a clean Petri dish to begin the dehydration process.

4.2.1 Step 4. Dehydration

Dehydrate agar-embedded cells to remove water from the sample by transfer through a graded ethanol series for at least 1 h each. The steps of the graded series can be used as holding points during processing if time does not permit proceeding through them directly. Only absolute ethanol (200 proof) should be used to make the solutions for the dehydration series. It is easiest to put samples into microfuge tubes so that they can be mixed throughout dehydration using a rotary mixer or shaking table. Start with samples in 35%–50% ethanol in water for at least 1 h. Move samples to the next solution 75% ethanol in water and then 90%–95% ethanol in water for at least 1 h. Complete the dehydration by three exchanges in 100% ethanol.

4.2.2 Step 5. En bloc staining

Depleted uranium is typically used as a heavy metal stain during sample preparation to enhance the inherently low contrast of biological materials. *Caution* in the use of uranium solutions: dispose of this staining solution and all contaminated dry waste generated with it appropriately in a radioactive waste stream. Uranyl acetate solutions are light sensitive and solutions with precipitate should not be used. During the dehydration in 75% ethanol in water add 1% uranyl acetate for en bloc staining (3 volumes of 100% ethanol to 1 volume of 4% uranyl acetate stock solution). Continue with subsequent dehydration steps (95% ethanol, etc.).

4.2.3 Step 6. Infiltration

During infiltration, the ethanol solvent is exchanged for another solvent compatible with an embedding resin. Multiple resins are available from commercial suppliers and compatible solvents may vary; Epon 812 is routinely used and is compatible with acetone or propylene oxide as a solvent.

Stock Epon can be stored in a freezer for intermediate time periods (approximately 1–2 months) but should be prepared as a graded series with the solvent of choice immediately before use. The infiltration can be performed in microfuge tubes, but they should be tightly sealed or clamped closed to prevent lids from popping open or leaking due to the high vapor pressure of acetone. Switch solution from ethanol to 100% acetone for at least 1 h for three exchanges of acetone. Transfer to a prepared 33% Epon:66% acetone solution for at least 6 h or overnight with constant movement in a rotary mixer or shaking table. Then transfer to 66% Epon:33% acetone for the same time period with constant movement of the sample within the solution. Lastly transfer the samples to 100% Epon resin for at least 8 h to (preferably) overnight.

4.2.4 Step 7. Polymerization

Transfer infiltrated samples from 100% Epon into silicon molds and cover with 100% Epon. Create labels in a word processor document containing sample description and date written in font size <3. Cut out labels and add to each mold. Degas them at room temperature under vacuum (few hours to overnight). This will minimize drift of samples and labels within the mold during the baking process. Bake Epon resin at 60°C under vacuum for at least 48 h. Solidified Epon will have a hard, glassy surface that is not tacky and will resist pressure applied to the surface.

4.2.5 Step 8. Sectioning

Reproducible preparation of sections having uniform thickness is the most critical step of sample preparation for TEM. Specialized tools such as ultramicrotome, glass knives, and a diamond knife are critical for quality sectioning of microbial samples. Several resources for proper sample sectioning are available from ultramicrotome and diamond knife manufacturers such as Leica. Remove samples from molds and begin sectioning by trimming excess Epon with a razor blade. Samples should be trimmed into square or trapezoidal faces having slightly pyramidal sides to support the sample. Using a glass knife, cut the face of the block down to 100 nm sections until the surface is level and reflective. Move to cutting on a diamond knife down to 65–100 nm sections for the final sections. The surface of the sample face should be mirror-like and highly reflective when viewed under the microtome. Sections can then be picked up using forceps by submerging a copper grid in the diamond knife water boat and picking up the sections from below. Blot excess water from the grid and place into a grid box.

4.2.6 Step 9. Section staining

Sections can be imaged after sectioning and drying but benefit from enhanced sample contrast generated by staining. The simplest combination of stains uses a uranium solution as previously used during en bloc staining as well as a lead solution. Stain samples postsectioning in 1% uranyl acetate for approximately 1 min. Overstaining is not a concern with uranium solutions. Briefly (5–10 s) rinse sections with deionized water and blot dry with filter paper. Stain sections in a 40 m*M* lead nitrate solution for 30 s to 1 min. Additional time in lead solutions can result in overstaining. Rinse in deionized water again and blot dry. Let grids dry in fume hood for approximately 1 h and store in a grid box until imaged.

4.2.7 Step 10. TEM imaging

Examine sections with TEM after drying under high vacuum at accelerating voltages of at least 100 keV and high magnification 10–270k times.

4.3 Protocol 5. Scanning electron microscopy

Scanning electron microscopy (SEM) allows the observation of surface features of methanotrophs. One of the unique applications is the investigation of cell surface layers (Fig. 4), or nanowires. Sample preparation involves many similar reagents to TEM sample preparations, but is simpler due to the preservation of exterior structures.

4.3.1 Step 1. Fixation

Caution: Aldehydes and osmium are toxic and carcinogenic; the arsenic buffer is also toxic and requires a dedicated waste stream. Bacterial culture fixation is performed in a single step using a solution of 2% glutaraldehyde and 0.5%–1% osmium tetroxide in 0.1 *M* cacodylate buffer (final concentrations) for 1 h on ice.

4.3.2 Step 2. Deposition

Wash cells by spinning down them gently in a tabletop centrifuge at low speed and resuspending in 1 mL of 0.2 μm filtered ddH_2O. Dilute an aliquot of fixed cells 1:100 in 0.2 μm filtered ddH_2O (i.e., 50 μL into 5 mL) and deposit cells onto a nucleopore membrane (Millipore) by passing the diluted cell solution from a Luer lock syringe into a Swinnex filter unit device (Millipore) holding the nucleopore membrane. These filters have a very clean background in SEM and come in various pore sizes. Bacterial samples should be deposited on 1 or 5 μm pore filters.

Dehydrate filters containing cells in a graded ethanol series (50%, 70%, 95%, 100% ethanol) for approximately 5 min at each concentration. Filters should not be allowed to be exposed to the air and must be submerged in ethanol at all times during dehydration. Ethanol can then be removed by CO_2 critical-point drying or by replacing the ethanol by three exchanges with hexadimethylsilazane and allowing it to evaporate in a chemical fume hood.

Dried filters must then be mounted on SEM stubs with adhesive carbon tape or colloidal graphite solution and made conductive by sputter coating. A HiRes sputter coater with a film thickness monitor should be used to apply at least 6 Snm of nonreactive metal such as Pt or Au to the filters.

4.3.3 Step 3. Imaging

View coated filters in SEM under high vacuum conditions ($<10^{-3}$ Pa) with low accelerating voltage (<10 keV) and at a short working distance (10 mm) to optimize determination of cell surface ultrastructure.

4.4 Protocol 6. Fluorescence microscopy

Bacterial cell biology has benefited by the use of epifluorescence to track the development of structures and localization of tagged proteins in the context of live cells. This technique involves many routine reagents that differentially stain cellular components such as DAPI for DNA, and use of lipophilic dyes such as FM4-64, FM1-43, and Mitotracker (Invitrogen, ThermoFisher) to stain cellular membranes. Consideration must be given to excitation and emission wavelengths for each stain used to determine accurate cellular localization. Fluorophores for protein fusions, such as GFP, mCherry, and mVenus, also can be used provided they maintain intracellular fluorescence. Below we describe the main steps for preparation of live, stained methanotrophic bacterial cells.

4.4.1 Step 1. Slide preparation

Prepare a 1% agarose solution in optically clear media such as $P_{3\%}$ or in phosphate-buffered saline. Pipette agarose solution onto a clean microscope slide and sandwich between another clean microscope slide to allow agarose to harden for at least 30–40 min. Do not allow agarose to dry longer than this as it may become brittle. Slowly remove the cover microscope slide by sliding it off of the agarose pad to maintain the smooth surface of the agarose.

4.4.2 Step 2. Cell preparation

Stain cells by mixing an aliquot of cells with each dye as directed by the manufacturer in culture media or cells resuspended in phosphate-buffered saline.

The working concentration for many dyes is generally between 1 and 5 μg/mL and can be used in combination to stain cells simultaneously. Cells of membrane-rich methanotrophs are stained with 0.5–1 μg/mL Fm4-64 for at least an hour along with 0.5–1 μg/mL of DAPI and incubated at optimal growth temperature.

4.4.3 Step 3. Cell loading

Seat cells onto the prepared agarose pad by pipetting 5–10 μL of stained cells onto the agarose and allowing to stand for approximately 10 min. Apply a clean coverslip to the agarose pad on top of the seated cells and proceed to imaging with a fluorescence microscope.

4.4.4 Step 4. Imaging

Locate deposited cells at low magnification using brightfield transmitted light before increasing objective magnification up to oil immersion and reflected light using an appropriate emission/excitation filter set. Capture images using each filter set for the dyes used to stain the cells (Fm4-64: 515 nm excitation, 640 nm emission; DAPI: 341 nm excitation, 452 nm emission; GFP: 488 nm excitation, 510 nm emission) and merge images for each channel using a software package such as ImageJ to create composites.

5. Systems level characterization

Nowadays, it is hard to imagine microbial physiology research without *-omics* data. Typical data sets include whole-genome transcriptomics, proteomics, and nontargeted metabolomics. The data enable further improvements of in silico tools, such as metabolic flux modeling. Below we summarize our approaches for the preparation of cell samples for large-scale *-omic* analyses. Many of the analyses can be outsourced and carried out by dedicated and centralized facilities, such as Metabolon (https://www.metabolon.com).

5.1 Protocol 7. Proteomic studies

Samples collected from steady-state cells generally provide a good overview of expressed functions which are essential for the given conditions. Typically, such samples also provide good reproducibility. Some additional insights might be obtained from transition experiments; however, those experiments tend to be less reproducible, and thus require large numbers of replicates, as well as multiple time points. We recommend at least three biological replicates (BR), with two technical replicates per BR.

5.1.1 Step 1. Cell harvesting

Grow cells in continuous culture under the desired test conditions (i.e., O_2 limitation, methanol, La vs Ca, without copper, etc.). Collect key parameters, such as growth rate and oxygen and methane consumption rates. Harvest biomass by centrifuging 50 mL of culture for each technical replicate at 4000 rpm for 20 min. Discard supernatant and freeze cell pellets at −80°C. Samples can be stored for 3–6 months.

5.1.2 Step 2. Cell lysis

Resuspend biomass in lysis buffer [4% sodium dodecyl sulfate (SDS), 100 m*M* Tris/HCl pH 7.6, 100 m*M* dithiothreitol (DTT)] by vortexing. Transfer 100 μL fractions into 1.5 mL centrifuge tubes and incubate at 95°C for 5 min to complete cell lysis and fully reduce and denature proteins. Cool samples at 4°C for 30 min and centrifuge at 15,000 × *g* and 4°C for 10 min to pellet the remaining cell debris. Use a Filter-Aided Sample Preparation (FASP) (Wiśniewskiet, Zougman, Nagaraj, & Mann, 2009) kit for protein digestion (Expedeon, San Diego, CA) according to the manufacturer's instructions. Add 400 μL of 8 *M* urea to each 500 μL 30k molecular-weight cut-off (MWCO) FASP spin column along with 50 μL of the sample in lysis buffer. Centrifuge at 14,000 × *g* for 30 min to reduce the sample to the dead volume. Remove waste from the collection tube, add another 400 μL of 8 *M* urea to the column, and centrifuge again at 14,000 × *g* for 30 min. Repeat the 8 *M* urea wash one additional time. Add 400 μL of 50 m*M* ammonium bicarbonate to each column and centrifuge for 30 min; repeat ammonium bicarbonate wash twice. Place column in a new, clean, and labeled collection tube. Prepare digestion solution by dissolving 4 μg trypsin in 75 μL 50 m*M* ammonium bicarbonate solution and incubate with the sample for 3 h at 37°C, using a Thermotop (Eppendorf, Hamburg, Germany). Centrifuge digested peptides through the filter unit and into the collection tube at 14,000 × *g* for 15 min. Snap freeze the collection tube in liquid N_2 and place the column into a new collection tube and digest again overnight with 150 μL of digestion solution. Centrifuge the peptides the following day at 14,000 × *g* for 15 min and add to the 3 h peptide collection tube and concentrate the sample to approximately 30 μL using a SpeedVac.

5.1.3 Step 3. Protein estimation

Determine final peptide concentrations using a bicinchoninic acid (BCA) assay (Thermo Scientific, Waltham, MA, USA). Dilute samples with water to 200 ng/μL for subsequent mass spectrometry analysis.

5.1.4 Step 4. LC–MS/MS analyses

Peptides dissolved in water with a total mass of 500 ng can then be analyzed by liquid chromatography–tandem mass spectrometry (LC–MS/MS). Process LC–MS/MS data with parameters: (1) tryptic digestion in at least one terminus of the peptide (Lys/Arg), (2) 20 ppm parent ion mass error tolerance, and (3) methionine oxidation and lysine trimethylation as variable modifications. Filter identifications with an appropriate probability score $\leq 1^{-9}$ with a false-discovery rate $\leq 1\%$ at the protein level. The number of observed peptides can then be determined using their relative abundances as a proxy. Average the number of peptides observed across replicates and consider a fold-change of greater than 2 as significant.

5.2 Protocol 8. Transcriptomics

Whole-genome gene expression studies are becoming a routine approach for investigating metabolic states. Furthermore, new generation sequencing provides massive amounts of information, which can be used for the identification of transcription start sites and promoter regions. When coupled with proteomics, the data can deliver additional insights into posttranscriptional regulation. Finally, we found that integration of transcriptomic data, metabolomic data, and flux balance analysis (FBA) enables identification of novel metabolic arrangements (summarized in Akberdin, Thompson, & Kalyuzhnaya, 2018; Kabimoldayev et al., 2018).

5.2.1 Step 1. Cell harvesting

Collect samples (45 mL) of bacterial cultures and immediately add 5 mL of stop solution (5% water-equilibrated phenol in ethanol). Pellet cells by centrifugation at 4700 rpm for 15 min, extract RNA using an RNeasy kit (Qiagen), and treat cells with PureLink DNaseI (ThermoFisher Scientific) to digest genomic DNA according to the manufacturer's instructions.

5.2.2 Step 2. Sequencing

Sequence samples on an appropriate next-generation sequencing platform such as an Illumina HiSeq2500 with at least two biological replicates per condition.

5.2.3 Step 3. Data analysis

Check the quality of the obtained raw Fastq files and analyze with FastQC (http://www.bioinformatics.babraham.ac.uk/projects/fastqc/). To improve raw read quality, use a software program tool such as Trimmomatic

(Bolger, Lohse, & Usadel, 2014) with these procedures: Remove a base from either the start or end position if the quality is low, and trim bases using a sliding window method. Remove any remaining reads that are <36 bases in length. Align trimmed reads to the organism's annotated genome retrieved from the NCBI using TopHat2 (Kim et al., 2013). Post-process alignments into sorted BAM files with SAMTools version 1.4 (Li et al., 2009). Assign reads to open reading frames using a count tool with coordinates of genes from the sequenced genome. Differential expression analysis and principal component analysis should be carried out using normalized logarithmic transformed read counts to determine replicate reproducibility. Analyze genes for differential expression based on average fold-change compared to the normalized counts as well as their adjusted *P* value for statistical significance (Anders & Huber, 2010).

6. Data Integration

Many of the high-throughput *-omics* and imaging methods have become routine in methanotroph labs. However, we still lack a universal computational tool which could help to integrate multiscale biological data. Below we describe some of our approaches to make predictions based on metabolomic and transcriptomic outputs, combined with computational modeling of methanotrophic cells. Validation of the prediction relies on additional enzymatic studies or in-depth investigation of the regulatory networks that control cell output under varying conditions. The enzymatic assays can be used for characterizing the central metabolic pathways enzymes.

6.1 Protocol 9. Flux balance model construction and validation

6.1.1 Step 1. Model construction

The availability of a whole-genome sequence is an essential starting point for genome-scale metabolic framework reconstruction. Previously published whole-genome metabolic reconstructions, such as *Methylomicrobium buryatense* 5G(B1) (de la Torre et al., 2015) or *M. alcaliphilum* $20Z^R$ (Akberdin, Thompson, Hamilton, et al., 2018; Akberdin, Thompson, & Kalyuzhnaya, 2018), can be used for initial input of central metabolic pathways as well as for the metabolic systems underlying nucleotide and amino acid biosynthesis, lipid biosynthesis, and fatty acid metabolism. Additional reactions must be manually curated, corrected, or reconstructed. Each reaction in these metabolic pathways should be checked for mass and charge balances. The model should be developed in a format compatible for FBAs (Bordbar, Monk, King, & Palsson, 2014; Cuevas et al., 2016; Orth, Thiele, & Palsson,

2010). Reaction and metabolite IDs should be standardized according to BiGG Models ID Specification and Guidelines (King et al., 2016). Genome sequence annotation can be applied to link genes with reactions. To mathematically model the methane utilization network and to solve FBA optimization problems use the GNU Linear Programming Kit (GLPK) (http://www.gnu.org/software/glpk/) solver in MATLAB employing the well-known COBRA toolbox designed for flux balance model reconstruction and analyses (Schellenberger et al., 2011). We found that the Escher web tool is simple to use and efficient to visualize metabolic networks and obtained fluxes.

6.1.2 Step 2. Model validation

Initial model construction should be navigated and/or validated by gene expression and or global proteomic data. The data provide additional support for the existence of cellular mechanisms (enzymes) predicted by computational annotation-based approaches. We found that the validity of metabolic models could be further assessed by metabolomics. In general to prepare cells for metabolomic studies, samples of cultures (50 mL, OD = 1–2) grown at the desired conditions should be collected by centrifugation, supernatant should be separated from cell pellets, and both (cell pellets and supernatant) should be flash frozen with liquid nitrogen and stored at −80°C. Sample analyses can be outsourced and carried out by dedicated and centralized facilities, such as Metabolon (https://www.metabolon.com).

6.1.3 Step 3. Data analyses

Transcriptomic and metabolomic data analyses are described above. Metabolomic data typically represent accurate relative quantification only for individual compounds across all the samples which contain them. The original scale data represent raw ion counts for the integrated peaks specifically representing each compound. The scaled data are generated by simply dividing each value for a compound by the median (among detected samples), then imputing any null values with the minimum detected value. Thus, the scaled data are centered around 1.0 (median) for all compounds, but the variation among the samples is not affected (Jolliffe, 2002; Welch, 1938). Statistical analyses are typically performed on natural log-transformed data.

6.1.4 Step 4. Statistical analysis

Statistical comparisons between methane-grown cells (methane batch) and each of the other culture conditions should be initially conducted to reveal key differences in metabolic signatures. The magnitudes of changes between two groups are then compared to predicted flux changes between the same two conditions. An enzymatic reaction with the highest value of flux for biosynthesis of corresponding metabolite is used to calculate flux ratio between two growth conditions. In order to estimate Spearman's correlation index, predicted flux ratios and metabolic signatures can be ranked according to the method described by Lehman, O'Rourke, Hatcher, and Stepanski (2005). 3D plots reflecting metabolomics and model ranks considering metabolic pathways for measured substances can be generated with the assistance of Plotly (Plotly Technologies Inc., Collaborative data science, Montréal, QC, 2015; https://plot.ly).

6.1.5 Step 5. Enzymatic profiling

The next step of model validation is enzymatic profiling for newly predicted functions. It can be carried out with whole-cell extracts. Cell lysates should be generated from biomass cultivated at the conditions for which the enzymatic reaction was predicted. Typically, cells are pelleted via centrifugation for 15 min at 13,000 × *g* and culture supernatants are discarded. Cell pellets are then quenched in liquid nitrogen, thawed, and solubilized on ice in 1 mL of lysis buffer (50 m*M* MOPS, pH 7.4, 50 m*M* NaCl, supplemented with 1 × Complete Protease Inhibitor Cocktail solution (Roche Diagnostics Corporation, Indianapolis, IN)). 50 m*M* Tris–HCl buffer or 50 m*M* K-phosphate buffer can be used instead of MOPS. The cells are then sonicated on ice at 4°C, at 90% power setting for 30 s × 3 cycles, with a 1-min cool-down period between sonication cycles using a Braun-Sonic-L ultrasonicator. Alternatively, cell extracts could be prepared by passing through a French Press mini-cell at 1000 psi pressure. Lysates are cleared by centrifugation at 16,000 × *g* at 4°C for 20 min. Resulting supernatants can then be used for enzymatic analysis.

6.2 Example. Model validation based on methanol vs methane metabolomic data

Metabolomic profiles were obtained for cells grown in continuous culture on methane, as well as for exponentially grown cells from batch cultures on methane and methanol. Two hundred and fifty-three biochemicals were identified in bacterial cell pellets. We conducted statistical comparisons between methane-grown cells (methane batch or optimal culture considered as controls) and other culture conditions to reveal key differences in metabolic signatures. The magnitude of changes between the two groups was compared to flux changes between the same two conditions predicted by the model. A positive and statistically significant correlation between predicted fluxes and experimental metabolomics data was observed (Fig. 5).

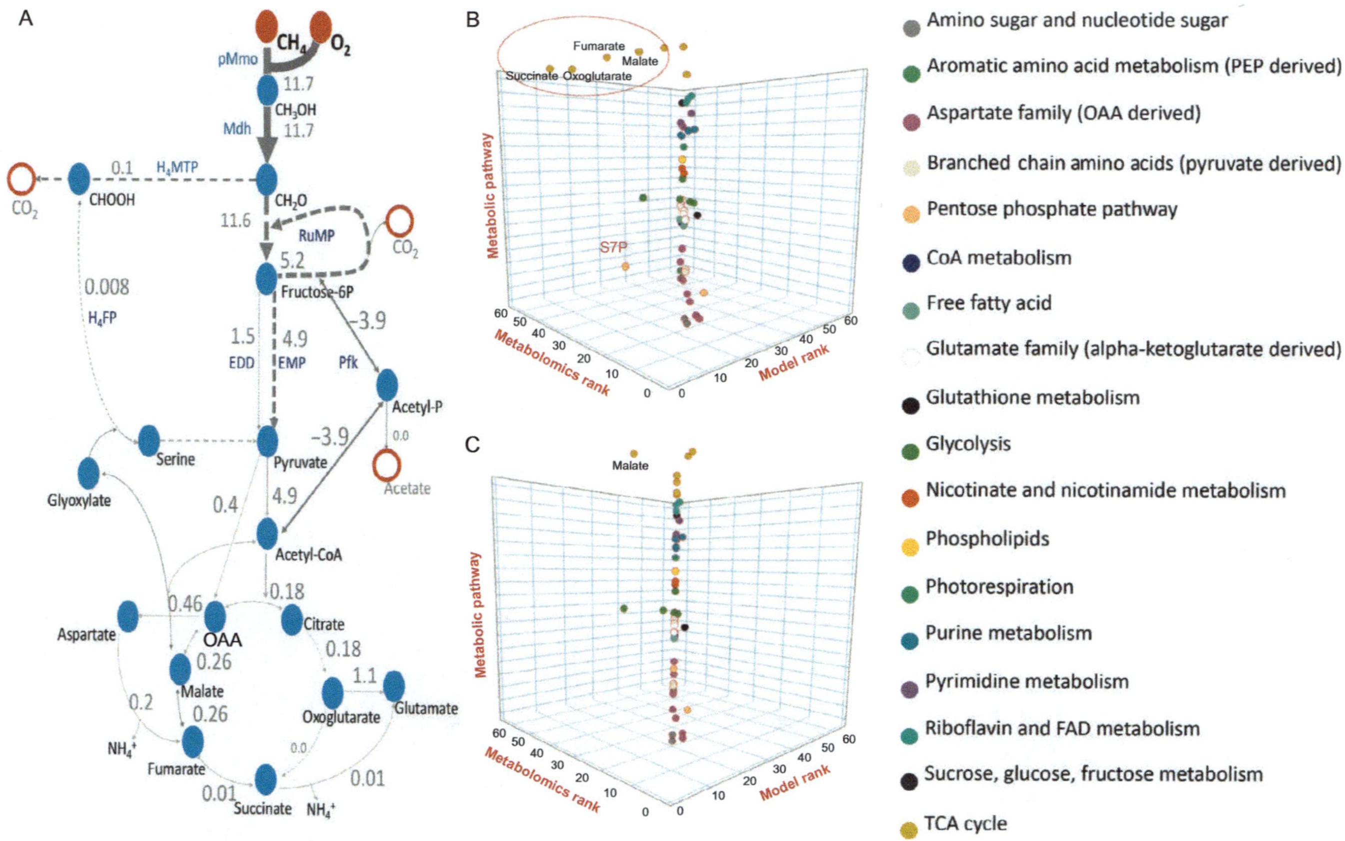

Fig. 5 See legend on next page.

Fig. 5 Flux balance model of *Methylomicrobium alcaliphilum* $20Z^R$. (A) Central metabolic pathways of methane utilization. *Red circles* indicate input substrates (CH_4, O_2, or CH_3OH), *white circles* with *red outlines* indicate excreted compounds (CO_2 and acetate), and *thickness of arrow* represents predicted value of the corresponding flux via the reaction (*solid arrow*) or metabolic pathway (*dotted arrows*). (B–C) 3D plots of comparative analysis of metabolic profiling vs predicted flux ratios between the same growth conditions (axis *X*—model rank; axis *Y*—metabolomics rank; axis *Z*—metabolic pathway that a compound belongs to). (B) Comparison of results for complete TCA cycle (Spearman's index equals to 0.9, P value$=2.7\times10^{-24}$). (C) Comparison of results for branched TCA cycle and reverse phosphoketolase reaction ($R=0.99$, P value$=2\times10^{-55}$). *The figure is adapted from Akberdin, I. R., Thompson, M., & Kalyuzhnaya, M. G. (2018). Systems biology and metabolic modeling of C1-metabolism. In M. G. Kalyuzhnaya & X. H. Xing (Eds.),* Methane biocatalysis: Paving the way to sustainability *(pp. 99–115). Switzerland: Springer International Publishing. Akberdin, I. R., Thompson, M., Hamilton, R., Desai, N., Alexander, D., Henard, C. A., et al. (2018). Methane utilization in* Methylomicrobium alcaliphilum *$20Z^R$: A systems approach.* Scientific Reports, *8, 2512.*

We found that the integration of nontargeted metabolomics data with metabolic flux predictions can help to refine the central metabolic pathways, by highlighting nodes that are poorly resolved by in silico reconstruction, such as the pentose phosphate pathways and tricarboxylic acid cycle. Further enzymatic studies confirmed the carbon flux from acetyl-CoA to xylulose-5-phosphate via phosphoketolase, from glutamate to succinate via succinate-semialdehyde dehydrogenase and glutamate carboxylase, and from aspartate to fumarate via aspartate lyase.

References

Akberdin, I. R., Thompson, M., Hamilton, R., Desai, N., Alexander, D., Henard, C. A., et al. (2018). Methane utilization in *Methylomicrobium alcaliphilum* $20Z^R$: A systems approach. *Scientific Reports*, *8*, 2512.

Akberdin, I. R., Thompson, M., & Kalyuzhnaya, M. G. (2018). Systems biology and metabolic modeling of C1-metabolism. In M. G. Kalyuzhnaya & X. H. Xing (Eds.), *Methane biocatalysis: Paving the way to sustainability* (pp. 99–115). Switzerland: Springer International Publishing.

Anders, S., & Huber, W. (2010). Differential expression analysis for sequence count data. *Genome Biology*, *11*, R106.

Bolger, A. M., Lohse, M., & Usadel, B. (2014). Trimmomatic: A flexible trimmer for Illumina sequence data. *Bioinformatics*, *30*, 2114–2120.

Bordbar, A., Monk, J. M., King, Z. A., & Palsson, B. O. (2014). Constraint-based models predict metabolic and associated cellular functions. *Nature Reviews. Genetics*, *15*, 107–120.

Chistoserdova, L., & Kalyuzhnaya, M. G. (2018). Current trends in methylotrophy. *Trends in Microbiology*, *26*, 703–714.

Chistoserdova, L., Laukel, M., Portais, J.-C., Vorholt, J. A., & Lidstrom, M. E. (2004). Multiple formate dehydrogenase enzymes in the facultative methylotroph *Methylobacterium extorquens* AM1 are dispensable for growth on methanol. *Journal of Bacteriology*, *186*, 22–28.

Chu, F., & Lidstrom, M. E. (2016). XoxF acts as the predominant methanol dehydrogenase in the Type I methanotroph *Methylomicrobium buryatense*. *Journal of Bacteriology, 198,* 1317–1325. https://doi.org/10.1128/JB.00959-15.

Crombie, A., & Murrell, J. C. (2011). Development of a system for genetic manipulation of the facultative methanotroph *Methylocella silvestris* BL2. *Methods in Enzymology, 495,* 119–133.

Cuevas, D. A., Edirisinghe, J., Henry, C. S., Overbeek, R., O'Connell, T. G., & Edwards, R. A. (2016). From DNA to FBA: How to build your own genome-scale metabolic model. *Frontiers in Microbiology*, 7, 907.

Dedysh, S. N., Panikov, N. S., & Tiedje, J. M. (1998). Acidophilic methanotrophic communities from *Spagnum* peat bogs. *Applied and Environmental Microbiology, 64,* 922–929.

de la Torre, A., Metivier, A., Chu, F., Laurens, L. M., Beck, D. A., Pienkos, P. T., et al. (2015). Genome-scale metabolic reconstructions and theoretical investigation of methane conversion in *Methylomicrobium buryatense* strain 5G(B1). *Microbial Cell Factories, 14,* 188.

Figurski, D. H., & Helinski, D. R. (1979). Replication of an origin-containing derivative of plasmid RK2 dependent on a plasmid function provided in trans. *Proceedings of the National Academy of Sciences of the United States of America, 76,* 1648–1652.

Gibson, D. G., Young, L., Chuang, R. Y., Venter, J. C., Hutchinson, C. A., III, & O'Smith, H. (2009). Enzymatic assembly of DNA molecules up to several hundred kilobases. *Nature Methods, 6,* 343–345.

Gilman, A., Fu, Y., Hendershott, M., Chu, F., Puri, A. W., Smith, A. L., et al. (2017). Oxygen-limited metabolism in the methanotroph *Methylomicrobium buryatense* 5GB1C. *PeerJ, 5,* e3945. https://doi.org/10.7717/peerj.3945.

Green, R., & Rogers, E. J. (2013). Transformation of chemically competent. *Methods in Enzymology, 529,* 329–336.

Gu, W., Haque, F. M. U. l., DiSpirito, A. A., & Semrau, J. D. (2016). Uptake and effect of rare earth elements on gene expression in *Methylosinus trichosporium* OB3b. *FEMS Microbiology Letters, 363.* pii: fnw129.

Gayazov, R. R., & Mshenskii, Y. N. (1991). Growth of *Methylococcus capsulatus* at various concentrations of ammonium nitrogen and inorganic phosphates. *Microbiologiya (Moscow, Russia), 60,* 171–174.

Haque, M. U. I., Gu, W., DiSpirito, A. A., & Semrau, J. D. (2016). Marker exchange mutagenesis of *mxaF,* encoding the large subunit of the Mxa methanol dehydrogenase, in *Methylosinus trichosporium* OB3b. *Applied and Environmental Microbiology, 82,* 1549–1555.

Hayat, M. A. (1981). *Fixation for electron microscopy*. Academic Press.

Henard, C. A., & Guarnieri, M. T. (2018). Metabolic engineering of methanotrophic bacteria for industrial biomanufacturing. In M. G. Kalyuzhnaya & X. Xing (Eds.), *Methane biocatalysis: Paving the way to sustainability* (pp. 117–132). Springer.

Henard, C. A., Smith, H., Dowe, N., Kalyuzhnaya, M. G., Pienkos, P. T., & Guarnieri, M. T. (2016). Bioconversion of methane to lactate by an obligate methanotrophic bacterium. *Scientific Reports, 6,* 21585.

Jolliffe, I. (2002). *Principal component analysis.* John Wiley & Sons, Ltd.

Kabimoldayev, I., Nguyen, A. D., Yang, L., Park, S., Lee, E. Y., & Kim, D. (2018). Basics of genome-scale metabolic modeling and applications on C1-utilization. *FEMS Microbiology Letters, 365.* https://doi.org/10.1093/femsle/fny241.

Kalyuzhnaya, M. G., Khmelenina, V., Eshinimaev, B., Suzina, N., Nikitin, D., Solonin, A., et al. (2001). Taxonomic characterization of new alkaliphilic and alkalitolerant methanotrophs from soda lakes of the Southeastern Transkaibakal region and description of the *Methylomicrobium buryatense* sp. nov. *Systematic and Applied Microbiology, 24,* 166–176.

Kim, D., Pertea, G., Trapnell, C., Pimentel, H., Kelley, R., & Salzberg, S. L. (2013). TopHat2: Accurate alignment of transcriptomes in the presence of insertions, deletions, and gene fusions. *Genome Biology, 14,* R36.

King, Z. A., Lu, J., Dräger, A., Miller, P., Federowicz, S., Lerman, J. A., et al. (2016). BiGG models: A platform for integrating, standardizing and sharing genome-scale models. *Nucleic Acids Research, 44*, 515–522.

Lehman, A., O'Rourke, N., Hatcher, L., & Stepanski, E. (2005). *JMP for basic univariate and multivariate statistics* (p. 481). Cary, NC: SAS Institute Inc.

Li, H., Handsaker, B., Wysoker, A., Fennell, T., Ruan, J., Homer, N., et al. (2009). 1000 Genome project data processing subgroup. The sequence alignment/map format and SAMtools. *Bioinformatics, 25*, 2078–2079. https://doi.org/10.1093/bioinformatics/btp352.

Ma, T. (2015). *Ignitability and explosibility of gases and vapors* (p. 212). New York: Springer Science. https://doi.org/10.1007/978-1-4939-2665-7.

Marx, C. J. (2008). Development of a broad-host-range *sacB*-based vector for unmarked allelic exchange. *BMC Research Notes, 1*, 1. https://doi.org/10.1186/1756-0500-1-1.

Marx, C. J., & Lidstrom, M. E. (2002). Broad-host range *cre-lox* system for antibiotic marker recycling in Gram-negative bacteria. *BioTechniques, 33*, 1062–1067.

Mshensky, Y. N. (1979). *PhD thesis: Growth of methanotrophic bacteria*. Pushchino: IBPM RAS.

Mühlemeier, M. I., Speight, R., & Strong, P. J. (2018). Biogas, bioreactors and bacterial methane oxidation. In M. G. Kalyuzhnaya & X. Xing (Eds.), *Methnane biocatalysis: Paving the way to sustainability* (pp. 213–235). Springer.

Ojala, D. S., Beck, D. A., & Kalyuzhnaya, M. G. (2011). Genetic systems for moderately halo (alkali) philic bacteria of the genus *Methylomicrobium*. *Methods in Enzymology, 495*, 99–118.

Orth, J. D., Thiele, I., & Palsson, B. Ø. (2010). What is flux balance analysis? *Nature Biotechnology, 28*, 245–248.

Pédelacq, J. D., Cabantous, S., Tran, T., Terwilliger, T. C., & Waldo, G. S. (2006). Engineering and characterization of a superfolder green fluorescent protein. *Nature Biotechnology, 24*, 79–88.

Pol, A., Heijmans, K., Harhangi, H. R., Tedesco, D., Jetten, M. S. M., & Op den Camp, H. J. M. (2007). Methanotrophy below pH 1 by a new *Verrucomicrobia* species. *Nature, 450*, 874–878.

Puri, A. W., Owen, S., Chu, F., Chavkin, T., Beck, D. A. C., Kalyuzhnaya, M. G., et al. (2015). Genetic tools for the industrially promising methanotroph *Methylomicrobium buryatense*. *Applied and Environmental Microbiology, 81*, 1775–1781.

Schellenberger, J., Que, R., Fleming, R. M., Thiele, I., Orth, J. D., Feist, A. M., et al. (2011). Quantitative prediction of cellular metabolism with constraint-based models: The COBRA Toolbox. v2. 0. *Nature Protocols, 6*, 1290–1307.

Semrau, J. D., DiSpirito, A. A., Gu, W., & Yoon, S. (2018). Metals and methanotrophy. *Applied and Environmental Microbiology, 84*, e02289-17. https://doi.org/10.1128/AEM.02289-17.

Semrau, J. D., Jagadevan, S., DiSpirito, A. A., Khalifa, A., Scanlan, J., Bergman, B. H., et al. (2013). Methanobactin and MmoD work in concert to act as the 'copper-switch' in methanotrophs. *Environmental Microbiology, 15*, 3077–3086.

Simon, R., Priefer, U., & Pühler, A. (1984). A broad host range mobilization system for in vivo genetic engineering: Transposon mutagenesis in Gram-negative bacteria. *Biotechnology, 1*, 784–791.

Sirajuddin, S., & Rosenzweig, A. C. (2015). Enzymatic oxidation of methane. *The Biochemistry, 54*, 2283–2294.

Welch, B. L. (1938). The significance of the difference between two means when the population variances are unequal. *Biometrika, 29*, 350–362.

Whittenbury, R., Phillips, K. C., & Wilkinson, J. F. (1970). Enrichment, isolation and some properties of methane-utilizing bacteria. *Journal of General Microbiology, 61*, 205–218.

Wiśniewskiet, J. R., Zougman, A., Nagaraj, N., & Mann, M. (2009). Universal sample preparation method for proteome analysis. *Nature Methods, 6*, 359–362.

Yan, X., Chu, F., Puri, A. W., Fu, Y., & Lidstrom, M. E. (2016). Electroporation-based genetic manipulations in type I methanotrophs. *Applied and Environmental Microbiology*, *82*, 2062–2069.
Zhivotchenko, A. G., Nikonova, E. S., & Jørgensen, M. H. (1995). Copper effect on the growth kinetics of *Methylococcus capsulatus (Bath)*. *Biotechnology Techniques*, *9*, 163.

Further reading

Galchenko, V. F. (2001). *Methanotrophic bacteria*. Moscow: GEOS. ISBN: 5-89-118-200-9500.
Heng, L., Handsaker, B., Wysoker, A., Fennell, T., Ruan, J., Homer, N., et al. (2009). The sequence alignment/map format and SAMtools. *Bioinformatics*, *25*, 2078–2079.
Kalyuzhnaya, M. G., Puri, A., & Lidstrom, M. E. (2015). Metabolic engineering in methanotrophic bacteria. *Metabolic Engineering*, *29*, 142–152.

MIX
Papier | Fördert gute Waldnutzung
FSC® C083411

Kolibri 360 GmbH
Ferdinand-Jühlke-Straße 7
99095 Erfurt, Deutschland
produktsicherheit@kolibri360.de